T0269186

CAMBRIDGE LIBRARY COLLECTION

Books of enduring scholarly value

Physical Sciences

From ancient times, humans have tried to understand the workings of the world around them. The roots of modern physical science go back to the very earliest mechanical devices such as levers and rollers, the mixing of paints and dyes, and the importance of the heavenly bodies in early religious observance and navigation. The physical sciences as we know them today began to emerge as independent academic subjects during the early modern period, in the work of Newton and other 'natural philosophers', and numerous sub-disciplines developed during the centuries that followed. This part of the Cambridge Library Collection is devoted to landmark publications in this area which will be of interest to historians of science concerned with individual scientists, particular discoveries, and advances in scientific method, or with the establishment and development of scientific institutions around the world.

Conferences Held in Connection with the Special Loan Collection of Scientific Apparatus, 1876

In 1876 the South Kensington Museum held a major international exhibition of scientific instruments and equipment, both historical and contemporary. Many of the items eventually formed the basis of collections now held at London's Science Museum. In May 1876, organisers arranged a series of conferences at which leading British and European scientists explained and demonstrated some of the items on display. The purpose was to emphasise the exhibition's goal not merely to preserve archaic treasures (such as Galileo's telescopes or Janssen's microscope) but to juxtapose them with current technology and so inspire future scientific developments. Volume 1 of the proceedings is devoted to physics and mechanics. The contributors include William Thomson (Baron Kelvin), John Tyndall, Francis Galton, James Clerk Maxwell, J.A. Froude and Thomas Stevenson, all of whom have other works reissued in the Cambridge Library Collection, which also includes the full catalogue of the exhibition itself.

Cambridge University Press has long been a pioneer in the reissuing of out-of-print titles from its own backlist, producing digital reprints of books that are still sought after by scholars and students but could not be reprinted economically using traditional technology. The Cambridge Library Collection extends this activity to a wider range of books which are still of importance to researchers and professionals, either for the source material they contain, or as landmarks in the history of their academic discipline.

Drawing from the world-renowned collections in the Cambridge University Library and other partner libraries, and guided by the advice of experts in each subject area, Cambridge University Press is using state-of-the-art scanning machines in its own Printing House to capture the content of each book selected for inclusion. The files are processed to give a consistently clear, crisp image, and the books finished to the high quality standard for which the Press is recognised around the world. The latest print-on-demand technology ensures that the books will remain available indefinitely, and that orders for single or multiple copies can quickly be supplied.

The Cambridge Library Collection brings back to life books of enduring scholarly value (including out-of-copyright works originally issued by other publishers) across a wide range of disciplines in the humanities and social sciences and in science and technology.

Conferences Held in Connection with the Special Loan Collection of Scientific Apparatus, 1876

Physics and Mechanics

VOLUME 1

VARIOUS

CAMBRIDGE
UNIVERSITY PRESS

University Printing House, Cambridge, CB2 8BS, United Kingdom

Cambridge University Press is part of the University of Cambridge.
It furthers the University's mission by disseminating knowledge in the pursuit of education, learning and research at the highest international levels of excellence.

www.cambridge.org
Information on this title: www.cambridge.org/9781108078139

This edition first published 1876
This digitally printed version 2015

ISBN 978-1-108-07813-9 Paperback

SOUTH KENSINGTON MUSEUM.

CONFERENCES.

SPECIAL LOAN COLLECTION OF SCIENTIFIC APPARATUS,

1876.

SOUTH KENSINGTON MUSEUM.

CONFERENCES

HELD IN CONNECTION WITH

THE

SPECIAL LOAN COLLECTION OF SCIENTIFIC APPARATUS.

1876.

PHYSICS AND MECHANICS.

Published for the Lords of the Committee of Council on Education

BY

CHAPMAN AND HALL, 193, PICCADILLY.

LONDON:

PRINTED BY VINCENT BROOKS, DAY AND SON,

GATE STREET, LINCOLN'S INN FIELDS.

CONTENTS.

SECTION—PHYSICS (including Astronomy).

SECTION—MECHANICS (including Pure and Applied Mathematics and Mechanical Drawing).

INTRODUCTION.

The Conferences in connection with the Special Loan Collection of Scientific Apparatus at South Kensington Museum, of which these volumes form a record, originated in a suggestion contained in a letter addressed by the Right Hon. Viscount Sandon, M.P., Vice-President of the Committee of Council on Education, to the Presidents of the various Learned Societies. In this letter his lordship stated that it had been represented to the Lords of the Committee of Council on Education that the utility of the approaching Loan Collection of Scientific Apparatus would be much enhanced, both to the English and to the Foreign men of Science who might be expected to visit it, if arrangements were made for explaining and demonstrating the method of using the various instruments, as well as for reading papers on, and discussing, scientific subjects; and he added that their lordships would be glad to afford every facility for carring out this proposal, and they desired to suggest, that with this view the various learned Societies should organise a series of Conferences similar to the sectional meetings of the British Association for the Advancement of Science.

In accordance with this suggestion a Sub-Committee was formed consisting of the Presidents and Vice-Presidents of each of the Learned Societies by whom the details of the scheme were considered, and it was finally decided to recommend to their Lordships that the Sub-Committees of the various sections should be requested to undertake the organization of the Conferences in those branches of science which came within their cognizance. It was also proposed that the Conferences should begin on the 16th of May, 1876, and last till the end of the month, and that the subject of the Conferences should be confined to the description and use of the instruments exhibited. The following are the Sub-Committees to whose labours the Lords of the Committee of Council on Education are so largely indebted for the successful carrying out of this proposal.

SECTION—PHYSICS (including Astronomy.)

President:

Mr. W. Spottiswoode, M.A., LL.D., F.R.S.

Vice-Presidents:

Il Commendatore Professore Blaserna.

Mr. De La Rue, D.C.L., F.R.S.

Professor Carey Foster, B.A., F.R.S.

Professor Guthrie, F.R.S.

Herr Professor Dr. Helmholtz.

Herr Professor Dr. Rijke.

M. le Professeur Soret.

Professor Tyndall, D.C.L., LL.D., F.R.S.

Professor Sir W. Thomson, LL.D., F.R.S.

M. le Professeur Wartmann.

SECTION—MECHANICS (including Pure and Applied Mathematics and Mechanical Drawing.)

President:

Mr. C. W. Siemens, D.C.L., F.R.S.

Vice-Presidents:

Mr. F. J. Bramwell, F.R.S.

Mr. W. Froude, M.A., F.R.S.

M. le Général Morin, Directeur du Conservatoire des Arts et Métiers

Dr. Werner Siemens

M. Tresca, Sous-Directeur du Conservatoire des Arts et Métiers.

Sir Joseph Whitworth, Bart., D.C.L., F.R.S.

SECTION—CHEMISTRY.

President:

Professor E. Frankland, Ph.D., D.C.L., F.R.S.

Vice-Presidents:

Professor Abel, F.R.S.

Herr Professor Dr. Von Babo.

M. le Professeur Beilstein.

Il Commendatore Professore Blaserna.

M. le Professeur Fremy.

Dr. Gilbert, F.R.S.

Professor Gladstone, Ph.D., F.R.S.

Herr Professor Heintz.

Herr Professor Himly.

Herr Professor Dr. Hofmann, F.R.S.

Herr Professor Dr. de Loos.

The Right Hon. Lyon Playfair, C.B., M.P., F.R.S.

Professor Roscoe, Ph.D., F.R.S.

Herr Professor Waage.

Professor Williamson, Ph.D., F.R.S.

Herr Professor Dr. Wöhler, F.R.S.

SECTION—BIOLOGY.

President:

Professor J. Burdon Sanderson, M.D., LL.D., F.R.S.

Vice-Presidents:

Dr. G. J. Allman, F.R.S.
M. le Professeur Van Beneden, F.R.S.
Herr Professor Cohn.
Herr Professor Dr. Donders, F.R.S.
Professor Michael Foster, M.A., M.D., F.R.S.
Colonel Lane Fox.
Herr Professor Ewald Hering.
Dr. Hooker, C.B., P.R.S.
M. le Professeur Marey.
Professor Rolleston, M.D., F.RS.

SECTION—PHYSICAL GEOGRAPHY, GEOLOGY, MINING, AND METEOROLOGY.

President:

Mr. John Evans, F.R.S.

Vice-Presidents:

Herr Professor Dr. Beyrich.
M. Daubrée, Directeur de l'Ecole des Mines, Paris.
His Excellency Dr. Von Dechen.
M. le Professeur Dewalque.
Mr. H. S. Eaton, President of the Meteorological Society.
M. le Professeur Dr. Forel.
Mr. N. Story-Maskelyne, M.A., F.R.S.
Professor Ramsey, LL.D., F.R.S.
Major-General Sir H. Rawlinson, K.C.B., F.R.S.
Mr. W. Warington Smyth, M.A., F.R.S.
The Baron Ferdinand Von Wrangell.

A central room in the range of buildings in which the collections were displayed was assigned for the purposes of the Conferences, and the first, that on Physics, was held on the 16th May. The Right Hon. Viscount Sandon, M.P., Vice-President of the Committee of Council on Education, took the chair at the commencement of the opening meeting.

His Lordship expressed his great regret that, owing to the special pressure of official business which the Education Bill entailed upon him, it would be impossible for him to attend the Conferences—and even on that occasion, to do more than convey the assurance of his own gratitude and that of Her Majesty's Government, to the men of science all over the world, for their invaluable assistance to this unique undertaking, which had made it, what he might now with confidence pronounce it to be, a great success, of which not only this country but the civilized world might be justly proud. He could not however refrain from bearing his special testimony on this, the first public occasion which presented itself, to the unwearied zeal which had been shown, and the extraordinary sacrifices of time which had been made by the first men of science in this country, working together as one brotherhood in endeavouring to make this collection as complete, as useful, and as widely instructive as possible. He would at the same time wish to express the warm appreciation of Her Majesty's Government of the exertions which had been made by gentlemen of the highest distinction in scientific pursuits on the Continent to promote the success of this work to which their Governments have given a most cordial and gratifying support. It was the earnest hope, both of the Lord President and of himself, that this collection would not be a mere gazing place where nothing but feelings of wonder and pride would be excited by the past triumphs of science, but that much instruction would be gained from it. With this view the preparation of handbooks had been committed by the Education Department to gentlemen of the highest capacity, and he had reason to believe that they would be found invaluable aids to such instruction, and would be considered to be in themselves of high intrinsic value. With this same view they had given their cordial consent to the Conferences, of which this was the opening meeting, and he trusted that at these gatherings many an old friendship between the workers in the fields of Science would be renewed, and many a new friendship between those who were labouring in the same cause in the different countries of the world would be formed, so that by the interchange of ideas and the comparison of their various researches and labours, the seeds might be sown on the occasion of this Loan Collection of Scientific Apparatus, of fresh achievements, to the general advantage of the human race.

Before they proceeded to the business of the day, he must be allowed to call their attention to the high services which had been rendered by the officers of the Science and Art Department, to the undertaking, the success of which they were now celebrating : he had a personal knowledge of the large extent to which that success was due to their highly cultivated intelligence, their zeal, their resource, and their ungrudging devotion of time, and, he feared he might say health. He could only say that he was proud to serve the Queen in his Department of the State, in concert with such officers.

In conclusion, he must express his disappointment that his special Parliamentry duties this Session would prevent his deriving that advantage from the Loan Collection which he would have wished, and would rob him of the power of profiting by the opportunity to make the acquaintance of the many eminent men who would be gathered together by this exhibition, in the preparation for which he had from the first taken a warm personal interest. It only remained for him now to bid a hearty welcome on the part of Her Majesty's Government to the men of Science who were, and who would be gathered in these galleries from all parts of the world, and to renew the expression of the hope that when these Conferences have come to an end, and when this collection is dispersed, it might be found that no unimportant assistance had been given to those who were labouring in the noble cause of scientific investigations.

The Conferences were held on the following days :—

Physics (including Astronomy), 16th, 19th, and 24th May; Mechanics (including Pure and Applied Mathematics and Measurement), 17th, 22nd, and 25th May; Chemistry, 18th and 23rd May; Biology, 26th and 29th May; Physical Geography, Geology, Mineralogy, and Meteorology, 30th May, 1st, and 2nd June.

SECTION—PHYSICS (including Astronomy).

President: Mr. W. SPOTTISWOODE, M.A., LL.D., F.R.S.

Vice-Presidents:

Il Commendatore Professore BLASERNA.
Mr. DE LA RUE, D.C.L., F.R.S.
Professor CAREY FOSTER, B.A., F.R.S.
Professor GUTHRIE, F.R.S.
Herr Professor Dr. HELMHOLTZ.
Herr Professor Dr. RIJKE.
M. le Professeur SORET.
Professor TYNDALL, D.C.L., LL.D., F.R.S.
Professor SIR W. THOMSON, LL.D., F.R.S.
M. le Professeur WARTMANN.

May 16th, 1876.

MR. SPOTTISWOODE: The opening of this Exhibition may prove an epoch in the science of Great Britain. We find here collected, for the first time within the walls of one building, a large number of the most remarkable instruments, gathered from all parts of the civilised world, and from almost every period of scientific research. These instruments, it must be remembered, are not merely masterpieces of constructive skill, but are the visible expression of the penetrative thought, the mechanical equivalent of the intellectual processes of the great minds whose outcome they are.

There have been in former years, both in this country and elsewhere, exhibitions including some of the then newest inventions of the day; but none have been so exclusively devoted to scientific objects, nor any so extensive in their range as this. There exist in most seats of learning museums of instruments accumulated from the laboratories in which the professors have worked; but these are, by their very nature, confined to local traditions. The present one is, I believe, the first serious, or at all events the first successful, attempt at a cosmopolitan collection.

To mention only a few among the many foreign institutions which have contributed to this undertaking, we are specially indebted to the authorities of the Conservatoire des Arts et Métiers of Paris, the Physical Museum of Leyden, the Tayler Foundation of Haarlem, the Royal Museum of Berlin, the Physical Observatory of St. Petersburg, the Tribune of Florence, and the University of Rome.

Among those in our own country, we have to thank the Royal Society, the Royal Institution, the Ordnance Survey, the Post Office, the Royal Mint, the Kew Observatory, besides various other institutions and colleges, which have freely contributed their quota.

To enumerate even the chief of the individual instruments of historical interest would be a task beyond the limits both of my powers and of your patience. But I cannot refrain from naming as especially worth notice among the astronomical treasures, a quadrant of Tycho Brahé, telescopes of Galileo, a telescope of Newton, some lenses by Huygens, one of Sir W. Herschel's grinding machines for specula, and a telescope made by himself in intervals between his music lessons during his early days at Bath, at a time when, to use her own words, his sister Caroline "was continually obliged to feed him by putting victuals by bits into his mouth." This also is probably the "mirror from which he did not take his hands for sixteen hours together," and with which he may have seen for the first time the Georgium Sidus. To come to later days, we have the original siderostat of Foucault, lent from the Observatory of Paris, a compound speculum by the late Lord Rosse, the photoheliograph from Kew, and from still more recent times a complete transit of Venus equipment, from the Royal Observatory at Woolwich.

Turning to other branches of physics, we have a "composed microscope," now nearly three centuries old, constructed in 1590 by one Zacharias Janssen, a spectacle maker, possibly a connection, or at all events a worthy predecessor, of M. Janssen, the celebrated astronomical spectroscopist. We have an air-pump and two "Magdeburg hemispheres," with the original rope traces by which horses were attached in the presence of the Emperor Charles V., in order, if possible, to tear them asunder, when exhausted by the air-pump. We have the air-pump of Boyle, the compressor of Pappin, Regnault's apparatus for determining the specific heat of gases, Dumas' lobe for the determination of vapour densities, Fizeau and Foucault's original revolving mirrors and toothed wheels, whereby the velocity of light was first determined independently of astronomical aid, Daguerre's first photograph on glass, and the earliest astronomical photographs ever taken. To these may be added De la Rive's instruments for statical electricity; the actual table and appurtenances at which Ampère worked; and some contrivances as if fresh from the hands of Faraday himself.

Yet rich as is this part of our collection, and interesting as it might be made in the hands of one versed in the history and anecdote of the past, we must not linger even about these pleasant places. Indeed, a museum of only the past, venerable though it might be, would be also grey with the melancholy of departing life. For science should be living, instinct with vigour and organic growth. Without a continuance into the present, and a promise for the future, it would be like a tree whose branches are broken, whose growth is stopped, and whose sap is dried. And if I may carry the simile a stage further, an exhibition of the present, with no elements of the past, would be like the gathered fruits to be found in the market-place, ready to hand, it is true, but artificially arranged. But when past and present are represented in combination, as has been attempted here, the very newest achievements will be found in their natural places as ripened and ever-ripening fruit in the garden from whence they have sprung.

In reviewing the series of ancient, or at least now disused, instruments, one thing can hardly fail to strike the attention of those who are accustomed to the use of the modern forms. It is this—how much our predecessors managed to achieve with the limited means at their disposal. If we compare the magnificent telescopes, the exquisite clockwork, the multiplicity of optical appliances, now to be found in almost every private, and still more in every public, observatory, with those of two centuries past; or, again, if we look at the instruments with which Arago and Brewster made their magnificent discoveries in polarised light, in contrast to those with which the adjoining room is literally teeming, we may well pause to reflect how much of their discoveries was due to the men themselves, and how comparatively little to the instruments at their command.

And yet we must not measure either the men or their results by this standard alone. The character of the problems which nature propounds, or which our predecessors leave as a legacy to our generation, varies greatly from time to time. First, we have some great striking question, the very conception and statement of which demands the very highest powers of the human mind ; unless indeed, the clear and distinct statement of every problem may be regarded as the first and most important step towards its solution. Next follow the first outlines of the solution sketched in bold outline by some master hand ; afterwards, the careful and often tedious working out of the details of the problem, the numerical valuation of the constants involved, and the reduction of all the quantities to strict measurement. It is in this part of the business that the more elaborate instruments are especially required. It is for bringing small differences to actual measurement, for detecting quantities otherwise inappreciable, that the complex refinements with which we are here surrounded become of the first importance. But happily this somewhat overwhelming complication is not of perennial growth, for, curiously enough, by a kind of natural compensation, it relieves itself. In reviewing from time to time the various aspects of a problem in connection with the instrumental appliances designed for its solution, the essential features come out by degrees more strongly in relief. One by one the unimportant parts are cast aside, and the apparatus becomes reduced to its essential elements. This simplification of parts, this cutting off of redundancies, must not, however, be understood as detracting from the merit of the original devisors of the instruments so simplified ; the first grand requisite is to effect what is necessary for the solution of the problem, then follows the question whether it can be done more simply or by some better process.

And this leads me in the next place to advert for a moment to the advantages which may accrue to the cultivators of science, and through them to the nation at large, from a national collection of scientific apparatus. Through the liberality of our foreign neighbours, and through the exertions of our own countrymen, we have here a magnificent specimen, an almost ideal exemplar, of what such a collection may be. By bringing together in one place, and by rendering accessible to men of science generally, the instrumental treasures already accumulated, and constantly accumulating, we should not only portray in, as it were, living colours the history of science, we should not only be paying just tribute to the memory of the great men who have gone before us, but we should afford opportunities of reverting to old lines of thought, of repeating with the identical instruments important but half-forgotten experiments, of weaving together threads of scattered researches, which could otherwise be taken up again only with difficulty, and after an expenditure of much and irretrievable time.

Let me now turn for a moment to the other side of the picture. If the collection in the midst of which we are here assembled is an evidence of the valuable relics which still remain to us of the great men who have passed away, the circumstances under which some of them have found their way hither, and the vacant places due to the absence of others, are no less evidence of how much the preservation of such objects would be promoted by the establishment of a museum such as I have ventured to suggest. Many circumstances contribute to thrust into oblivion, or to put absolutely out of reach of future recovery, original apparatus. First, the paramount importance and immediate uses of an improved instrument or a new invention; next, in Government departments such as the Survey, the Post Office, &c., the imperative demands of the public service, which leave little or no time for a retrospect of the past; and if I may add a word from the experience of private individuals, the pressing calls of space and expense lead the possessors to throw away, or to utilise, by conversion of the materials to new purposes, apparatus which has done its work. I venture to particularise one or two considerations, which will probably have occurred to many of you, but which appear to me to illustrate the above remarks. In the case of the Ordnance Survey it is almost certain that the current work of the department would never have required, and it is doubtful whether any private interposition would have brought about the removal of the disused instruments, here exhibited, from the cellars at Southampton. Again, the Post Office would hardly have been justified in devoting valuable time to the arrangement, or valuable space to the storage, of instruments no longer on active service, except at the call of a public department, or for a public purpose. And surely it would be a matter of serious regret that the time already spent upon the collections now before us should have no issue beyond the purposes of the present exhibition. To take another instance; we have here fragments, but only fragments, of Baily's apparatus for repeating Cavendish's experiments; but of Cavendish's own apparatus we have simply nothing. Again, Wheatstone's instrumental remains must inevitably have been broken up and scattered or destroyed, if there had not been found at King's College a resting-place, and authorities intelligent enough to appreciate and willing to receive them. Of other individuals from whom apparatus, now of historical interest, has been received, some from sheer lack of space have been breaking up old instruments, while others, from a modesty commendable in itself, were with difficulty persuaded of, and even now are only beginning to perceive the value, in a national and cosmopolitan point of view, of their own contributions. Lastly, there is, I think, little doubt that, if the objects in question were to go a-begging, they would be gladly received in some of the foreign museums which have so liberally contributed on the present occasion.

To put the suggestion in a more tangible form I would venture to suggest that, in the first instance, instruments whose immediate use has gone by, but which are nevertheless of historical interest, lent either by public departments or by private individuals, might remain here on permanent loan; further, that other instruments as they pass out of active service—for example, from the Admiralty, from the Board of Trade, from the Ordnance Survey, or from the other departments—should similarly find a place in this museum. In such a category also might be included the scientific outfit of the "Challenger," and of the Arctic Expeditions, and likewise those of expeditions for the observations of the transit of Venus or of solar eclipses. To these might be added apparatus purchased for special investigations through the parliamentary grant annually administered by the Royal Society. And further if, as I would suggest, this deposit of instruments be made without alienation of ownership, then private societies or even individuals might be glad to avail themselves of such a depository of instruments not actually in use.

In making such a suggestion, it must of course be assumed that the custody of property so valuable in itself, and so delicate in its nature, would be confided to a curator thoroughly competent for such a charge, but I abstain from entering prematurely into further details.

And now let me turn in conclusion to one more aspect of this great undertaking. We have here collected not only the instruments which represent the most advanced posts of modern science, but we have not a few of the men whose genius and perseverance have led the way thither; men who stand in the forefront of our battle against ignorance and prejudice and against the host of evils which a better scientific education must certainly dispel; we have men whose powers are competent for, and whose very presence is an inspiration to further progress. But, while taking this first opportunity of offering them a hearty welcome, I shall however best consult their feelings and your wishes by abstaining from any panegyric upon them in their presence, and by giving them an opportunity of speaking, and you of hearing them, upon some of their own subjects in illustration of the remarkable instruments which they have with so much pains and trouble brought under our view.

Mr. WARREN DE LA RUE, in proposing thanks to the President, strongly expressed a hope that the collection might be the nucleus of a permanent one.

The addresses arranged for the day were then delivered as follows:

ON SPECTROSCOPY APPLIED TO THE HEAVENLY BODIES OTHER THAN THE SUN. BY WILLIAM HUGGINS, D.C.L., LL.D., F.R.S., *Corresponding Member of the Institute of France.*

There is not much that is very new in the subject on which I have been asked to speak. I will give, therefore, a short summary of the methods and present state of this part of science. It is, too, not inappropriate in this great collection of instruments to say something of the methods and instruments by which the results have been obtained.

When the spectroscope is to be applied to the heavenly bodies, the first consideration which presents itself is as to the method by which the image of the apparently moving heavenly body shall be made to remain steadily under observation. This effect of the moving platform on which the astronomer finds himself may be counteracted in two ways, either by mounting the telescope on an axis parallel to the earth's axis of rotation, so that the telescope will follow the star by a uniform motion given to it by a driving clock, or by giving suitable motions by clockwork to a large plane mirror placed in front of the telescope, which may then be immoveable, so that the light of the star as it travels from east to west will be always reflected in the same direction, and the star's image consequently remain stationary at one place. This latter instrument, called a heliostat or sidereostat according as it is applied to the sun or to the stars, is well represented in this collection by a beautiful apparatus designed by Colonel Campbell and constructed for him by Mr. Hilger.

When by either of these methods we have obtained a brilliant and stationary image of the heavenly body, it then becomes necessary to interpose somewhere in the path of the light, a prism or prisms by which it may be separated into the different kinds of light which come to us bound up together in the star's light. A prism of small angle may be placed before the object-glass of the telescope. This plan possesses some advantages, but the size of the telescope is limited by the size that can be given to the prism. The other and more usual method consists in subjecting the light to the separating power of the prism after it has been condensed by the object-glass or mirror. If a cursory general examination of the spectra of *stars* only (the images of which are points of light) is required, a very excellent form of apparatus consists in placing a direct vision prism immediately in front of the field lens of a negative eye-piece. When the eye-piece is withdrawn for a short distance, the spectrum is seen well defined and of sufficient width without the use of a cylindrical lens.

If measures, and especially comparisons with terrestrial spectra are desired, and if the objects, as the nebulæ and planets, have images of sensible size, it is best to employ a complete spectroscope. Several forms of this instrument by Mr. Browning are in the exhibition. The slit of the instrument is placed exactly at the spot where the image of the heavenly body is formed. The diverging rays are brought parallel in the usual way and the spectrum is viewed in the small telescope of the instrument.

Having by one of the above methods obtained the spectrum of a heavenly body, we have to seek the best method of applying exact measurement to the spectrum. This may be accomplished by some form of micrometer, or by simultaneous comparison with a known terrestrial spectrum. If the spectrum is sufficiently bright, webs or wires may be moved along it by a screw, and the relative positions of the lines obtained. If the spectra are faint, some method of illuminating the wires may be employed, or the continental method of forming by the side of the spectrum the image of an illuminated scale. Zöllner has applied the principle of the divided object-glass to the small telescope, with the addition of a prism, by which one spectrum is inverted relatively to the other, and the distance to be measured doubled. Perhaps the most convenient form of bright line micrometer is one recently invented by Mr. Hilger.

For the purpose of accurate comparison with terrestrial spectra the use of the ordinary little reflecting prism, before the slit, is not sufficiently trustworthy, unless some special arrangements are adopted. It is obvious that, unless the light from the terrestrial source comes upon the slit in precisely the same direction as that from the stars, the method fails in accuracy. Practically, the introduction of the spark of vacuum tube in the axis of the telescope, at a distance of about two feet from the slit, gives sufficient accuracy of identity of position of the two spectra. The most perfect method consists in causing the image of the spark, by a suitable arrangement of lenses, to fall precisely at the spot where the image of the celestial body is formed.

It is now time to state in a few words the principal results of spectrum analysis applied to the heavenly bodies other than the sun.

We have learned that the planets Mars, Jupiter, and Saturn, have atmospheres not very dissimilar probably from our own. The spectra of the more distant planets Uranus and Neptune indicate atmospheres of a wholly different constitution.

This analysis has taught us that the fixed stars are truly suns after the order of our own. The spectra of the stars are not precisely similar to the solar spectrum. They differ too, greatly, from each other. Roughly, the spectra of the stars may be arranged in some four or five divisions from the brilliant Sirius to telescopic red stars. These differences of the spectra point to different degrees of temperature, and it may be, also, to some differences of chemical constitution. It is evident that great differences of temperature would be sufficient to give rise to different chemical conditions of the investing atmospheres of the stars, even on the supposition that the stars do not really differ greatly as to the substances which compose them.

The spectroscope has revealed to us the true nature of the nebulæ. This diagram represents the latest results of a comparison of the lines of the nebula with terrestrial spectra.

Of the four bright lines, the two more refrangible ones appear certainly to show the presence of hydrogen. The brightest line appears in a small instrument to be coincident with the brightest line in the spectrum of nitrogen. A more critical examination shows the nebular line to be single, while the nitrogen line is double. The line of the nebula is sensibly coincident with the less refrangible of the components of the double line of nitrogen.

Spectrum of the Nebula in Orion.

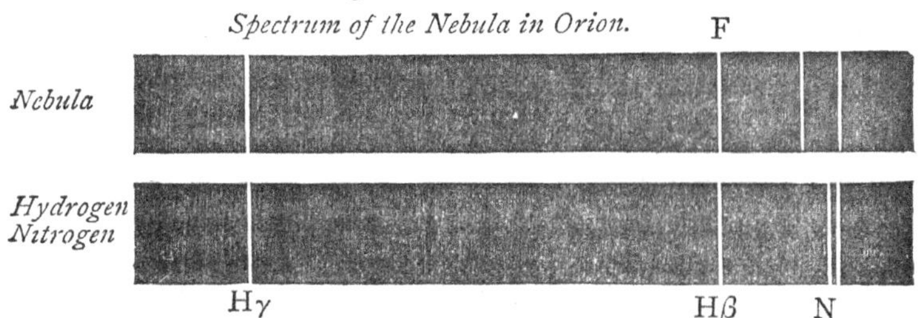

The same method of research has told us, that, a part at least of the light of comets is emitted by the cometary matter in a state of gas. Further, that this matter contains carbon in some form, probably combined with hydrogen.

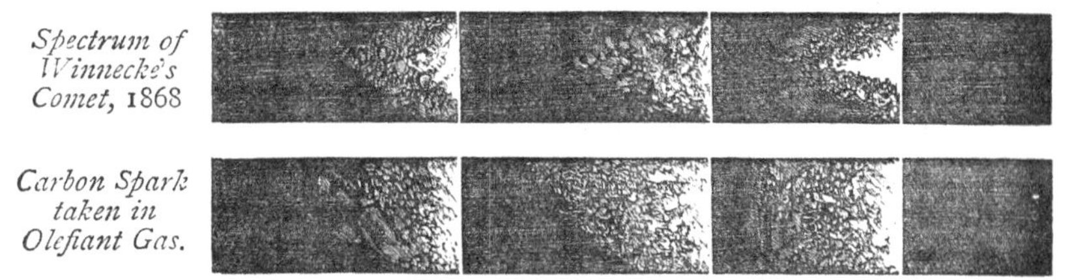

When the presence of a terrestrial substance has been established in a star, then it is possible by a more critical comparison with greater dispersive power to find out whether the star is in motion relatively to the earth in the line of sight.

If a small shift in the stellar lines is seen towards the more refrangible end of the spectrum, then we know that the star is approaching the earth. If the lines of the star are shifted towards the red, we learn that the star and the earth are receding from each other. Further, an exact determination of the amount of alteration of wave-length of the lines, will tell us the velocity with which the star is moving in approach or recession relatively to the earth.

The line F in Sirius compared with the line of hydrogen shows that this star is receding with a velocity of 25 miles per second.

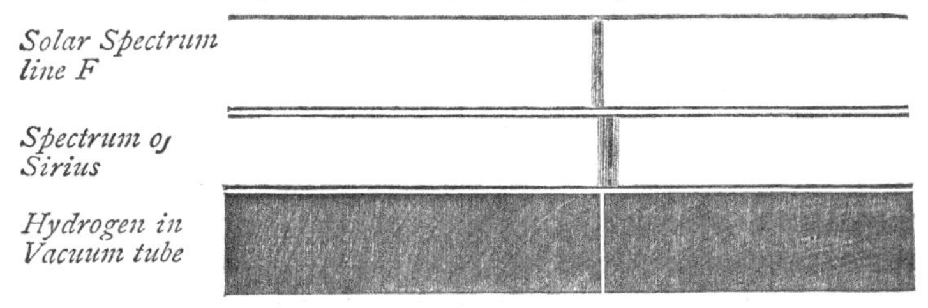

In the case of Arcturus independent comparisons were made with lines of magnesium, hydrogen and sodium. The lines in the star corresponding to "B," "F" and "D" of the solar spectrum were all found to have a small shift towards the blue end of the spectrum as compared with the bright lines of the substances to which they are due. This shift of the spectrum shows that the star has a motion of approach. These comparisons are represented in the accompanying diagram.

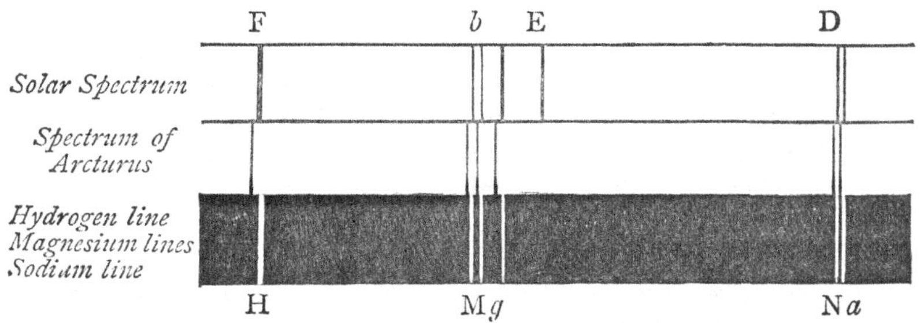

In a similar way, the motions in the line of sight of about 21 stars were ascertained. It is obvious that by combining these motions with those at right angles to the line of sight obtained, by ordinary astronomical observation, the true motions of the stars may be discovered. The latest contribution to this part of science was made by Mr.

Christie at the meeting of the Royal Astronomical Society on Friday, May 12th. His paper contains the following table of comparison of the recent application of this method at Greenwich.

A COMPARISON OF STAR MOTIONS IN LINE OF SIGHT.

+ = RECESSION. — = APPROACH.

STAR.	HUGGINS.	GREENWICH.	STAR.	HUGGINS.	GREENWICH.
α *Andromedæ*	—	—	α *Ursæ Majoris*	—46-60	—40
Aldebaran	+ ?	—	β *Leonis*	+ ?	—
Capella	+	+20	*Spica*	+	+
Rigel	+	+	γ *Ursæ Majoris*	+ ?	—
Betelgeuze	+22	+	*Arcturus*	—55	—35
Sirius	+18-22	+25	*Boötes*	— ?	— 8
Castor	+23-28	+25	α *Coronæ*	+	+ ?
Procyon	+	+43	*Vega*	—44-54	—37
Pollux	—49	—	α *Cygni*	—39	—50
Regulus	+12-17	+30	α *Pegasi*	—	—27
β *Ursæ Majoris*	+17-21	+20			

Mr. Christie concludes his paper by saying—"Notwithstanding the difficulties (connected with the methods of observation) it is gratifying to find that out of the list of 21 stars, which have been observed both by Dr. Huggins and Mr. Maunder (the observer at Greenwich), there are only two cases of discordance, as will be seen in the table ; and for both of these stars Dr. Huggins has expressed himself dissatisfied with his observations, whilst the Greenwich results rest on too few observations at present."

Mr. Christie informs me that since his paper was written, the motion of the planet Venus has been observed by means of a small displacement of the Fraunhofer lines of the sun's light reflected from that planet.

I am now engaged in applying photography to the spectra of the stars. This investigation may throw much light on the relative temperature of the suns and stars, and on some other important points of astronomical physics. The investigation is at present too incomplete for me to give any statement of the methods, and of the results obtained.

Mr. J. NORMAN LOCKYER, F.R.S., gave some account of the present state of spectroscopic research as applied to Solar and Molecular Physics. He confined his remarks to a few points in connection with the instruments exhibited relating to the construction of a new Normal map of the solar spectrum, together with a perfectly purified map of the metallic spectra, showing which lines of these latter are coincident with dark lines in the solar spectrum and which are not, and thus confirming the presence of elements suspected to exist in the sun's reversing layer by previous observers, and proving the presence of numerous others. Moreover, there are considerations which point to the existence of solar elements which are quite foreign to, or have not yet been discovered in this world. Mr. Lockyer exhibited the maps which have been constructed for this purpose, and, up to the present time new maps have been constructed for the region of the solar spectrum comprised between wave-lengths 39.000 and 43.000. The metallic spectra for the whole of this region are not yet completed, as the amount of labour involved in their production is very considerable. However, the spectra of about one-half of the metals for the first two sections, viz., wave-lengths 39.00-40.00 and 40.00-41.00, are nearly completed, and these are, in all probability, *absolutely pure.* The method of purification was then stated. Here the lecturer observed that he was glad to have that opportunity of expressing his obligations to Corporal Murray of the Royal Engineers, who had most efficiently prepared the enlarged maps, which were finally photographed down to the size desired. The enlarged maps are drawn on twelve times the scale of Ångström, and the final photograph will be reduced to four times, this being none too large to ensure the necessary amount of detail. To show how vast an improvement the photographic map exhibits over that of Ångström's, Mr. Lockyer stated that whereas in Ångström's map there were only *three* lines between the H_1 and H_2 lines (of the solar spectrum), the present maps showed *ninety-nine.* The lecturer proceeded to give some details of the manner in which photography had been utilized in the research. The instrument used consists of a spectroscope constructed on the model of Bunsen and Kirchoff's, with a train of four flint-glass prisms, 3 of 45° and 1 of 60°. The observing telescope is replaced by a camera provided with a simple quartz lens of about 6 feet focus. By these means the image of a tolerably large portion of the spectrum is thrown upon the photographic plate at the end of the camera, and a good focus for a part of the spectrum, at any rate, ensured. The novel feature of this spectroscope consists in the adaptation of a sliding shutter to the slit. It is well known that the width of a spectrum depends upon the extent of the slit exposed to the light, and this shutter admits light only through a very small portion of the whole slit. By successively exposing adjoining portions of the slit to the light emitted from the incandescent vapours of metals whose spectra it is necessary to confront with one another for the purpose of purification, the final effect on the sensidized plate is that two or more spectra successively record themselves thereon. All superposition of spectra and gaps between them are guarded against by the following means. The brass shutter with the square opening, slides in grooves in front of, and up and down the slit. On one side of the shutter in one of the grooves, holes are bored, the distance between each hole being the same as the height in the opening of the shutter. A short pin fixed to a spring falls into each hole in succession. The scale of measurement adopted is a wave-length scale. In the first instance the solar spectrum as obtained by the means here employed, being a refraction spectrum, is reduced to a wave-length spectrum by means of curves of graphical interpolation; then the spectra of the metals being photographed side by side with that of the solar spectrum, the measurements of the metallic lines may be deduced from their position in relation to the dark lines of the solar spectrum. Mr. Lockyer then mentioned that observations made during his investigations led him to the conclusion that different molecular aggregations produce five different kinds of spectra, viz.:

1. Line spectrum.
2. Channelled-space spectrum.
3. Continuous absorption at the blue end.
4. Continuous absorption at the red end.
5. Continuous absorption.

In fact modifications in the molecular construction of bodies would produce a corresponding modification in the spectra themselves. Mr. Lockyer also referred to some interesting experiments which he has made in connection with the spectrum of calcium. When, for instance, the chloride of calcium is subjected to a low temperature, we obtain a line in the blue part of the spectrum, together with a nearly complete spectrum of the chloride. This blue line is the blue line of calcium. As the dissociation of the chloride progresses, the blue ray becomes more brilliant and the chloride spectrum gradually disappears. If, instead of the comparatively low temperature hitherto employed, we now make use of that of the electric arc, the [result is that this blue line becomes extremely developed, and two additional lines in the violet make their appearance, corresponding in position with the two H lines in the solar spectrum. Now, in the sun, these two H lines are the thickest in the whole spectrum, and the blue calcium line before named is comparatively thin. The conditions in the spectrum of calcium in the arc are just the reverse of this, the blue line being very much thicker than the two violet lines. These facts suggested to Mr. Lockyer that between the temperature here employed and that of the sun, there should be a difference which affects the spectrum of calcium in a corresponding manner, as the difference in temperatures at our command affects the spectrum of the chloride of calcium. To verify this, Mr. Lockyer made direct experiments upon calcium at different temperatures, using for this purpose a large induction coil and interpolating Leyden jars of different sizes, and by these means he obtained at the lowest temperature the blue line without any trace of the two violet lines, whilst at the highest the two violet were present and the blue entirely absent. From these results, Mr. Lockyer argues that here we may have a dissociation of calcium itself, seeing that a parallel exists between them and the changes taking place during the dissociation of the chloride. In conclusion, Mr. Lockyer justly observed that this work was one which would have taken a single individual very many years to accomplish,but as portions had been taken up by other workers at Owens College, Manchester, and at Potsdam, a completion of this important work will be brought about in much less time.

Professor SORET (of Geneva) recalled that he gave, two years ago, the description of a spectroscope with a fluorescent ocular glass (*Bibliotheque Universelle Archives*, Sc. phys. et nat. 1874 t. xlix., p. 338.) The essential disposition consists in placing at the focus of the telescope of an ordinary spectroscope, a thin plate of a transparent and fluorescent substance, such as glass of uranium or a layer of a solution of esculine between two lamels of glass. The ultra-violet spectrum is formed upon this plate, as in the celebrated experiments of Professor Stokes, and it can be perceived with an ordinary positive ocular glass, inclined towards the axis of the telescope.

This ocular glass was first applied to a spectroscope of which the prisms and the lens were made of ordinary optical glasses. Under these conditions it is hardly possible to perceive rays more refrangible than N.

It remained to be seen whether the apparatus would work satisfactorily if use were made of prisms and lenses of quartz or of Iceland spar, neither of which absorb rays of great refrangibility.

In one first arrangement, M. Soret followed the system of direct vision employed by Herschel and Browning, the two prisms made of Iceland spar being so cut that the crystallographic axis should be parallel to the edges. The lenses are made of quartz. By this apparatus the ultra-violet solar spectrum can be seen to the line R; but as much light is diffused (proceeding partly from extraordinary rays, which are not included in the field of the spectrum) diaphragms must be adapted to the tube for the observation to be easy. And besides this, combination is somewhat difficult to carry out practically: it is necessary that the prisms should be extremely well cut and adjusted with the greatest precision.

The second arrangement simply consists in fitting a prism of Iceland spar, in which the edges are parallel to the axis, to an ordinary spectroscope furnished with lenses of quartz.

By concentrating the light of the sun upon the slit of this spectroscope by means of a quartz lens with a long focus, the solar spectrum can easily be distinguished as far as S; if use is made of the light of the Voltaic arc between different metals, the very refrangible brilliant lines can be most easily perceived in it. Thus with cadmium all the lines pointed out by Mr. Mascart can be seen up to the 25th.

With regard to the nature of the fluorescent substance on which the spectrum presents itself at the focus of the lens, a solution of esculine in water seems to be the most favourable for the observation of the part between H and O. For radiations of greater refrangibility, glass of uranium is preferable.

The spectroscope with a fluorescent ocular glass, without having, perhaps, the precision of the photographic method, has nevertheless great advantages for rapid observations. It allows of angular measurement being taken, if, beforehand, two lines, in the shape of a cross, have been drawn on the fluorescent plate, which would fulfil the functions of a reticule. It can be used for a great number of researches. MM. Soret and Sarasin have employed it for measuring the rotatory power of quartz ; their observations—which are not yet ended—go as far as the line R.

But it is chiefly to the use of this apparatus for astronomical observations that M. Soret wishes to draw attention: the study of the ultra-violet spectrum, in the centre or on the edges of the solar disc, in the protuberances and in the spots, would be most easy with this instrument, especially by adapting it to reflecting telescopes ; it would no doubt lead to interesting results.

On Spectrum Microscopes and the Measuring Apparatus used with them. By H. C. Sorby, F.R.S.; Pres. R.M.S., &c.

The object of the author was to exhibit and explain the various kinds of apparatus shown in the Exhibition, that had been contrived to examine and measure the spectra of small coloured objects seen under the microscope. The first form was that described by the author in the *Quarterly Journal of Science* for 1865, vol. ii., p. 198, in which the slit was placed some distance from the microscope, and the prism under the achromatic condenser, so that a small image of a spectrum was seen in focus at the same time as the object on the stage. The characteristic spectrum of the object was then shown by the absorption of particular rays. In this form of apparatus the measurement of the spectra was effected by means of a micrometer placed in the eye-piece. The original apparatus is exhibited. It served very well to prove that a wide field for research would be opened out by the further use of the instrument, but it was soon found to be very inconvenient not to be able to observe the spectra of very imperfectly transparent or opaque substances. This led to the adoption of a spectrum eye-piece, in which the slit is placed in the focus of the eye lens, and compound direct-vision prisms are placed over it. They can thus be taken off, and the object seen through the opened slit, and then on closing up the slit and placing on the prisms the spectrum of the object can be seen. A reflecting prism and side stage enable the observer to comparé two spectra together. Eye-pieces of this kind are exhibited by Mr. Browning and Messrs. Becks, and another by Mr.

Pillischer, in which the prisms can be pushed on one side, and the object seen through the slit without taking off the prisms. These spectrum eye-pieces can be very conveniently used with a binocular microscope, since the natural object can be found and examined by means of one tube, and its spectrum seen by the other tube.

For the sake of compactness in a travelling microscope, a moveable slit fitting into the eye-piece like a micrometer may be used, as exhibited by Messrs. R. and J. Beck.

All these eye-piece methods have the advantage of enabling us to examine the spectra of very minute objects, but the author has found that when they are of moderate size, it is more convenient and less trying to the eyes to make use of the binocular spectrum apparatus, described in his paper in the Proceedings of the Royal Society for 1867, vol. xv., p. 433. The slit and small reflecting prism are placed at the focus of a special object glass of low power, and the direct vision prisms are fixed between the slit and the lens. With this arrangement, it is, however, necessary to insert a cylindrical lens to make the spectrum in focus at the same time as the line of division between the two spectra which are compared side by side. An apparatus of this kind is exhibited by Messrs. R. and J. Beck.

For the measurement of the position of the absorption bands seen in spectra, the author has contrived and regularly used for many years a standard spectrum produced by means of a plate of quartz, cut parallel to the principal axis, placed between two Nicol's prisms, as exhibited by Messrs. Becks, which gives twelve well-defined bands, spread regularly over the spectrum, as described in the paper in the Proceedings of the Royal Society already named. For quick and moderately accurate measurements this plan is very convenient. The chief objection is the difficulty of so preparing the plate of quartz, as to give the bands exactly in one uniform position in every scale. Mr. Sorby, however, proposes that in all cases the position of bands should be expressed in wave-lengths; and if each observer made a table of wave-lengths for the quartz scale used by him, perfect identity in the position of the bands is of very little importance.

Mr. Browning exhibits the bright slit micrometer made by him to measure the position of absorption bands, as described in the *Monthly Microscopical Journal* for 1870, vol. iii., p. 68. For certain purposes this is very convenient, and with some modifications suggested by the author may probably be much improved. Each observer should construct a table applicable to his own apparatus, and describe the results in wave-lengths.

The chief objection to such a form of instrument is the absence of a fixed datum, and the necessity of verifying the adjustment by reference to the sodium D-line, and also the fact that the measurements vary with the focal adjustment. In order to overcome these difficulties, and also to be able to measure with greater accuracy than is possible with the above-named quartz scale, in which the bands are fixed, the author has contrived an instrument, exhibited by Messrs. R. and J. Beck, fully described in a paper in the *Monthly Microscopical Journal* for 1875, vol. xiv., p. 269. This consists of a piece of quartz, 1½ inch long, cut and mounted so that the light passes in the line of the principal axis of the crystal, along which there is no double refraction, but circular polarisation. This is mounted between two Nicol's prisms, one of which can be rotated along with a graduated circle, and the other turned for adjustment. This gives a spectrum with seven black bands, each of which moves into exactly the position of the one above or below for each half revolution of the circle, which is so graduated that it can easily be read off to one one-hundredth of this half revolution. The author has constructed a table of wave-lengths, as ascertained by using a diffraction spectroscope. By means of this new apparatus it is easy to measure the position of the centre of well-marked absorption bands to within one-millionth of a millemetre. The chief objection to the apparatus is that it is somewhat large, but by placing it under the stage in the fitting made for the condenser, using the binocular spectrum arrangement, and placing the object in front of the reflecting prism, very excellent results can be obtained. It must, however, be admitted that it will be of use more for ascertaining the general laws of spectra and the value of minute differences, than for carrying out the ordinary kind of qualitative analysis of colouring matters for which the spectrum microscope is so well adapted. For this kind of research various branches of biology furnish an almost boundless field, and promise most valuable results.

Professor CLIFTON, F.R.S., after giving a sketch of the history of the discovery of phenomena of interference of light, and of the progress made in the investigation of this branch of optics, drew attention to some of the instruments in the Exhibition, intended to facilitate the illustration of this class of phenomena, and to enable the conclusions of the undulatory theory of light to be compared with the results of observation. He further described the method of adjusting the *optical bench*, as arranged by himself, for repeating accurately the investigations of Fresnel, Young, and others, and mentioned some results obtained with this instrument in illustration of the close agreement between observation and the deductions from the undulatory theory. He also drew attention to the assistance which photography may render in the study of interference phenomena, and exhibited some photographs in illustration of his remarks.

The PRESIDENT: I will now ask M. Pictet to give an account of the Sulphurous Acid Ice-Machine upon R. Pictet's Anhydrous System.

M. RAOUL PICTET: In order clearly to understand the working of ice-making machines, it is first of all necessary to explain the theory upon which they are founded. Cold is produced by the evaporation of a volatile liquid; all liquids—without a single exception—in passing from a liquid to a gaseous state, absorb a considerable quantity of heat, which is consumed by the molecular work; the constituent particles of the liquid are strongly drawn together by the attraction of cohesion, heat separates them, and thus does a great work. If, on the other hand, vapours be compressed in a receiver, these vapours will transform themselves into liquid, and will give out, during the process of condensation, the same quantity of heat as that absorbed in their first change of state. Thus, in theory, all liquids indiscriminately can, by their passage through liquid or gaseous forms, be used for the artificial production of ice. It is merely necessary to cause a volatile liquid to boil in a closed vessel, surrounded by the water which one wishes to freeze; a pump continually draws away the vapours which are formed and compresses them in a condenser, in which they are condensed by means of the temperature of a stream of water, and of pressure.

The mechanical theory of heat allows the exact relation to be established which exists between the work expended by the compression-pump and the cold produced in the shape of ice.

The great differences which exist in the products of various ice-machines are due solely to the practical considerations of which we will now mention the chiet.

Ether machines have to exhaust very rarified vapours, which have a tension of but a few centimetres of mercury. They consequently require immense cylinders for the pump. And, moreover, the relative vacuum of the refrigerator allows the exterior air to penetrate, by the slightest opening, into the interior of the apparatus, and thus completely frustrates the work of the machine. It is necessary to grease the piston of the pump, and this grease mingles thoroughly with the ether; this has the effect of diminishing the volatilizing power of this liquid. Finally, after many volatilizations the chemical state of the ether transforms itself into acid substances, which differ materially from sulphuric ether.

Accordingly, these machines do not work regularly in the factories where they are used.

The ammonia machine of Caré avoids many of these defects, inasmuch as ammonia is much more volatile than ether, but there is a greater difficulty in the use of the machine—namely, the great pressure which exists in the condenser and in the boiler. This pressure can reach from eighteen to twenty atmospheres in hot countries, and thus renders escape of gas inevitable and explosions much to be feared.

The grease, also, saponifies itself with the ammonia, and quickly transforms itself into soap.

It can be perceived from these few words that the practical manufacture of cold for industrial purposes requires certain particular conditions, which are not satisfied either by the ammonia or by the ether machines.

It was to fulfil the various requirements of this question that I introduced anhydrous sulphurous acid into the manufacture of ice.

This liquid plainly satisfies all the conditions requisite for carrying on the process of manufacture regularly.

1. This liquid boils, under atmospheric pressure, at a temperature of 10° of cold 15° Fahrenheit.
2. It never gives—even in the tropics—pressure greater than three

and a-half atmospheres. This is very feeble to that given by the ammonia.

3. This liquid cannot be decomposed, and has no action on metals or grease.
4. This liquid, in a gaseous form, is so good a lubricator that all grease is unnecessary.
5. By using this liquid all danger of fire or explosion is avoided.
6. This liquid can be produced at a lower price than that of ether or ammonia.
7. The price of a ton of ice, when sulphurous acid is used, is about seven shillings, all told.

These are the principal advantages gained by the use of anhydrous sulphurous acid in the manufacture of cold. I thank the Committee of the Exhibition for having allowed me to explain my process.

The PRESIDENT: It is perhaps already known to most present that the machine, which has been so well described by M. Pictet, is exhibited downstairs, and is frequently in operation, so that you can all see not only the results, but the actual *modus operandi.* I believe I may say M. Pictet will have much pleasure in explaining it again on the spot to any who may be desirous of understanding it.

I will now call on Sir William Thomson to speak on

THE PRINCIPLES OF COMPASS CORRECTION IN IRON SHIPS.

Sir W. THOMSON, LL.D., F.R.S.: Mr. President, Ladies, and Gentlemen, the principles by which a compass in an iron ship may be corrected so that it shall point to the true magnetic north in every position of the ship, were pointed out about forty years ago by the Astronomer Royal. He then shewed that by the application of masses of soft iron, and also of permanent magnets, in the neighbourhood of the compass, the whole disturbance produced by the iron of the ship could be annulled and the compass brought to indicate the true north in whatever position the ship's head might lie. In the Astronomer Royal's investigation, however, there was a certain assumption made regarding the magnetic induction, which could only be approximately true for substances of exceedingly small inductive capacity, and which was very far from being true for iron. The nature of this assumption was such that according to it a long bar of iron would have the same degree of magnetisation by induction whether its length is held along or across the lines of magnetic force. It seems strange that the widely divergent character of the actual phenomena from those which would result from this assertion did not strike him, his sole ground for making the assumption being that there were no means of calculating the difference between the amounts of effect in the two cases, and that, therefore, they might be taken, for the sake of the investigation, as being the same. This theoretical error was pointed out by Mr. Archibald Smith, and an exceedingly curious result as regards the effect of ships magnetism on the compass was also pointed out, when the deviation from Mr. Smith's theory was also taken into account. One point as regards the accuracy of the Astronomer Royal's method of calculation, which would touch on this very important theoretical point noticed by Mr. Archibald Smith, was that whereas, according to the Astronomer Royal's theory, the compass would point correctly on every course if corrected by magnets and soft iron placed in a certain manner, which he points out; according to Archibald Smith's theory there might be a constant deviation left uncorrected, so that for positions of the ship all round, the compass would point at a certain angle from the true north and always in the same direction. This is an exceedingly interesting result, and the consequences have been followed out with great care and worked out by an exceedingly curious and beautiful analysis, the method of doing which was pointed out by Mr. Archibald Smith, which has been followed now most ably for about thirty years in the Hydrographic Department of the Admiralty. The nature of the disposition of iron which should produce that peculiar effect discovered by Archibald Smith is essentially one in which the masses of iron shall have a certain skewness or want of symmetry in relation to the compass, so that when the compass is placed amidships, and when the iron of the ship is symmetrical, as it usually is, this peculiar term does not exist. Practically, as the compass is placed in most ordinary ships, the amount of this effect is so small that it may be regarded as practically insensible, and the indications of it, which are found in the records of the Hydrographic Department, may be considered as being in some degree due to inevitable errors of observation. Still, I think I may be borne out in saying that there is perfectly distinct evidence in the Hydrographic Office of the existence of this very remarkable result.

Having said so much, and having pointed out that, in the theoretical point of view, an exceedingly important correction was required in the Astronomer Royal's theory, I must also point out distinctly that the Astronomer Royal himself was conscious of the fact that his theory made an assumption which was not rigorously true. He justified it by the consideration that the results did not produce effects of considerable importance with respect to the practical problem. But there is just one thing in which this way of putting the Astronomer Royal's case must be accepted with considerable reserve. That is, that if any attempt were made on his theory to calculate beforehand what would be the effect of particular masses of iron, such as bars or globes, the theory would be found to be greatly at fault. But if we merely take the integral result of the whole iron of the ship, then we get precisely the same effect when the iron is distributed almost symmetrically, as it is in all ordinary ships, as if the particular assumption referred to had not been made. So that unless we wish to calculate beforehand the effect of particular masses such as bars and spheres, we may accept the Astronomer Royal's conclusion without reserve. In respect to the estimation of the nature of the effects produced by particular masses of iron in the ship, Archibald Smith's more complete theory has been very valuable. The Astronomer Royal's theory for instance could not have accounted for the great effects that may be due to magnetic induction upon a vertical mass of iron such as the stern post of a ship or a vertical staunchion. But setting that aside altogether, and leaving what no doubt the Astronomer Royal was ready to admit was a subject for special investigation, the effect of bars and so on, and the fact that they differ enormously in virtue of the mutual action of the different portions of the iron from what they would be were such mutual action insensible, then I have no more to say, but that the Astronomer Royal has given a complete and practical method for correcting the compass. When we come to some of the modern ironclads, such as the Inflexible, in which I believe there are turrets unsymmetrically distributed, that is to say a forward turret on one side, say the starboard, and an after turret on the port side, then the want of symmetry must produce considerable effects, and may even produce that remarkably curious effect, discovered by Archibald Smith. I need not say any more on this point, but I would remind those who have followed the investigations on this subject that the coefficient called by the letter A. in Mr. Smith's theory may be looked for as being of more practical importance in some of the modern ironclads with an unsymmetrical disposition of iron than it ever has been in any other ships hitherto. Before leaving the subject of this particular result, I may say it would be altogether corrected, and permanently, on the same ship, by shifting the needles round a little in the compass card, so that the magnetic axis of the needles should be 1°, 1½°, or 2°, or whatever it might be on one side or other of the true north. Then that same compass card would always shew precisely the same result as if this peculiar term called A. in Smith's theory did not exist. It has sometimes been supposed that correction in the lubber line would correct this error, but that is a mistake, it can only be done by a correction of the position of the needles in the compass card.

In respect to the principles of correction pointed out by the Astronomer Royal, in the first place, the actual magnetism of the ship's iron at any moment depends on two influences, the first being the magnetisation which the iron has acquired somehow or other in the process of manufacture. The hammering of bars of iron, if their lengths be in any other direction than perpendicular to the lines of magnetic north, tends to make them magnetic; hammering masses of iron when they are red-hot, and allowing them to become cold under the hammer, produces this effect to a very marked degree, but it is also produced even in hammering cold iron. The very hammering in of the rivets in fastening the cold plates of the ship's sides shakes the plates themselves in such a manner as to shake in a great deal of magnetism. It shakes up the molecular structure, and causes them to take magnetism due to the position in which they are when the rivets are being hammered in. It appears that there is a great deal of evidence of a very curious kind which has been collected by various writers on this subject, to which I shall refer presently, shewing the influence of the position in which the ship's head was when being built upon the magnetism which she is found to have when launched. Part of the magnetism thus hammered into the ship, as it were, is held loosely, and is shaken out again, but part of it remains as long as the plates remain attached to the frames, and the ship for twenty, thirty, or fifty years, or whatever may be the life of an iron ship, retains a very large part of the magnetism originally imparted to her by terrestrial magnetic influence under the blows and shocks which the metal experienced in the building. But besides the magnetism which I have thus referred to, and which has been called by the Astronomer Royal permanent magnetism and sub-permanent magnetism, and which has been called by Dr. Scoresby retentive magnetism, which differs from induced magnetism which comes and goes with the inducing force, there is a certain amount of magnetism depending upon the position of the ship at the moment. The magnetic influence of the earth may be resolved into two components, a vertical and a horizontal component. Whichever way the ship turns the vertical component remains unchanged, and hence, as long as she remains in one neighbourhood, or does not go to a different part of the world, she experiences a magnetisation due to the constant vertical component of the earth's magnetism. But, besides that, there is a variable magnetism due to the horizontal component which differs according to the position of the ship. Thus, when the ship's bow is pointing to the magnetic north, magnetism becomes induced throughout the whole of the ship, which makes the bow become a true south pole and the stern a true north pole. When she is turned broadside to the magnetic north, so as to point east and west, she experiences magnetism in which the side next the magnetic north has a true south magnetic polarity. Whatever permanent and sub-permanent magnetism there may be, this inductive magnetism comes and goes so far as we know independently of it, so that the change of magnetism due to turning the ship round is sensibly the same as if she had no permanent magnetism. It seems to me this is a subject which wants experimenting upon, and one reason for my wishing to bring this question before you is, that it seems to be one in which an impulse is just now wanting for experimental investigation. Among other points, it is very important to find whether the inductive magnetism is quite independent of the permanent magnetism retained by the position of the metal in question—to find whether a compass will, for instance, experience the same inductive magnetism when it is turned into different positions relatively to the earth's magnetising forces as it would if unmagnetised. We are almost without experimental information upon this subject. Some most valuable experiments are being made, and were communicated to the British Association Meeting at Bristol last September, and they are still being continued by Rowland in the New John Hopkins University at Baltimore, which promises to be the greatest institution for experimental investigation the world has ever seen. He is making exceedingly valuable investigations, the result of which will be most important with respect to the problem of correcting compasses at sea. In the meantime, so far as we know, inductive magnetism takes place quite independently of any permanent or sub-permanent magnetism of the ship.

The Astronomer Royal shewed that placing steel magnets in proper positions in the neighbourhood of the compass corrects perfectly the effect of the permanent and sub-permanent magnetism of the ship. You will readily understand how that happens. Suppose the ship itself to be a permanent magnet without change of magnetism, then as the ship turns round it carries, so to speak, its magnetic force with it, and if you apply a steel magnet, or set of steel magnets in the neighbourhood of the compass whose magnetic force is equal and opposite to the ship, the position being found by turning the ship round and adjusting the distance until you find that the magnetic force of the ship is exactly compensated by the steel magnets, then as you turn the ship round into all positions the effect of the permament magnetism of the ship will be annulled, and the compass will point just the same as if the ship were non-magnetic. The ship and the magnets exercise, in fact, a zero magnetic force upon the compass in its actual position. But there remains the effect due to the induction. At first, let us suppose that somehow the effect of the permament magnetism of the ship has been annulled, and that we have nothing but the effect of the magnetism induced according to the different positions of the ship to deal with. Now, it appears, that if the ship is symmetrical, the effect of the induced magnetism on the compass will be zero when the ship's head is north or south, and when it is east or west. You will see that readily. If you imagine the ship to be perfectly symmetrical, and the compass to be placed amidships, when the ship points due north it is made magnetic by its inductive influence, with its true south pole towards the north, and its true north pole towards the south. That magnet being perfectly symmetrical will exercise a force in the direction of the length of the ship, and will either augment or diminish the force that the needle experiences from the earth, but it will not alter the direction of that force. Consequently, the needle will point due north, the magnetism induced in the ship nothwithstanding. When the ship's head is east and west, notwithstanding the inductive magnetism of the ship, the perfect symmetry will still cause the needle to point due north. This leads to the simple way of correcting the errors in a symmetrical ship. We know that the error due to inductive magnetism is zero when the ship's head is north or south, or when it is due east or west; then let the ship be placed north or south, and then let the compass be made to point accurately by fixed magnets, then place the ship east or west, and by another fixed magnet so arranged as not to alter the effect of the first, let the compass be again made to point correctly. Thus, by correcting the compass when the ship is north and south, and again when it is east and west, we are perfectly sure that we have annulled the effect of the permanent and sub-permanent magnetism of the ship, provided always the iron of the ship is symmetrical.

It remains to correct the effect of the inductive magnetism of the ship's iron. Now, I must introduce two names. The first is "quadrantal error"—that name was introduced by the Astronomer Royal when he brought forward the subject forty years ago—and secondly: "semi-circular error," which was the name given by Archibald Smith according to the analogy of that used by the Astronomer Royal. The semi-circular error, is that due to the permanent and sub-permanent magnetism, whilst the quadrantal is that due to the inductive magnetism of the ship by the horizontal component of the earth's magnetism. It is called quadrantal, because it has a maximum value in the middle of each of the four quadrants, from north to east, east to south, south to west, and west to north. Thus, the quadrantal error is at its maximum when the ship's head is north-east, and is zero when the ship's head is east. Then it has a maximum in the other direction when the ship's head is south-east, it is zero when the ship's head is south, at the maximum again when the ship's head is south-west, but in the same direction as when it was north-east.

Then again, when the ship's head is north-west the quadrantal error is at the maximum in the same direction as when the ship's head was south-east. Let the quadrantal error be corrected in any one of the maximum points, and the compass will point accurately for any position of the ship. Get, first a north or south, then an east or west, and then any one of the four quadrantal points correct by the Astronomer Royal's method, and the compass will be correct for all positions of the ship. This exceedingly beautiful and simple method might seem at once to settle the whole of the problem of ship's magnetism and allow us to send ships to sea with confidence that the compass would always point correctly. But it would be a most dangerous confidence, because there are a number of points which in a short plausible statement such as that I have brought before you are altogether overlooked, and points which are of paramount importance in the practical problem.

First of all, can the correction be made perfect for any one place of the ship? If the process I have indicated be carried out, and the ship is sent to sea, will the compass be right within half-an-hour after the correction is made? It will not be right. It will be far wrong in actual compasses and modes of adjustment to which the method has been frequently applied, and is still very often applied. There is one little assumption at the beginning of the Astronomer Royal's method in his mathematical paper, which he makes in common with all mathematicians—that is, that the needles are infinitely small. But what are they in reality? From about 7½ inches long in the Admiralty standard compass, to 14 or 15 in some of the compasses to which this method has been applied. The consequence is, as discovered by Captain Evans, that, although the compass is quite correct in the north and north-east and east positions, and so on, there is an error amounting, if I remember rightly, sometimes to about 5° in the intervening points. In one of the compasses of the "Great Eastern," for instance, Captain Evans found in turning the ship round to all points, that, although the quadrantal error had been corrected perfectly enough for practical purposes, for the north-east point there were errors in the intermediate points having a maximum ⅛ part of the way round from the earth in one direction, and another error when another ⅛ part of the way round. These errors he called octantal errors. In conjunction with

Mr. Archibald Smith, the theory of the octantal errors was worked out and fully described in a paper communicated to the transactions of the Royal Society of England, and published about the year 1861, the title of the paper being *Effect produced in the deviations of the compass by the length and arrangement of the compass needles, and a new mode of correcting the Quadrantal Deviation.* This paper, by Messrs. Smith and Evans, contains the mathematical investigation of the theory of these octantal errors, and a very curious and interesting result is arrived at, according to which these errors are very much less with such a compass as the Admiralty standard, in which there are four needles instead of only two or one, as in many common compasses, and with the needles arranged in a particular way, the two ends nearest the north being 30° apart, and those on each side of them being, I think, 30° more, so that the four ends were placed at intervals of 30° from one another. With this particular disposition of the needles it was proved that the octantal error was theoretically very much less, and, by Captain Evans's experiments on the "Great Eastern," it seems that, when an Admiralty compass was substituted for the great compass with one needle which had been in the binnacle before, the error was corrected nearly enough at all points by the correctors as actually applied. That very dangerous kind of error then may be considered as being rectifiable, and thus, although we cannot have infinitely small needles, we can at all events have a correction which will be perfect, or practically perfect, for the ship in all positions by the correctors to which I have referred.

The quadrantal error has to be corrected by placing masses of soft iron on each side of the compass, in such masses and in such positions as shall, by the magnetism induced in them, counteract the effect of the magnetism induced in the ship. The natural history of ships of all classes, which has been worked out most admirably by the Admiralty Compass Department, in respect of magnetism, worked by the fullest development of the mathematical theory of Archibald Smith, with the most thoroughly business-like perfection of detail, and with great scientific accuracy of observation, and which has been going on for thirty years, has given us knowledge which is of inestimable value on this subject. Among other things it shows us the amount of quadrantal error we have to meet with in different kinds of ships. One fact it brings out is, that the quadrantal error is of such a kind that masses of soft iron placed not before and behind the compass, but on the two sides of the binnacle (the starboard and port side), will correct it. Where you have such an arrangement of iron in such a ship, you could find a place (not the usual place for a compass) in which the quadrantal error would be zero, or in which it would be of the opposite kind, so that a mass of iron placed before it and another behind it would be suitable for correcting it. But one point brought out in the natural history of ships is, that the quadrantal correctors must be applied on each side for every ship in any position of the compass.

There is one other point of considerable theoretical interest with respect to quadrantal corrections which has hitherto escaped investigation, and which has not yet been published. It is this, that even with the Admiralty compass and its perfectness with respect to escaping the danger of octantal errors there is another action which vitiates the perfect completeness of that correctness, viz., that the induction produced by the compass needles themselves upon the soft iron correctors produces a very sensible disturbance. I will tell you what I have in my mind about that. I found—to my surprise rather—some years ago, that out of $12\frac{1}{2}°$ of quadrantal error, corrected with the Admiralty standard compass by cylinders of soft iron of the dimensions recommended by the Compass Committee, only $6\frac{1}{2}°$ were genuine, and $7°$ degrees depended on the influence of the magnets of the compass card upon the correctors. That would not in the slightest degree vitiate the correction as long as the same compass is used and the ship remains in the same place. But if a compass card with weaker magnets were used, then the correctors would not be so efficacious, and if a compass card with more powerful needles were introduced the correction would be overdone.

Now I have to refer to one most important and valuable quality of the method of correcting quadrantal error, introduced by the Astronomer Royal, and that is, that when once made for a ship in any latitude, it remains perfect for that ship as long as the iron of the ship remains unchanged, to whatever part of the world she may go; but that is on the assumption that the needles are infinitely small, and of infinitely small magnetic moment, that last condition having never been stated. The state of the case shortly would be this: with regard to quadrantal corrections with an Admiralty compass, that, that which would be perfect and quite satisfactory, and quite useful practically, as long as the ship remains about the British Channel, or even about the British Islands; if she goes away to Labrador, where the horizontal component of the magnetic force of the earth is much smaller, the quadrantal error would be found to be greatly over-corrected; or if she went to the magnetic equator, where the horizontal force is much greater, the quadrantal error would be under-corrected. Thus, the most valuable quality of the quadrantal error is vitiated by the greatness of the magnetic moment of the compass card in the best of the compasses hitherto used at sea. Notwithstanding the fallacy in this respect, I should say that the quadrantal error, if it were not possible to get compass cards in which the defect would not exist, still the quadrantal error ought to be corrected; but there would be considerable awkwardness, and the possibility of being liable to error, by being obliged to vary the distance of the correctors according to the different latitudes. Thus, the octantal corrector would want to be brought nearer the compass at the equator, and moved farther away in high north or south magnetic latitudes. But, if we can get over the defect, and have the quadrantal error, as has always been stated hitherto in books and papers on the subject, perfect in all latitudes, it would be a very great advantage. It was on this account, and also to avoid the octantal errors that I was allowed a good many years ago to attempt to produce practical working compasses with very small needles, and I have now succeeded in obtaining an instrument which works well, as I may say, from actual experience at sea with needles, the largest of which is 3½ inches, and of which the magnetic moment is utterly inadequate to produce in any sensible degree the effect I have alluded to. This is the largest kind of compass card that I have as yet made, and it is really large enough for all practical purposes. I do not think that any sailor could say that these divisions are not big enough for him to see them, and certainly no navigating officer can say that he cannot see them when he is using a sextant every day, and reading the scale off to minutes. With this compass, correctors cannot be brought nearer than within about 6 inches from the centre of the compass.

The only reason for ever using a smaller compass than this, provided this works well, except for purposes of convenience, would be that if the quadrantal error to be corrected is very large, we may want to bring the correctors nearer than the dimensions of the compass card will allow. I have therefore, constructed a smaller one with an 8-inch compass card, the length of the needle being rather less than 3 inches. When there is a very great quadrantal error to be corrected, as in some of the heavy ironclad ships, the masses of iron required to correct it would be incoveniently great, and therefore, I have taken as a convenient size for such correctors a globe of iron 6 inches diameter which weighs 31 lbs. That is easily handled and put in position, and does not sensibly increase the cumbrousness of the binnacle. But if this is not sufficient I would rather diminish the size of the compass then increase the size of the globe, and therefore my practical rule would be this: if you want as large a compass as this, use it, unless it requires more than a 6-inch globe for quadrantal correcton. If the quadrantal error does not exceed 5° a pair of 6-inch globes will correct it, and with a compass of this size allows two such globes to be placed at the distance of 7 inches on each side; so that for any error less than 5° this sized compass is perfectly convenient. When the quadrantal error is anything between 6° and 11° I take a smaller size compass, allowing the correctors to be brought within 7 inches of the centre of the card, and if the error exceeds 11° and does not exceed 22° then 6-inch correctors are still perfectly available for a compass of this smaller size which has a 6-inch card. It really is not incoveniently small. It is as large as the steering compass in many ships in the Navy, and therefore there is not the slightest occasion for using incoveniently large correctors by having a larger compass than that. If you imagine a ship with about forty inches of iron sheathing such as the ships of the future will probably be, and with 200 tons guns, then perhaps even with this compass you might want a ton or two of iron on each side, but in such a ship perhaps even that would not be much thought of. It may be considered desirable however to have a somewhat smaller compass, and I have therefore, made a miniature one. Although so small it can be tossed about or rougly handled without being in the slightest degree injured. I will now briefly explain the compass card. I have an aluminium rim which is covered with a piece of paper gummed to it in order that the compass card itself may be gummed on to that. The card consists simply of a single thickness of paper, and I find it necessary in the larger sizes to cut the paper radially, because when it becomes heated by the sun it shrinks and tends to warp the rim into a saddle shaped surface. That is obviated by radial slits in the paper. The magnets are hung direct to the rim, and this specimen before you is the first in which there is an arrangement for a perfect adjustment of the magnets. I have got them out of position by throwing it about, but I can adjust them in a quarter of a minute, and before going to sea a little shellac could be applied to make the position permanent. The magnets are placed in a row like a rope ladder, being fastened by half hitches of silk thread around each magnet, like the ratlins of a rope ladder. By the aid of a straight edge I can very easily adjust the magnets exactly to the north and south line. The magnets are hung on the rim by a silk thread and lastly the rim is supported on the central boss by a silk thread also, that being put on under equal tension in such a way as to secure exact and constant centreing. The object is to get needles in a compass which will work well at sea, and for this purpose it is found that a certain slowness of period is necessary, accordingly I have a forty second period with this, and this will be as steady at sea in air as any compass can be, so that liquid compasses will be unnecessary. The other one has a period of thirty three seconds, and the smallest has the same period as the Admiralty compass, and therefore will have approximately the same steadiness at sea.

With regard to the other correctors, I shall only say that a perfect system of adjustment for different latitudes must be carried out. The Astronomer Royal long ago pointed out this, and many patents have been taken out for various methods of adjustment which have fulfilled the conditions laid down by the Astronomer Royal with more or less practical availability. I have endeavoured to introduce a system of adjustment which would be more absolutely and perfectly simple and ready in its use. I should have had great pleasure in shewing you a binnacle with this compass in it, but by some mischance it is not here. It is, however, in London, I believe, and you will have an opportunity of seeing it from to-day forward. The variable effect due to sub-permanent magnetism, varying, and to the different induction produced by the different amount of the vertical component of the earth's magnetic force in different parts of the world may also be corrected. The rule will simply be to keep the compass correct on whatever course you are steering, and once every three or four days, or a week, or a month, according to circumstances, put the ship on some other course, keep her on that course for a few minutes, correct the error on that other course, and then the quadrantal error having been once corrected originally, you may be perfectly certain that the compass will be correct, and will remain correct until either the sub-permanent magnetism of the ship changes or until she goes to a different magnetic latitude. It may be supposed that it is using too strong a term to say it will be perfectly correct, and I admit that mathematically there is no such thing as perfection. I mean as correct as it is possible for a compass to be with the correctors which we have in the ship. How small in fact can we get the practical error by applying the table of corrections? how near can we be sure of our result? Can we apply the table with any accuracy as supplied by the Admiralty? Can we do it within a quarter of a degree, within 1°, or within 2°? We can sometimes and sometimes we cannot, according to circumstances. Then we can be neither more nor less near, and neither more nor less sure, of the degree of nearness with which we obtain our result by the application of the correctors on the system which I am advocating, as by the application of the table of errors on the principles laid down by Mr. Archibald Smith, and practised thoroughly in the Admiralty, with the aid of the magnetic correctors since introduced by Captain Evans. I must apologise for having detained you so long, but I thought the importance of the subject warranted me in bringing it before you.

The PRESIDENT: I am sure we are all much obliged to Professor Thomson for the able manner in which he has brought this important subject before us, and I have no doubt it is well known to many of you that he is a bold British navigator himself, and therefore competent to speak on both the theory and practice of the matter; I beg to offer our best thanks to Professor Thomson for his communication.

CAPTAIN EVANS: In case an erroneous impression should be left by Sir W. Thomson's very able remarks upon this new error which he has found, arising from induction of the compass needle upon soft iron correctors, amounting in some cases to 7° or 8°, I should wish it to be understood that the Admiralty compass department have long recognized the difficulty of applying these correctors for the quadrantal deviation, and to avoid that source of error, or indeed any source of error that may arise, they have inculcated the practice, and I think it has been so far a very wholesome practice, of constantly observing the amount of error of Ships' compasses. Sir W. Thomson rightly gave it as a theoretical principle that when once this quadrantal deviation was corrected it remained permanent in all parts of the world. That which he described as the natural history of the ships of the navy we have found very useful, and we know from it the quadrantal deviation of every ship, and we also know this deviation remains constant in all parts of the world. By not correcting the quadrantal deviation we are enabled to see, knowing its constancy, that the compass has been properly attended to. In attending to its values as given by the officers in charge of the navigation from time to time as they progress over the globe there is a great check upon them. We have found it very useful in practice never to allow an officer to suppose that his compass is correct, or that it can be practically accurately corrected. Therefore from day to day they make careful observations, and it is something like a man with his watch, if he knows it is five minutes in error, he simply allows for the difference, and in the same way the quadrantal deviation is allowed for instead of being corrected. With reference to these beautiful compass cards which Sir W. Thomson has brought before us, I think I can see where they are likely to be of far greater use than in the ordinary navigation to which he wishes apparently to confine them. They appear to me to be likely to be of use in those great iron-clads which he so well described, where there are several apparently abnormal conditions, but which are in fact obedient to law. I think they might be introduced there, and probably we should have less trouble in correcting our compasses with the very small needle that he employs than with the ordinary compass. That leads me to another point. I consider Sir W. Thomson has done good in calling attention to the advantage of using short needles. When long ships were introduced both into the navy and the merchant service, the idea prevailed that you should have large compasses, but really it should be just the reverse, and practically for accuracy the larger the ship the smaller should be the compass. If I had my will a ship like the Great Eastern instead of having, (as she originally had,) a compass with a needle fourteen inches long, should have one of three or four inches, as being more conducive to her safety. It would take too long to enter upon either the theoretical or mathematical considerations bearing on this subject; but they are perfectly sound, and my opinion is confirmed by the views that Sir W. Thomson has expressed in introducing these cards to your notice.

The PRESIDENT: Dr. Wartmann, Professor of Natural Philosophy at the University of Geneva, will now give you an account of his experiments with the Radiometer.

M. ELIE WARTMANN, Professor of Natural Philosophy in the University of Geneva: In the address which he delivered at the opening of this conference, Mr. W. Spottiswoode recalled to mind the invention of the air-pump by Otto von Guericke. As soon as scientific men knew how to produce a vacuum, they set to work to observe what phenomena might manifest themselves in it. About the beginning of the 19th century, Sir Humphrey Davy proved that it is permeable to heat, which influenced a thermometer placed in the middle of the chamber of a barometer. Three years ago, Mr. W. Crookes, F.R.S., excited general attention by the curious instrument to which he has given the name of the Radiometer. It is known that it consists of a delicate mill, revolving freely in a glass vessel, in which the best possible vacuum has been produced. This mill has four arms supporting as many paddles (palettes), each one of which is painted black on one side and white on the other.

This pretty instrument has, in England, often been called the "Light Mill," and many persons are endeavouring to make use of it as a photometer to determine the intensity of the light produced by gas and by other means. This is the result of an erroneous impression which ought to be removed. It is by no means light, as such, which causes the instrument to revolve, but solely heat, either luminous or obscure. By exposing one of the blackened paddles to the focus of a large lens, or of a concave mirror, the light of the full moon passing the meridian may be concentrated on it without producing the slightest motion. And it can likewise be proved that when revolving under the influence of heat, given out by a properly arranged Bunsen's burner, the mill will keep up exactly the same rate of speed whether the flame be rendered luminous or not.

There has been much difference of opinion as to the cause of the rotation of the arms. To solve the difficulty I have made a great number of experiments, of which I shall mention but a few. The Radiometer is not affected by electrical sparks, produced outside and quite close to the globe in which it stands, however intense or often repeated they may be. But under a magnetic current its metallic arms are influenced, and its motion is either partially or completely arrested. If, by means of lenses of equal power (ouverture), the radiation of exactly similar moderator lamps be concentrated on both surfaces of the same paddle, a perfect equilibrium can be obtained by placing the white surface always much closer than the blackened one. *This equilibrium is established when the intensity of the radiations on each surface is in inverse ratio to their absorbing power.* The slightest change in the distance of, or in the degree ot power displayed by the lamps, is sufficient to set the machine in motion.

In a horizontal plane, and concentrically to the axle of the radiometer, let an iron ring, thirty centimeters in diameter, and heated red hot, be presented to the mill. The arms will immediately revolve with a velocity which becomes very great indeed, when the plane of the branches which support them is mingled with the plane of the ring. *The speed produced is the same, at equal distances from the ring, either above or below this plane.*

The radiometer, placed either in the air or in water, does not move when heat is applied equally in all directions. And this is the case even at the most elevated temperatures.

If strong solar heat be concentrated on the paddles of the mill, the mica, of which they are made, is partly split, and the instrument pervaded by a grayish smoke; from that moment it loses a very great part of its delicacy.

When the same radiometer is repeatedly exposed to the focus of an optical instrument, which collects the rays of a powerful lamp, an occasional deposit can be seen gradually forming inside, which is composed of a material that has been vaporized, and which at times assumes the appearance of sublimated crystals. I suppose it is produced by the impurities contained in the smoke-black, or in the alcohol used to lay it with a brush on the surfaces of the paddles; and as pulverized coal condenses vapours with an energy that is well-known, it is impossible to avoid coming to the conclusion that the vacuum produced by Sprengel's pump is by no means perfect. The pressure of the gases which have not been removed varies according to circumstances, and it is to the different degrees of this pressure that we must attribute the various motions of the instrument invented by Mr. Crookes.

The PRESIDENT: I do not know whether Mr. Crookes is present, but we should have been very happy to have heard anything from him on this subject, which has excited so much interest both here and on the Continent. If there is any other gentleman present who has any remarks to offer we shall be very pleased for him to do so.

I am afraid Lord Rosse, who had promised to take a part in this Conference, has been prevented by some means from coming, but Mr. Fletcher is here, and he will give us an account of his Anemometers.

Mr. FLETCHER: When an effort was made to examine chemical works with a view to carry out the provisions of the new Alkali Act it was found necessary to measure the amount of vapour, gas or smoke which was passing along the flues which entered the chimneys of the works, and as a first step it was necessary to find out the speed at which the gases were passing along. As there was corrosive vapour and soot, and in many cases flame passing with the air, it was impossible to use the ordinary Anemometer consisting of delicate mechanism of various kinds, a light fan wheel passing round and the revolutions being counted by delicately poised wheels. Of course, such an instrument once introduced into a flue which was red hot would not shew any very accurate indications when it came out again, therefore this Anemometer was devised to answer the purpose which was wanted. It consists of two tubes of ordinary gas piping which may be of any length. One is bent at right angles, and the other is cut of short. They are passed through a hole in the brick work and thus introduced in the current of gas. If the vapour is passing upwards the bent one would be turned down so as to receive the pressure into its open end, whilst the other one is crossed by the current. The action in the open one is to experience a slight vacuum or exhaustion, and a slight pressure is experienced in the bent one. By flexible tubing the other ends of these tubes are attached to a simple U tube, mounted with some little care, such as you see here, by which the pressure is measured. As the exhaustion of the chimney operates on each lime equally, the ether in the U tube will not be moved by it, but as one is drawn by virtue of the slight vacuum experienced, and the other pushed by virtue of the slight pressure, they both act in the same direction, and so the ether used in this U tube is moved, one limb being depressed and the other elevated. Affixed to the flexible tubes is a small switch or button, by reversing which, the limb which was first depressed is elevated, and *vice versa.* In using the instrument the operation will be to place the scales level with the surface of the ether in the first instance, then to reverse the button, and then move the scale to the new position and read again, and by deducting the one from the other, you get a reading which is double the amount of the elevation. Thus if the difference of level is $\frac{1}{10}$th of an inch you get $\frac{2}{10}$ths, and this double reading of course halves the error. The question then was to find out how this was to be connected with the speed, and for that purpose I made a mathematioal calculation which gave this formula, $V = \sqrt{p} \times 28.55$. I made a number of experiments with currents of air of known velocity for the purpose of checking this result. There was some difficulty to get currents of air of known velocity, but I accomplished it by getting a flue of 100 feet in length, connected with a chimney letting off a small flash of gunpowder at one end, and by means of a pane of glass, by which I could see through the flue, I counted the number of seconds which the smoke took in coming from the starting point to the place where I stood at 100 feet distance. Estimating the exact speed in that way, and testing it with the instrument, I got the constant 28.50., being only 5 from the figure given by calculation. I then calculated a table accordingly. I also made a table of corrections for temperatures. The first table gives the speed, supposing the air to be at 60°, but as the air in these flues is often as high as 500°, or 1000°, that also required a careful calculation, which I have given. I take this opportunity of shewing this instrument, thinking it may be useful to those who have to measure gases where ordinary Anemometers of the mechanical kind cannot be introduced. In this there is no friction whatever, as there are no moving parts. The column of liquid in the Anemometer attains its position, and is there kept by the constant pressure of the current of air. There is nothing to move backwards and forwards, and therefore friction is out of the question. I can measure the rate of air, supposing it is going at the rate of 4 feet a second, and can tell it is not going at the rate of 4 feet 1 inch a second. When it gets down to a very low speed, below 12 inches a second, its action is very feeble, and that is the weak part of the instrument, but when it gets up to anything like 4 feet per second you can tell with great accuracy.

THE PRESIDENT: I have now to convey the thanks of the meeting to all those gentlemen who have favoured us with communications this afternoon. It is remarkable to observe how subjects connected with light have come forward. They were dealt with this morning and have again reappeared for our consideration this afternoon. The mode in which the different branches of optics which turned up this morning, in the first place dispersion, then fluorescence, then interference and polarisation, keep cropping up here and there, shows how very much all these different branches of the subject, notwithstanding the care with which they are kept apart, as subjects of actual research, do flow into one another, and by their mutual interference throw light upon one another. The arrangement for our next Physical Conference, which takes place on the 19th, are nearly complete, but I hope the audience will always remember that we are indebted very much to the voluntary assistance of those who are good enough to make communications, and that our arrangements are, at all times, liable to some kind of disturbance. Nevertheless, I feel quite sure that we shall never lack for interesting matter during the Conferences.

SECTION—PHYSICS (including Astronomy).

May 19th, 1876.

The PRESIDENT: Ladies and Gentlemen, as it is now the hour of commencement I think I need make no preface whatever in the case of our first communication, but have only to mention the name of Professor Tyndall to insure an attentive audience. I therefore call upon Dr. Tyndall to give us his communication upon his remarkable experiments upon the Reflection of Sound.

Professor TYNDALL, D.C.L., LL.D., F.R.S.: If gas be permitted to issue from a nipple with a circular orifice, such as I now hold in my hand, and if it issue under a very low pressure and in a very calm atmosphere the column of unignited gas will rise to a height, or can be made to rise to a height of from eight inches to a foot or more. Great calmness on the part of the atmosphere is required to accomplish this, and if with the gas, smoke, or some other floating matter be associated you can see the column rising through the air to this height. If you augment the pressure a little very soon that column breaks up and you have the gas issuing from the nipple thrown into *tourbillons* or vortices. If you ignite that gas then those vortices instantly disappear, and a flame more or less tall arises from the burner. You can then go on augmenting your pressure, doubling it, trebling it, or even quadrupling it, and it may be even more than this, the flame at the same time becoming taller and taller. You can in this way, with appropriate gas, raise your flame to a height of eighteen inches or two feet. But if you go on augmenting the pressure you come to a point where the ignited jet also breaks up into those vortices, or *tourbillons;* the gas flares, to use a common term, but if, before the gas jet or flame reaches this point, you bring it just to the verge of flaring, then by an appropriate musical sound that gas is utterly broken up, and the flame which would burn silently without the action of the musical vibration at a height of two feet, will fall suddenly and break up and roar at a height of only eight or nine inches. Here is a flame issuing from a nipple such as I have described. The pressure is so arranged as to bring it near the point of flaring, and a very slight action of an appropriate sound upon the flame will bring it down, abolishing the light altogether. Thus, when I take this bunch of keys and shake them, the noise of the keys brings the flame down. In such a flame we have a re-agent of extraordinary delicacy as regards musical vibrations. We are indebted for the discovery of this action of sound upon flame to one of the most remarkable men in the United States, a man whom I am sorry to say was cut off by the war from active work in science for some time, but he is now engaged upon it again; I mean Professor John Le Comte. He observed this action of sound, and, moreover, he gave us the distinct intimation that the flame required to be brought to the edge of flaring in order to get this effect. You bring the flame to the edge of a precipice as it were, and the musical sound pushes it over. Subsequently, when I had the honour of his assistance at the Royal Institution, Professor Barrett observed this effect also upon a flame, and afterwards made various interesting experiments upon the subject. It has also been experimented upon by Mr. Philip Barry and by my present excellent assistant, and I have done something towards exalting the sensitiveness of the flame myself. It is to be our re-agent at the present time. Many of my continental friends who desire to repeat these experiments find some difficulty in doing so, because a full exposition of the proper conditions necessary to success has never yet been given. Hence, my reason for introducing the subject here. It is advisable to have a gas-holder that will enable you to apply considerable pressure, and the gas-holder before you is loaded in the manner you perceive, in order to bring the flame to the required proximity to its point of flaring. It is also desirable that all passages between the burner and the gas-holder should be fully open. This is a point of considerable importance. With regard to the seat of sensitiveness in this flame, the action is not caused by the impinging of the sonorous waves upon the nipple, it is not caused by the impinging of the sonorous waves on the flame higher up. The action occurs in the orifice. Converging the sonorous waves of a small vibrating reed upon the nipple, a little below the orifice, they produce no effect. Causing the waves to converge higher up, there is no action upon the flame. Converging the waves on the orifice itself, violent action occurs on the flame. We are to make this flame the test of the reflection of sound. When my excellent friend, your Chairman, mentioned the reflection of sound, he meant it to be from the limiting surfaces of gaseous layers, not the ordinary reflection of sound from the surfaces of solids or liquids. This is the peculiar feature of the experiments that these echoes will take place at the limiting surfaces of layers of gas. And here, I think, we may confine ourselves, inasmuch as I know there are many gentlemen coming after me, to one illustrative experiment. For the purpose of making that experiment we have this apparatus devised, I may say, by Mr. Cotterell, and executed by Messrs. Tisley and Spiller. Through a series of apertures, we might throw into this apparatus air saturated with various vapours; such, for example, as the vapour of ether. We should, in this way, destroy the homogeneous character of the air in this horizontal tube. But instead of making an experiment upon vapours, I pass on to one representative experiment which will enable you to infer the character of all other experiments made with this apparatus. Underneath this tube are apertures through which, when this series of gas flames are turned on, the columns of heated air from the gas flames will enter the tube. In that way, instead of having a homogeneous horizontal column of air, we obtain a column of air wherein heated layers alternate with cooler layers. The action of this upon the sound will be made manifest to you. At every passage of the sound wave, from the heated layer to the non-heated layer, there is a slight echo, and those echoes occur even in that short distance so frequently as entirely to waste the direct sound in echoes. It is precisely analagous to the action of a cloud or of foam upon light. You have there the mixture of two transparent substances—air and water—and in virtue of those repeated reflections at the limiting surfaces of both, the mixture of those two transparent substances becomes opaque. Foam becomes opaque, not in virtue of absorption, but of these repeated internal reflections. My assistant will now start his reed and regulate the sensitiveness of his flame. The flame, you observe, is steadied by the interposition of my hand between it and the tube. And now, instead of my hand, my assistant turns the series of gas flames round, the heated columns of air enter the tube, and that moment the flame is stilled as effectually as when I intercepted the waves with my solid hand. This experiment illustrates all of a similar class. We have operated with a great number of gases. We have here, and can use, if necessary, carbonic acid gas, hydrogen gas—anything, in fact, that will render the column of air non-homogeneous. I said, subsequently, to my assistant that we ought to be able, not only to shew the interposition of the sound waves by these different layers of gaseous matter, but by proper experiment to be able to make the echoes evident by their action upon a flame. Mr. Cotterell, in an exceedingly ingenious and beautiful manner, devised the means of doing so, and he will now make his experiment himself. Two tubes are placed so as to form a V; at the end of one of them is placed a vibrating reed, and opposite the end of the other a sensitive flame. The two ends here referred to are those widest apart. The sound from the reed passes down one tube, impinges on the broad flame of a batswing burner, and is reflected by it through the other tube to the sensitive flame, which is thrown into violent agitation by the waves reflected from the hot gaseous layer. As to the transmission of these sounds through different media, it is very desirable that we should be clear upon this subject, inasmuch as various theoretical notions have been entertained which had a very important practical bearing as regards the establishment of sound signals at sea. It was necessary to test these notions with the utmost strictness, to examine the permeability of various kinds of air and various states of the atmosphere for the waves of sound. Here is a vibrating reed, from which the waves of sound will pass through the tube, and act upon the flame. The interposition of my hand completely stops the action, and causes the flame to become quiescent. I will next try a pocket-handkerchief. The sound waves go through the handkerchief as if it were not there. I take a flannel, and you will observe that it has hardly any influence on the sound. I double it so as to make a thick blanket of four layers, and that is practically transparent to the waves of sound. Here is a thick woollen shawl, and you see that it has hardly any effect upon the waves of sound, they go through it as if it were not there. I have here a piece of close felt, impervious to the noonday sun; it has hardly any effect upon the waves of sound. They go through it almost as if it were not there. Here are a hundred layers of cotton net all sewn together, and you will see it has no sensible effect upon the waves of sound. Here is another hundred which I add to them, and you see the thing is sensibly transparent to the waves of sound. A piece of card-board, on the contrary, immediately cuts off the sound. A piece of oilskin, also, stops the waves of sound. Here is a piece of cambric, which is really like nothing as regards its action upon the sound waves. But the whole secret of it is, that these waves of sound possess the power in a most astonishing manner of getting through any substance in whose interstices the air is continuous. If you can suck the air freely through a body, the waves of sound will go through it. This piece of cambric, for example, has no sensible effect upon the waves of sound. I dip it into water, so as to cause it to be thoroughly wetted by the water, and when that is done you will find that its perviousness to sound ceases. We have here a film of water filling up the interstices of the cambric, and the result is that the sound waves are entirely cut off. Let us repeat this experiment the converse way. I put this cambric between bibulous paper and rub it, so as to take away the water from the interstices, and you see its transparency to sound is restored. I dip it again, and then it stops the sound. Therefore, as long as there is no solution of continuity in the air, the waves of sound possess this extraordinary power of permeation. You have seen the action of these layers of differently heated air upon the waves of sound. Sometimes the whole atmosphere is filled with these layers of air of different densities produced in part by heat and in part by different degrees of saturation of aqueous vapour, so that you have here what we may fairly call an acoustic fog in the air, uniformly diffused through it. You can go beyond that. These masses of non-homogeneous air sometimes drift through the air with definite boundaries, just as the clouds of the ordinary atmosphere drift over the blue sky; and if you simply exercise the patience of observing a bell or a clock strike with a certain definite force for a single week, sometimes for a single day, you are able to see with the mind's eye, those invisible accoustic clouds drifting through the atmosphere just as certainly as the clouds that you see floating before your eyes in the blue heavens. Standing, for instance at the end of the Serpentine, and listening to Big Ben, let it be twelve o'clock in the day, or twelve at night, you hear the first stroke of the clock with powerful force—the second with power, the third absolutely quenched, the fourth quenched, the fifth quenched, a feeble sound at the sixth, the seventh comes out with sudden and extraordinary power, the eighth with great power, the ninth and tenth are again silent ; and in this way by observing a clock, or bell, struck with a definite mechanical power you can realise to your minds the drifting of these acoustic clouds through the atmosphere just as vividly as the clouds you see with your eyes. Not only is this the case, but these accoustic clouds are also superposed upon ordinary clouds, and upon ordinary fog. You find this action upon bells during foggy weather, the fog itself having no sensible influence on the sound. Not by the most accurate measurements can any sensible influence be established with regard to fog, nor any sensible interception of the sound waves by the particles of fog. To follow up the parallel a little further, you see a cloud shone upon by the sun, and you sometimes receive dazzling white light from the surfaces of that cloud; those are the echoes, so to say, of the light reflected back from the cloud. In precisely the same way, with regard to these acoustic clouds, if you have your source of sound strong enough, you get an echo of a most extraordinary character. Put yourself in front of an acoustic cloud, and you are able to detect by reflection the very sound that had been refused transmission. I am sure you will find no difficulty in believing, after having seen these experiments on bodies impervious to light, but transmitting the waves of sound, that there is no connection whatever between optical transparency and acoustical transparency. What has been alleged regarding fog is also true of hail, rain, and snow, and we find the same thing both by observation and by experiment. I had a chamber at the Royal Institution, which I filled with fumes, so dense that a layer of a couple of feet was sufficient entirely to efface the strongest electric light. This fog was sometimes produced by the combustion of gunpowder, sometimes by the combustion of phosphorous, sometimes by the combustion of resin, and sometimes by the precipitation of a real cloud denser than any fog you have ever seen in London, and still in no case did the fog or fumes exert any sensible influence upon the waves of sound. The sensitive flame was as much affected by the sound passing through these smoky media as when the smoke was absent, whereas a couple of candles or a couple of gas burners, placed within the chamber, in half a minute stilled the waves of sound by virtue of a state of things perfectly invisible to the eye, that is to say, layers of air of different densities produced by these burning bodies. We have made artificial showers of snow, artificial showers of hail, and artificial showers of rain. We have had showers of bran, showers of grain of various kinds, and showers of sand; showers of real water and showers of pieces of paper to imitate flakes of snow, and you will be prepared from what you have seen to acknowledge the reasonableness of my statement—that none of them had the slightest influence upon the waves of sound. You told me, Mr. President, that I had half-an-hour, and I do not think I have exceeded the time which you allotted to me.

The PRESIDENT: Ladies and Gentlemen, I will express once more on your behalf our very sincere thanks to Professor Tyndall for his, in the first place, extremely clear and interesting account of these remarkable experiments, which he terminated some little time ago. We have secondly to thank him for bringing the apparatus here and arranging them so successfully as he has done, a thing certainly which could not be done without great labour and great thought, especially at the distance which his usual laboratory and usual lecture room are from this place. We have to thank him for filling this room with an atmosphere of such intellectual transparency, that I feel there is not one trace of fog, not one intellectually opaque cloud, if I may judge at all from the intelligence and attention which has been shown by the audience, remaining in any corner of this room. Professor Tyndall has shown himself to-day not only a successful experimental philosopher, but also a great moral philosopher in restricting his remarks, of which we should have been glad to have heard more if time had permitted; to the time which I had ventured to indicate to him, and thereby giving other gentlemen who are kind enough to offer communications, an opportunity of making them, and us hearing them. I beg to move our sincere thanks to Professor Tyndall, and to call upon Dr. Stone for his communication

ON JUST INTONATION.

DR. STONE: Mr. President, Ladies and Gentlemen—In an international exhibition of scientific instruments, the object of conferences must necessarily be to draw the attention of persons present to those particular points in that collection which deserve our study, and to this I propose in the section which has been kindly committed to me, to confine myself. We have in the acoustical department of the present exhibition, an exceedingly fine collection of instruments brought together, illustrating the debatable point of just or tempered intonation. Indeed I may say that we have, with one exception, a perfect collection of such contrivances. In entering upon just intonation, I am aware however, that *incedo perignes* and that I am liable to touch upon a little warm controversial matter, which I shall endeavour to my utmost to exclude. If we take for instance the written words of the father of just intonation, Perronet Thomson, whose original instrument stands in the corridor, we find him simply saying this—"The temptation to the old systematic teaching to play out of tune was, that performers might play with perfect freedom in all keys by playing in none. Hence the rivalry in magnitude of organs, and the sleight of hand and foot to conceal. But a reaction is setting in, and the world is finding out that music is not a noise, but the concord of sweet sounds." This was written in 1830—the original edition. Yet in 1876 we find Dr. Stainer, an equally able acoustician and musician saying, as regards equal temperament—"When musical mathematicians shall have agreed among themselves on the exact number of divisions necessary in the octave, and when mathematicians shall have constructed instruments upon which the new scale can be played; when the practical musician shall have framed a new notation which shall point out to the performer the ratio of the note he is to sound to the generator,—when genius shall have used all this new material to the glory of art; then"—and here comes the bathos—"then it will be time enough to found a new theory of harmony on a mathematical basis." I must confess, that this putting off to the Greek Kalends, of scientific improvement for practical purposes, does not seem to me, although I generally agree with much that Dr. Stainer says—scientific, or indeed artistic—and the more so as the discrepancy and difficulty lie not in the unfortunate mathematician whom he so sarcastically alludes to, but in a law of nature. As well may the surveyor say, "when mathematicians have made the diameter of the circle commensurate with the circumference, when they have established a simple formula for the circular measurement of an angle, then we will proceed to mensuration and go into the art of surveying." In both instances we might reply that there is no time but now; and we must struggle in spite of the difficulties which the accidental or possibly the intended conformation of nature, at present hidden from us, puts in our way, to produce as great an amount of perfection and as great a union between science and practice, as we can possibly obtain. This incommensurability of nature it would take too long to explain at length, but it is so admirably given in a few words by Mr. Alexander Ellis, in his translation of the great work of Helmholtz, that I may venture to quote the single sentence which contains its exposition. "It is impossible," says Mr. Ellis, in his appendix, "to form octaves by just fifths or just thirds, or both combined, or to form just thirds by just fifths, because it is impossible by multiplying any one of the numbers $\frac{3}{2}$, or $\frac{5}{4}$, or $\frac{2}{3}$, each by itself, or one by the other any number of times, to produce the same result as by multiplying any other one of these numbers by itself any number of times." Thus, having this initial difficulty, how has it been met, and how may it be met? The number of plans brought forward for meeting it are numerous—almost numberless. I will only venture to mention, on this occasion, one, two, or three which stand out beyond the rest. The first is the old unequal temperament. In former times players and composers of music were content to restrict themselves to a certain number of keys. Modulation was not so rapid nor so extensive as it has become of late years. The early effects of music had to be developed before the more complicated. Indeed, Handel's and Bach's music keep in one key. Even Mozart in his early works keeps very much in one or two keys, but with Mozart and Beethoven there came a time when equal temperament became known, and immediately these giants began to throw their gigantic arms about and travel into all the irrelevant parts of the scale. Thus the difficulty soon became manifest. But the older composers, as I was saying, being content to keep to a certain limited number of keys, were also content to exclude themselves voluntarily from certain other keys, and those keys were very graphically called wolves, because they howled. They threw all the howling into certain unfortunate keys, and they promised never to use them. That was the first system. Then came the second, in which the howling, instead of coming from a few individuals, was spread over the whole community, and all the keys were allowed to howl to a certain extent. The howling was distributed over all the keys, and it became, some people say, so small that you do not notice it. Other people think otherwise. That is the second system, termed the system of equal temperament. Then there is a third, and this is the arrangement of which I have to speak most, which was, that by increased mechanism and by greater contrivance you might combine the two. For this purpose there were obviously two ways open. You might increase the number of the keys on the key-board; you might have more digitals to put your fingers upon, each speaking to a more true note, or you might increase the mechanism beyond the digitals in such a way that by drawing combination stops, as they are termed, the proper number of sounds for the particular scale should be brought into action to the exclusion of the others. The first attempt of this kind I have already noticed in speaking of Perronet Thomson's organ. It is not only the first in that direction; it is also a very admirable contrivance; and, I believe, it remains, as yet, without a fellow. His idea was to complicate the key-board to any extent, so that the true sounds were obtained. However, finding that this would be almost impracticable, he limited himself to forty sounds; these forty sounds are distributed, over three key-boards, each key-board containing, besides the ordinary digitals, black and white red ones, what he termed quarrels, what he termed flutals, and what he termed buttons. There are three key-boards, therefore, furnished with these various kinds of keys. No doubt the idea is very good. Anybody can judge for himself who likes to play a simple melody on those key-boards. But the fingering is extremely difficult; and I fear the difficulty is so great that it will not be very largely adopted by practical musicians. They would have to re-learn all the mechanical work of music, generally learned in childhood, and the work is too great. Since then we have had several excellent contrivances, and the most perfect of all, perhaps, is the key-board of Mr. Bosanquet. He uses fifty-three sounds to the octave, and I believe he has them all there present for use. But he is here to-day, and has kindly undertaken to explain the subject, in which he is much better informed than I am; he has also undertaken to mention to you the key-board of Colin Browne, Ewing Professor in the Andersonian Institution of Glasgow, of which, by Mr. Browne's kindness, I have a model. The real instrument is only just patented, and is not completed. As soon as one is completed, Mr. Browne tells me he will kindly send it to this Exhibition for inspection by visitors. Between these elaborate contrivances of mechanism and the simple unaltered key-board there lie two others. The first is that proposed by Professor Helmholtz, and the second, which I hoped to be able to exhibit, but am unfortunately prevented, is that of Mr. Ellis, the translator of Helmholtz, whom I have just mentioned. He uses a combination stop system, and it may be some consolation to us to know that the harmonium would have shown very little; that is to say, only the ordinary twelve notes. But there are stops added, each of which draws out the proper combination for the particular key. This, of course, lays you open to the difficulty that the combination stops cannot be always drawn rapidly during performance, and in extemporising or playing complicated music you do not always know what key you are going to travel into, and therefore what stop to draw. If the key-board is before you, you can in a moment choose the right one, your very ear directs you to it. Of this I have a specimen sent from Paris. It is termed Gueroult's Harmonium, but it is practically, I believe, equivalent to the system suggested by Helmholtz. There are twenty-four notes to the octave, simply arranged on two key-boards, one above the other, like the key-board of an organ, but close together so as to enable the finger of the hand to touch the proper key of either range. The description of it I have before me is as follows, if you will pardon a rough translation: The two key-boards are tuned each in true fifths; but the posterior key-board is tuned a comma lower than the anterior key-board, which is, in the diapason normal, the French pitch. I need not go into the account closely, but you can consider some notes as flats instead of sharps, and then considered as flats, the keys on the second key-board represents the sharps on a third key-board, which would be tuned a comma still lower than the second. We have not time I fear to play that harmonium; but I think we can use it as a means of demonstrating the fact that real just intonation is not inaudible. I think anybody who tried that key-board would be able to tell without looking at the keys whether I am using the common chord on the one key-board, or putting in the more accurate note on the other. The second is much more true than the first. I will take another chord. I leave the exposition of the mathematical principle to Mr. Bosanquet. But, as a practical conclusion, it seems to me that we have here to strike the balance between mathematical and mechanical difficulty. Absolute truth should be aimed at. Don't give it up as Dr. Stainer says, because surely it is within hope that some mechanism may simplify the contrivance so as to make it at any rate possible. It seems to me, however, speaking as a medical man, that there is a physiological condition involved, which has not been sufficiently adverted to in these rather acrimonious debates about true and untrue intonation. It appears that the ear gets deadened—spoilt, if you like, as Perronet Thomson calls it. I confess that my ear is spoilt, to some extent, by the habitual use of equal temperament. It has ceased to give me pain—at least, in keyed instruments. And, singularly enough, the real mathematical sixth seems to me what organists call keen. Whether we have a right to call this a vitiation, which seems to be a natural compensation for what we cannot help, I will not say. I have only one other thing to speak about, and hope to imitate Professor Tyndall's eloquent brevity. The question of intonation has hitherto been entirely dealt with by persons playing on keyed instruments. It is the more difficult problem, no doubt. There are many other instruments which require intonation as much, or even more, but that question has not been so much studied. Players generally trust in orchestral instruments to the lip. The lip will do a great deal, but it is obviously unfair to throw on nature's mechanism, beautiful as it is, what can be made more simple by man's contrivance. As Helmholtz says in another passage, in the whole question of equal temperament, too much has been sacrificed to the instrument, and too little attention paid to God's handiwork, which is the voice. That, I am sorry to say, is a very pregnant truth. Merely taking your ordinary orchestral instrument, there is still much to be done. This clarionet, which would not strike a musician as being different from others, has most of the important quarter tones on it. Simply because, by doubling a key in one place, and by utilising the different fingerings which are upon the instrument, if they could only be attended to, you can get nearly all the quarter tones. I shall be happy to show it afterwards to anyone who knows the mechanism of a clarionet. Lastly, there comes the large department of brass instruments. And here it seemed, for a long time, to be rather a hopeless question, especially as to valve instruments. They are very powerful, certainly a great addition to our sum total of means for producing musical sounds. But they were always considered to be very incorrect. This, which is a recent invention of my friend, Mr. Bassett, is a trumpet which possesses, to all intents and purposes, the ordinary mechanism of a trumpet, but not quite the same. The third valve, instead of lowering the pitch, as it does in the ordinary trumpet, has a separate function. The first valve lowers the pitch a major tone; the second valve lowers the pitch a diatonic semitone. The third valve raises the pitch of any note, produced by the first by the interval of a comma; therefore it has been termed the comma trumpet. In other words, the first and third valves together, lower the pitch a minor tone, and when the first is used together with the second valve or alone, it gives, of course, other modified intervals, resulting in the production of a more correct musical scale than has yet been obtained on any valve instrument, with very little alteration of the usual fingering. It appears to me that this excellent system, which has hardly yet attracted the notice of professional musicians, ought to be applied to other instruments. Mr. Bassett is here, and no doubt will, during the recess, play some notes and exhibit the system. Here I beg to conclude for the moment, and can only hope that the President will kindly allow Mr. Bosanquet to supplement those parts which, as mathematician and acoustician, I am less competent than he to undertake.

The PRESIDENT: I beg to return your thanks to Dr. Stone for his communication, and I will now call on Mr. Bosanquet to give us a short account of his instruments, on the theory and practice of which he has been for some time experimenting.

ON INSTRUMENTS OF JUST INTONATION.

Mr. BOSANQUET : As the combinations which are to be dealt with are complex, forms of arrangement have been employed which reduce all the phenomena in practice and theory to a few simple types. These I have called symmetrical arrangements. They were not employed by General Thompson.

There are two forms of symmetrical arrangements. Those which I employ I shall distinguish co-ordinate symmetrical arrangements; other forms employed by Mr. Poole and Mr. Brown may be called key-relationship symmetrical arrangements. Into these latter I shall not enter, but only remark that those which depend on co-ordinates possess all symmetrical properties in a more extended form than the others; *i.e.*, they include the key-relationship and more besides.

Mr. Brown's position relations are exactly the same as Mr. Poole's, with omission of the two series of sevenths.

In my co-ordinate symmetrical arrangements a number of equal temperament semi-tones, or twelfth parts of an octave, is taken as abscissa, and deviation or departure from the note thus arrived at as ordinate. Thus, the exact pitch of notes can be expressed by reference to co-ordinates in a plane.

To express a series of equal temperament (E.T.) fifths in this manner:

The E.T. fifth is $\frac{7}{12}$ of an octave. Taking twelve fifths up, we have seven octaves exactly. The result is expressed by abscissœ only.

To express a series of perfect fifths in this manner:

The perfect fifth is seven semi-tones and a departure = 01955 E.T.S. = $\frac{1}{51.15}$ · .·. twelve fifths are seven octaves and a departure = 23460 E.T.S., the Pythagorean comma. The resulting notes have ordinates which increase uniformly as we pass along the series of fifths; and the position arrived at after the twelve fifths has an ordinate which represents the Pythagorean comma when the fifths are perfect, and abscissa $7 \times 12 = 84$ or seven octaves.

The form of symmetrical arrangement employed in the generalized key-board, is arrived at by arranging a series of notes in the order of the scale as far as the abscissæ are concerned, and taking for ordinates distances proportional to those thus arrived at. That is to say, the ordinate of any note is proportional to its distance from a fixed point in the series of fifths.

The series of fifths is selected for the application of the conditions, because it is the most convenient; but the variations of all the concords in any system are linked together in such a manner that it is indifferent which is taken as independent variable, so to speak; the results would be always the same.

The generalized key-board, of which the harmonium exhibited offers an example, may be conveniently described with reference to abscissæ from left to right, and horizontal ordinates on plan from back to front. The vertical ordinates are one-third of the horizontal ones.

In the abscissæ, half an inch corresponds to an E.T. semi-tone. Twelve semi-tones make an octave. The octave measures 6 inches in abscissa, and nothing in ordinate.

In the ordinates on plan, 3 inches correspond to the Pythagorean comma or departure of twelve fifths. Thus the difference between the ordinates of two notes on same abscissa, between which one series of twelve fifths lies, is 3 inches.

The ordinates of the intermediate fifths are increased by $\frac{1}{4}$ inch at each step upwards in the series of fifths, so that twelve steps upwards in the series correspond to 3 inches.

I have described the key-board as connected with the system of perfect fifths; and it is so in this harmonium to all intents and purposes. But it is clear that if each fifth have any departure from E.T. whatever, this may be equally represented by the ordinates in question, as no use has been made of the amount of the departure; and we can say that a key-board, constructed in the form of a co-ordinate symmetrical arrangement, forms a graphical representation of the interval relations of any set of notes belonging to a regular succession of fifths.

Thirds can always be referred to fifths. In systems such as that of perfect fifths—which we are dealing with here—by means of a theorem brought into notice by Helmholtz: in other cases, in other ways.

The most important property of key arrangements which form graphical representations of their intervals is, that any combination of intervals has the same form to the finger on whatever notes or in whatever key it may be taken. Thus a common chord always has the same form.

Non-co-ordinate or key-relationship symmetrical arrangements, such as those of Mr. Poole and Mr. Brown, possess a similar property of more limited extent. In these it is, for instance, possible that a common chord may assume different forms to the finger in cases where the key relationship is differently assumed: not so in co-ordinate arrangements.

I will only allude to one property of the division of the octave into 53 equal intervals, according to which the harmonium exhibited is tuned.

The mode in which the number 53 is arrived at has been explained by me, as part of a general theory. But we can verify its properties independently by noticing, that if we take 31 units for the fifth of the system, then $12 \times 31 = 372$ and $7 \times 53 = 371$; so that we see directly without formulæ, that the departure of twelve fifths $= \frac{1}{53}$ of an octave $= \frac{12}{53}$ of a semi-tone; and departure of one fifth $= \frac{1}{53}$ of a semitone. Now, the departure of a perfect fifth is $\frac{1}{51\cdot151}$, and the difference is about $\frac{2}{2700} = \frac{1}{1350}$ of a semitone, which is the error of the fifth of the system. Hence, we may say that the system of fifty-three is sensibly identical with a system of perfect fifths.

In the enharmonic organ recently constructed, I have applied to a generalized key-board of forty-eight notes per octave, Helmholtz's approximately just intonation, and also the mean tone system, which is of historical interest. Each system is brought on to the key-board separately by a draw-stop. In the same way all systems of interest are accessible; it is this employment of the key-board that I would at present commend to those who inquire into its utility.

Considering the facilities that we see about us for manipulating just and approximately just systems, it is difficult to see why mere book knowledge should continue to be alone regarded in the study of this portion of the elements of music. When it is taught, for instance, that certain vibration ratios correspond to certain musical effects, the lesson should be taught experimentally; as it is, musicians for the most part only know what consonances are from descriptions in books. As illustrations, I may point out that we have, in the harmonium now exhibited, the means of distinguishing three different kinds of minor thirds, whose ratios are 6: 5, 32 : 27, 7 : 6; and these sound quite different to the ear. Again, Pythagorean thirds can be contrasted with exact thirds; the harmonic seventh compared with other forms of minor seventh, and numerous other theoretical results reduced to practical knowledge.

Of the applications of the various systems, I will only say that in my opinion it is a mistake to apply ordinary music to them indiscriminately. Just systems especially, which have both thirds and fifths nearly perfect, must be studied and written for before they can be used with advantage. I need hardly say that I think, when this is done, the advantage will be great.

Dr. Guthrie here took the chair.

The CHAIRMAN: I am sure you will all thank Mr. Bosanquet for his communication, and I am sure we are very glad to hear that there is some practical prospect of this very desirable end being brought about. We have now the subject of the limits of audible sounds, and if Dr. Stone will introduce the subject, I hope Mr. Galton will illustrate it by one or two practical results.

THE LIMITS OF AUDIBLE SOUND.

Dr. STONE: At the risk of the accusation of irrepressibility perhaps you will allow me to occupy one or two minutes in the way of the Roman nomenclator of old, to start a subject rather than to complete it. I wish to mention the limits of audible and musical sound. Of course we have them both above and below, at either extreme. We have, also, I am proud to say, a very fine collection of illustrative instruments in the exhibition, and I would name three at the upper limit and three at the lower. The first is a curious instrument of Mr. Griesbach's which is interesting, not only on account of the perfect way in which he illustrates the upper limits of audibility, but because he contrived it so as to show many principles of musical sound which are considered to be of recent discovery. This is the instrument. It is to all intents and purposes a small organ with a key-board, the pipes of which are exceedingly small. They look rather long, but the greater part of the pipe is foot, and the real acting part is very short indeed. Here is a note which would astonish modern pianists to play. It is so fine that I am afraid you will not hear it. Here is the semitone below it on the same scale, and although many persons are not able to hear these pipes separately, yet the resulting tone produced by blowing two together is perfectly audible to most people. This was made many years ago, and it is historically interesting as being an anticipation of modern research on the subject. Among the many discoveries of Professor Wheatstone was one of this kind. He produced the same effect by using two very small harmonium reeds; reeds of the same kind are here in this small box. These are made by Mr. Griesbach, but as to priority between the two I am unable to speak. Unluckily, that is the condition of these small reeds also: they are unable to speak, but there they are. Time has damaged them, but the same things were certainly contrived as early, if not earlier, by Professor Wheatstone. Then as to the lower limit of sound. Helmholtz may have erred from erroneous information given to him, and perhaps the appreciation of musical sound might be rather different in other persons than those he had to experiment with. I believe he is a violin player, but I do not think he pretends to be a musician of a practical kind. He says the deepest tone of musical character which can be heard is about forty-one vibrations in the second, in the upper half of the thirty-two feet octave he says the perception of the separate pulse is clear, but practically he does not admit that you can get any musical note below E or F on the common German double bass. Now what seems to be wanting, if I may use the term, in these investigations, is that the mass upon which he experimented was rather small. He used pianoforte strings weighted with a kreutzer in the middle. That is a very feeble source of sound. If we are to produce these low tones, the amplitude of the vibrations must be enormously increased. Whether he has quite utilized the effect of a consonant body or of a resonant case, on a vibrating string, to the extent to which it might be done, is a point on which I have some little doubt, because it has been done here. Here is Elliott's apparatus, originally invented by Chladni, a sort of wand passed through a slit. This Helmholtz alludes to, and declares it produces a false result, because the upper partial tones are very strong compared with the fundamental. No doubt they are. I should be delighted to find him correct, and he is probably speaking only of simple pendular vibrations such as you get in the lowest stopped diapason pipes of an organ; but if this is what he means by saying that you cannot hear simple pendular vibrations below forty-one, he is simply saying that you cannot carry the stopped diapason down below forty-one vibrations, a much less extensive statement than to say that the ear cannot distinguish sounds below that. As a matter of fact, all the extreme bass instruments we make use of, perhaps unfortunately, have the upper partials very strong; strong compared to the foundation note, but you can intensify the fundamental note by certain contrivances, especially those of consonance. On this I have spent a great deal of time myself. I have tried on a double bass. The double bass has been often before made to produce very low notes, and there is one, quite gigantic, on the other side of the building, which requires giant to play it; the present race of pigmies had to stand on a a table. It produces a very fine tone in the low notes, but it would not suit this generation; it needs sons of Anak to play upon it. Other attempts have been made to produce the low tone, by making the strings thicker; but then you cannot get at the centre of gravity of a cylindrical string; you must strike at the outer circumference, and these large strings rotate and produce false notes, not at all the tone that is wanted. There remained one thing which had not been done, and that is to work by weight. By covering the string very heavily with copper wire, and placing it on the double bass, I succeeded in getting a sound which I thought satisfactory. The tone contains a great predominance of 16-foot vibration. I had to strengthen the double bass very considerably by what I term elliptical tension bars, so as to give it, in the first place, force enough to resist the enormous pull of this heavy string, and secondly to give it a sound-conductor. There was already a bar from end to end which tended to counteract the dumbing effect of the "S" holes, which are necessary, however, for letting out the air vibrations. If you will permit me to add presently to my unmusical illustrations, I think I can obtain a note from the double bass, which you will say is musical. Another attempt was in the wind department. We have here an instrument which some of you may know I habitually play, especially with Sir Michael Costa's orchestra. I played it the other night at the Albert Hall. It is a reed instrument of 16 feet in length, the octave of a bassoon. It is not a new instrument, but it is on a new scale This, as I hope to demonstrate afterwards, brings out the CCC, the lowest note of the 16 feet octave, with a tone which you people may call musical or not, but I think it is. At any rate it is not wanting in power or in intensity. I have intentionally omitted to speak of one excellent set of investigations on the upper limit of musical sound, because Mr. Galton, the author, is here present himself, and will explain them.

The CHAIRMAN: I have to ask you to again express your thanks to Dr. Stone for these few supplementary remarks; but I think before we have any discussion on these communications, we had better complete this branch of the subject, and I will therefore call on Mr. F. Galton, F.R.S.

Mr. GALTON: I thought it would be of convenience to experimenters, that I should exhibit some little instruments I have combined for ascertaining what the upper limits of audible sound may be in different persons of the same race, and in individuals of different races, and in different kinds of animals. It is, of course, a matter of great interest to know whether insects and such small creatures can hear sounds, and can in any sense of the word, converse in language which to our ears is utterly inaudible. When I first devised to make experiments, I was checked by the great difficulty of finding instruments that vibrated with sufficient rapidity for the purpose in question. Dr. Wollaston (to whom we are indebted for the first experiments ever made on this subject, and for the fact that vibrations exist which the ear is incompetent to seize and render into sound) found very great difficulty in making his small pipes. I tried several plans for obtaining acute notes, and the one I finally adopted was this: I made a very small whistle, whose internal diameter was much less than one-tenth of an inch—I have many such here, made for me by Massrs. Tisley and Spiller, Opticians, 172, Brompton-road,—with a plug at the bottom, which plug is screwed up by a graduated screw. The graduations are marked on the side, so that when you use the instrument you know the depth of the tube, and knowing what that is, it is a matter of calculation to learn the rate of vibration. There is, however, a good deal of uncertainty in the matter, because there must be some fair proportion between the length and width of the tube in order that the calculations should give a correct result. A short whistle with a diameter exceeding two-thirds of its length, will certainly not give a note whose shrillness is governed wholly by its shortness. Therefore in some of my experiments I was driven to use very fine tubes indeed, not wider than those little glass tubes that hold the smallest leads for Mordan's pencils. It occurred to me, in order to produce a note that should be both shrill and powerful, and so correspond to a battery of small whistles, that a simple plan would be to take a piece of brass tube and flatten it, and pass another sheet of brass up it, and thus form a whistle the whole width of the sheet, but of very small diameter from front to back. I have such a whistle here, it makes a powerful note, but not a very pure one. I also made an annular whistle by means of three cylinders, one sliding within the other two, and graduated as before. I find that when the limits of audibility are approached, the sound becomes much fainter, and when that limit is reached, the sound usually gives place to a peculiar sensation, which is not sound but more like dizziness, and which some persons experience to a high degree. I am afraid it is of little use attempting to make the audience hear these small instruments; but I will try, beginning by making rather a low note. It was found that there was great variability in the audience, in their powers of hearing high notes, some few persons who were in no way deaf in the ordinary meaning of the word, being wholly insensible to shrill sounds that were piercingly heard by others. I find that young people hear shriller sounds than older people, and I am told there is a proverb in Dorsetshire, that no agricultural labourer who is more than forty years old, can hear a bat squeak. The power of hearing shrill notes has nothing to do with sharpness of hearing, any more than a wide range of the key-board of a piano has to do with the goodness of the sound of the individual strings. We all have our limits, and that limit may be quickly found in every case. The facility of hearing shrill sounds depends in some degree on the position of the whistle, for it is highest when the whistle is held exactly opposite the opening of the ear. Any roughness of the lining of the auditory canal appears to have a marked effect in checking rapid vibrations of the ear. For my part, I feel this in a marked degree, and I have long noted the effects in respect to the buzz of a mosquito. I do not hear the mosquito much as it flies about, but when it passes close by my ear I hear a sudden "ping," which is very striking. Mr. Dalby, the aurist, to whom I gave one of these instruments, tells me he uses it for diagnoses. When the power of hearing high notes is lost, the loss is commonly owing to failure in the nerves. On the other hand we may find very deaf people who can hear shrill notes, in which case the nerves are usually all right, but the fault is in the auditory canal. I have tried experiments with all kinds of animals on their powers of hearing shrill notes. I have gone through the whole of the Zoological Gardens using a machine of the kind that I hold in my hand. It consists of one of my little whistles at the end of a walking stick, that is in reality a long tube; it has a bit of india-rubber pipe under the handle, a sudden squeeze upon which forces a little air into the whistle and makes it sound. I hold it, as near as is safe, to the ears of the animals, and when they are quite accustomed to its presence and heedless of it, I make it sound, then if they prick their ears it shows that they hear the whistle, if they do not, it is probably inaudible to them. Still, it is very possible that in some cases they may hear but not heed the sound. Of all creatures, I have found none superior to cats in the power of hearing sharp sounds. It is perfectly remarkable what a faculty they have in this way. Cats, of course, have to deal in the dark with mice, and to find them out by their squealing. Many people cannot hear any notes in the squeal of a mouse. Some time ago, singing mice were exhibited in London, and of the people who went to hear them, some could hear nothing, whilst others could hear a little, and others again could hear much. Cats are differentiated by natural selection until they have a power of hearing all the high notes made by mice and other little creatures that they have to catch. You can make a cat, who is at a very considerable distance, turn its ear round by sounding a note that is too shrill to be audible by any human ear. Small dogs also hear very shrill notes, but large ones do not. You may pass through the streets of a town with an instrument like that which I used in the Zoological Gardens, and make nearly all the little dogs turn round, but not the large ones. At Berne, where there are more large dogs lying idly about the streets than in any other town in Europe, I tried this method for hours together, on a great many large dogs, but could not find one that heard it. Ponies and cattle, too, are sometimes able to hear very high notes—much more than horses; I once frightened a pony with one of these whistles in the middle of a large field. I can produce no effect on ants, nor on the great majority of insects, though there are some apparent exceptions about which, however, I am not yet prepared to speak.

The CHAIRMAN: I am sure I need not ask you to thank Mr. Galton for his very interesting remarks. Time is getting on, and I will now invite those who have anything to say upon the general subject of Just Intonation and the Limits of Audible Sound to do so. There was only one remark which occurred to me specially during Dr. Stone's last communication, and that was the able way in which he pointed out the bearing which the mass of the air which is set in motion has upon audibility. It has nothing to do with wave-length or wave amplitude but with the quantity of air, which you may compare to the end-length of sea waves. That seems to have great power and effect on the auditory nerves. Perhaps the lowest audible note which we have heard, at least, is that which you hear when you are outside a tunnel, and you hear the actual throbbing of the piston of the engine. You have an immense mass of air in the tunnel set in vibration. It has nothing whatever to do with the length of the tunnel; it does not act as an organ pipe establishing stationary waves, but you have an immense mass of air set in deliberate motion with slower vibrations even than sixteen in a second, and the sound is perfectly audible.

Mr. ALEX. J. ELLIS, F.R.S.: I only wish to say a few words with regard to the experiment of Helmholtz which was alluded to, for trying to determine the lowest limit of tone. His object in operating on a piano string loaded with a kreutzer was to render the upper partials inharmonic to the fundamental, so that he should be quite sure that he heard a simple pendular vibration. His object was to determine the lowest audible limits of such vibrations. The subject is one which has been very recently investigated, and accounts of the experiments have been given by Professor Preyer, of Jena, one of the members of the general committee. There are the pipes here which he experimented upon. He found that the best way to hear these sounds, was to obtain them first from a reed attached to a pipe, and then to shut oft the air, and listen to the tone as it vanished, when he heard the fundamental or lowest tone quite distinct from all the others, as a simple pendular vibration. As the result of a great number of experiments he found that ears differed very much as to what was really a musical tone, defining that to be one in which we hear no throbs, but only a continuity of sensation. He has also published a work on the limits of sensational power, and, in fact, that was his great point, to determine the limit of continuity of sensation. He found that he himself could hear continuous tone from as few as 14 vibrations to the second, but that most ears perceived sensation to be continuous when the number of vibrations reached 23. Therefore, somewhere between 14 and 23 vibrations must be fixed as the lowest limit of continuous simple pendular vibrational tone produced in the human ear, as far as it has yet been investigated. He also has gone very much into the question of the upper limits of tone, but his especial investigations were to determine the smallest amount of error in a melodic interval, which the most practised ears could hear. For this purpose he made use of some of these instruments of Herr Appun, of Hanau. There is one which gives a complete series of partial tones up to the 32nd, and there is another one in the next room which gives tones from 128 to 256 vibrations, proceeding by two beats at a time. With that instrument Herr Preyer experimented, and the investigation has a very important bearing on the method of representing just intonation by means such as that of Mr. Bosanquet, who uses an approximative scale, obtained by dividing the octave into 53 equal parts, which I consider to be really perfect enough. I have not calculated all Herr Preyer's results out completely, but I may state that no ear seems to detect an error in an interval melodically—not in a chord—which amounts to the hundredth part of an equal semitone, but that the fiftieth part (double that) may be detected by very fine ears indeed. With regard to just intonation and key-boards, I may say that key-boards like Mr. Brown's and Mr. Poole's (which is very good if we reject the natural sevenths), go upon the principle of carrying out a series of tones proceeding by perfect fifths, in three columns, so to speak, each column being a comma lower than the preceding. Guéroult uses only two. Guéroult and Helmholtz's plan is the one Dr. Stone said I wanted to simplify, using a single key-board by means of compound stops, and that was the instrument which was to have been exhibited; but he was a little in error. That instrument was one with single, not compound, stops, and was intended to exhibit the old organ tuning continued so as to play 21 notes to the octave. It was invented by Mr. T. Saunders, who subsequently declined to exhibit it. But the principle of Helmholtz's and Guéroult's instruments is to have two rows of 12 notes forming perfect fifths, one row being a comma flatter than the other. This would give the full succession of major keys, but only five minor keys complete. The other minor keys are quite imperfect, whereas with Mr. Bosanquet's instrument which, although it has only 53 tones to the octave, has actually 84 finger-keys to the octave, so as to be able to go round and round, all the keys, minor as well as major, are practically perfect. This instrument is really almost as simple to play as an ordinary harmonium when you understand that the major thirds are taken in a series below, and the minor thirds in a series above; whilst the oblique arrangement of the finger-keys obviates the necessity of jumping from one row to another and allows of playing each scale in one line. I consider Mr. Bosanquet's arrangement to be the acme of perfection in this respect, and I do not think that we are likely to arrive at anything which is simpler. I hope Mr. Bosanquet will give us an opportunity of hearing some of the effects of it afterwards, because until persons have heard music played in just intonation they cannot at all appreciate what it is that persons want to obtain as contra-distinguished from that which we are generally obliged to hear. I had an opportunity only last Christmas of hearing a well trained and educated choir of the Tonic Sol-fa College, accustomed to sing in perfect intonation, and, when they were singing unaccompanied, the chords in just intonation were perfectly divine, but when they sang immediately afterwards to a pianoforte which was almost inaudible, the chords were all torn to pieces in such an extraordinary way by the accommodation of the voices to the instrument that it was perfectly painful to listen to them. With regard to the upper limits of audible tone, although I am rather an old boy, I may say that I heard all the high tones produced by Captain Douglas Galton perfectly.

The late Sir Charles Wheatstone's Acoustical Discoveries.

Professor W. G. Adams, M.A., F.R.S.: If I were to speak of all the instruments—or even of all the musical instruments—which may be connected in some way or other with the name of Sir Charles Wheatstone, I am afraid I should occupy a very considerable time, weary you, and shut out those who have to come after me; but I propose to draw your attention to three classes of instruments with which Sir Charles Wheatstone was specially connected. First of all, if we consider the vibration of reeds, we may start from the very ancient instrument the Marimba, which has iron rods fixed into a sounding body in the same way as the iron fiddle, which consists of rods fixed at one end to a sounding board; from the iron fiddle, by lengthening the rods, we get to the kaleidophone, which is so well known, and the figures traced out by which are so familiar, that it will not be necessary for me to describe them in detail. If we take a cylindrical rod with one end fixed, and cause it to vibrate, being cylindrical, it will vibrate transversely at the same rate in all directions, but it may be put in vibration so as to give not only a simple figure, the ellipse, circle, or straight line, but by dividing it by nodes or points of rest into separate vibrating segments we may get also the super-position of the partial vibration figures combined with the original simple figures.

The simple figures are obtained by causing the rod to vibrate as a whole, and the partial vibrations are obtained by producing one or more nodes on the rod. The ratio of the number of partial vibrations to the number of fundamental vibrations is given by the number of indentations produced in the original figure traced out by the free end of the rod. The number of vibrations when there is one node on the rod is about $6\frac{1}{4}$ times the original number of vibrations of the rod when it vibrates as a whole. With one, two, three or more nodes the number of vibrations is as the squares of the second, third, or higher odd numbers. No. of nodes, 1, 2, 3, 4, &c.; No. of vibrations, 9, 25, 49, 81, &c. With a rectangular rod, when its section is a square, the curves traced out are the circle, the ellipse, or the straight line, exactly as in the case of cylindrical rods, and the partial vibrations bear the same relation to the fundamental vibrations with rectangular rods, which are not square in section, the number of vibrations will depend on and be proportional to the thickness of the rod in the direction in which the vibrations take place. If we increase that thickness, we shall increase the number of vibrations in a given time; if we double the thickness, the number of vibrations will be twice as many, and so, causing the rod to oscillate in one plane or in the other, we shall get in the direction of the greatest thickness, the greatest number of vibrations; and if the length of the rod is such as would produce musical tones, then the tone corresponding to the greatest thickness will be the octave of the tone corresponding to the least thickness, when the thickness in one direction is twice as great as in the other.

We may pass on from an ordinary rod of this kind, gradually thinning away in one direction, and if necessary, thickening a little in the other, and we pass to a thin reed or vibrating strip of metal fixed at one end, and placing that in an opening, we get the "free reed." It is only necessary to mention the free reed to recall again the name of Sir Charles Wheatstone, who developed it so much, and applied it, or caused it to be applied, to so many instruments. The figures traced out by these kaleidophones, when they are of different diameter in different directions, are very well known, and may be produced in various ways. There are many pieces of apparatus in the Exhibition which will show this. If the point of suspension of one pendulum be attached freely to the bob of another pendulum, so that they can swing in planes at right angles to one another, and the two are set in vibration, then the bob of the lower pendulum will trace out a curve which results from the two motions: the same curves may be traced out by Mr. Tisley's beautiful apparatus, having two pendulums vibrating in planes at right angles to one another, which are placed at the corner of a table, one at the end and one at the side, so that the oscillations take place in two planes, at right angles to one another, a point connected with both pendulums will trace out the curves. By altering the length of the pendulum or the times of oscillation, we may get a variety of different curves, and may make the times of oscillation so nearly coincident as to produce what corresponds to beats in music. Here is a very pretty instrument for showing the same thing. In this case we may consider that we have two free reeds, one attached to the end of the other, on the same principle as the pendulums attached one to the end of the other. There are two thin strips of metal soldered together end to end with their planes at right angles to one another. Fastening one of these in a vice, we may set them vibrating ; the motion of the free end will be the result of the combination of the motions of the two strips taken separately. On lengthening the lower strip, it will make a smaller number of vibrations in a given time, and in this way any combination of two rectangular motions may be obtained.

I must also call your attention to another instrument invented by Sir Charles Wheatstone for producing these figures by the motion of two cranks, in two planes, at right angles to one another. The ends of the cranks are fixed by a hinge to one end of a rod, the middle of the rod turns in a fixed socket, so that the free end describes the same curves as the end to which the crank arms are attached. There are several points worthy of attention in this instrument, especially the arrangement by which the number of revolutions of the two wheels which drive the cranks may be made to bear any given ratio to one another. I must now pass to another subject which is well illustrated by apparatus in this exhibition, and which was worked at and developed by Sir Charles Wheatstone, viz. :—The production of sound by exciting vibrations in tubes by means of gas flames.

If a small gas-jet of suitable form be placed just within the lower end of a tube open at both ends, vibrations will be excited in the tube, and those vibrations which correspond to the length of the tube will be reinforced by it, and a musical note will be heard. On raising the flame into the tube the sound ceases, but on making the flame larger, the sound is again produced and is louder than before. The sound gets louder as the flame is raised, and the most intense sound is produced when the flame approaches to the position of the node.

Working at this subject, Sir Charles Wheatstone produced an instrument which is called a chemical harmonicon or gas-jet organ. It consists of a key-board, to the keys of which are attached small gas burners coming from a tube containing hydrogen gas. Each burner is placed just within the lower end of a gas tube, and the tubes are of such a length as to resound to the successive notes of the diatonic scale, when a burning gas-jet is raised to the proper point of the tube by pressing down the keys. This instrument is from the Wheatstone Collection of Physical Apparatus at King's College. A photograph of a similar instrument is exhibited by Professor Oppel of Frankfort.

Wheatstone investigated experimentally the laws of vibration in conical tubes, and showed the agreement between the calculations of Bernouilli and experiment. For the first mode of vibration, *i.e.*, for the lowest note of a conical pipe, all the air in the pipe moves backwards and forwards in the same direction at the same time. The particles alternately approach to and recede from the apex of the cone. In the second mode of vibration there is a ventral section in the middle of the pipe, and the pipe is divided by it into two parts of equal length. In the third mode of vibration there are three equal lengths separated by two ventral sections.

Taking conical tubes, the different notes which may be produced from them correspond precisely to those which may be produced from open cylindrical pipes of the same length. Taking this conical tube, two feet long, the resonance corresponds to a middle C tuning fork, and if from any cone I take a part of the same length and open at both ends, I shall get the same note, so that with a conical pipe, either closed at the apex or open at both ends, we get the same note as from a cylindrical open pipe of the same length. The harmonics produced in the conical pipe are the same as those in the open cylindrical pipe, which are of course different from those produced by closed cylindrical pipes; the difference being seen in two musical instruments, the clarionet and the oboe. The oboe being conical the harmonics are those of an open cylindrical pipe. Cutting up the cone into equal lengths, we get the same note from each, so that if these short pipes, which were Sir Charles Wheatstone's, are sounded, the same musical note will be produced from each of them, but in each of these open pipes there is a node which is not at the centre of the pipe. In a cylindrical pipe the column of air will be divided equally into two vibrating parts at a node in the centre, but in the conical tube the node is not in the centre but will be nearer to the smaller end of the tube. As the pipe tapers more and more, the distance of the node from the centre of it will be more and more increased. I have here two conical tubes each about six inches long: one of them is open at both ends, the other I have divided into two parts at the node. If I stop these two tubes at the section where they have been divided by placing them on the palm of the hand, and sound the note, we shall find that they will produce the same note, and that this is the note given by the other conical tube open at both ends. I have found by trial the position of the node in these tubes, so that by stopping the tube at that point, the note produced from that short pipe when closed shall be the same as from the longer one open at both ends. If that is the case, we may expect also that by stopping the small end of this pipe we shall get the same note from it, which is the case, so that from those pipes, both closed, one at the larger end and the other at the smaller end, we get the same note produced, and this note is the same as that from an open cylindrical pipe equal in length to the sum of the two closed conical pipes. Taking tubes of the same length, but with different degrees of taper, we pass from a very low note, which is produced with the tube of greatest taper when the largest end is closed, through a succession of notes to the note of a closed cylindrical tube, then by inverting the tubes, taking them in reverse order, and closing the smaller ends, we may produce a succession of notes still increasing in pitch up to the note of an open cylindrical tube of the same length.

The CHAIRMAN: I will now ask you to record your indebtedness to Professor Adams for his very able exposition of one of the chapters of Sir Charles Wheatstone's great scientific career. It must have struck all those who have been working in science, ever and anon, that when they fancied they had found something new, they find it was done by Sir Charles Wheatstone years ago. That has happened scores of times, but I am happy to say there is every prospect soon of Sir Charles Wheatstone's published and unpublished papers being collected and presented to the public in a recognized form, so that that danger will in future be avoided. I will now call on Mr. Chappell to give us an account of ancient musical science.

ANCIENT MUSICAL SCIENCE.

MR. W. CHAPPELL: The limit of time imposed upon me makes it necessary that I should say very few words indeed. In this exhibition there are some models of ancient Egyptian pipes, and of a Greek, or rather of an Egyptian, hydraulic organ. It was the custom of the Egyptians, in the early dynasties of the empire to leave a pipe in the tomb of a deceased person, and to lay by it a straw of barley, by which the player might, as we assume, make fresh reeds when he awoke. This was upon the assumption that, having been a very good man, his soul would resume the human form. Here is an example of the sort of pipe which was deposited. This pipe could only be played with a double reed, such as that of the hautboy, the bassoon, or the ancient shepherds' pipe, because there is no notch in it, as in the flageolet or the diapason pipe of an organ, to excite the tone. Through the kind assistance of my friend Dr. Stone these pipes are fitted with reeds. The diameter of the pipe may not be exactly copied, but that is of minor importance because increase of diameter does but increase the volume of sound. It is the length of the pipe and the distance between the holes which are essential. There is no example of a bass pipe, thus deposited, but we have here two tenor pipes and two treble pipes, and from these we ascertain at least some of the scales of the ancient Egyptians. The Greeks constructed their minor scale by joining together two tetrachords, which we call fourths, but they are fourths with the semitone at the bottom, as in B, C, D, E—not as in C, D, E, F, where the semitone comes at the top. It was desirable to know how ancient that scale was in Egypt, and it is evident that these pipes were anterior to it, for they are all on the major scale. Greek writers upon music inform us that they joined together two tetrachords of four notes each, by commencing the upper tetrachord upon the highest note of the lower one, thus reducing the eight notes to seven, because there were seven planets. There are other associations connected with the number seven which might have influenced them to do so. In these pipes we discover two musical principles which we are not aware to have been known to the Egyptians. One pipe has a hole bored through it within an inch of the top, and it could not possibly have been sounded with those two holes left open. This is copied from a pipe in the British Museum, and it was necessary to cover the apertures with thin gutta percha to elicit the sound. It proves to us that the ancient Egyptians knew the principle of the famous pipe mentioned by Shakespeare, called the Recorder, which differed only from the soft English flute, played at the end, in having a little piece of bladder or something of that kind fastened over openings of this sort, the object being to give a tremulousness to the tone and to make it more like the human voice. The tone of the pipe would otherwise be perfectly pure and steady, as in an organ. In the case of a second Egyptian pipe we find it necessary to sink the reed down three inches within the tube to elicit any sound, and that is the principle of the drone of the bag-pipe. It serves to protect the fragile reed from injury. There is no perfect octave scale upon any of the four which have been tried. They are very limited in compass, but we find that the Egyptians often played in concert. There is a representation of three Egyptian pipers playing in concert with pipes of about one, two, and four feet in length, in the tomb of an Egyptian, named Tebhen in the hieroglyphics, and this tomb is of the fourth dynasty, or about the time of building the great pyramid. Three such pipes must necessarily be playing in three different octaves—treble, tenor, and bass. One of these pipes has six notes in the major scale, another has a diatesseron with the semitone at the top, but no tetrachord—which is very curious. The scale of two tetrachords which was borrowed by the Greeks must therefore be of later date. The tetrachord was the best possible arrangement for recitations limited to four notes. Suppose the tetrachord to be B C D E, C would be the reciting or key note, having B, the semitone below as the true seventh for drawing to a close. The chief part of the recitation would be upon the key note, and there was the power of rising to a major third above it, which was quite enough for eastern recitation. By joining two such tetrachords together they reduced the eight to seven, thus B, C, D, E, and beginning with E again, E, F, G, A, and then putting the octave A at the bottom they made our minor scale A, B, C, D, E, F, G, A. We had that minor scale earlier in England than in France, having received it through the Greek organ. We had a Greek as archbishop about the year 668, when Theodore was made Archbishop of Canterbury, and Adrian was sent with him to watch him, because he was a Greek, and might introduce some practices of the Eastern church which were not sanctioned in Rome. We find that Saint Aldhelm, who died in 705, began his "Praise of Virginity" with these words :—

"Maxima millennis auscultans organa flabris."

(Listening to the greatest organs with a thousand bellows), Wolstan fully describes an organ of the 10th century in the cathedral of Winchester, with 400 pipes, and it required many men to blow it. I am indebted to Dr. Stone for having made this model of the action of the hydraulic organ from my description in the *History of Music.* Here are two glass vessels, one inverted inside the other, and when we inject air into the inner one, water is expelled, and rises in the outer; and when we sound the pipe, the air escaping, the water returns to seek its own level. There are many erroneous descriptions of the hydraulic organ. Some say it was worked with boiling water, but that idea arose from the bubbling of the water. Thus when the cork which now floats sinks to the bottom, it shows that the vessel is filled with air, and if we continue to blow, the surplus air will ascend in bubbles outside, and thus give it the appearance of boiling.

The object of this invention of the Egyptian Ctesibius was to prevent the possibility of overblowing the instrument, to which the pneumatic organ was then subject. It proves that he understood the law that "Liquids transmit pressure equally in all directions, and the pressure they produce by their own weight is proportionate to the depth."

The CHAIRMAN: I am sorry that we have not time to discuss this interesting subject, but I must now call on Mr. Baillie Hamilton.

ÆOLIAN INSTRUMENTS.

Mr. J. BAILLIE HAMILTON: I am going to bring before your notice the last result of three years' work, and I select this particular one because it to a great extent embodies all the rest. I hold in my hand a small object, consisting of the following parts. There is a vibrator, consisting of the tongue of a reed, and above it there is a double circle, which can be made of any form to afford an elastic constraint. There is here a rod which forms a prolongation of the vibrator, and upon the reed a weight which can be turned upon a screw. By means of this weight the pitch of the note can be regulated to the utmost nicety, nor is there any chance of that being deranged; and by the constraint of this ring can be given any tone which I may desire. This little object embodies all the capabilities and all the phenomena characteristic of the æolian sounds; and when I speak of an æolian sound, I do not mean necessarily that of the æolian harp, but all that class of phenomena which occur whenever wind is applied to strings directly or indirectly. Here is a board illustrating the progress of wind and string amongst civilized nations. First of all, there is the earliest form of the æolian harp, in which the string was exposed to the action of the wind. Then, next, Professor Robison used a flattened reed, which was exposed to the wind along its whole length—a ribbon lying in a long narrow slit, and by that means the fitful sound of the æolian harp was reduced to one steady note. The next notable change was that which took place under Sir Charles Wheatstone, and I have here the original apparatus which was used by him, by which the wind was concentrated on one part of the string's length. There is another form in which the wire of a pianoforte was used, but in other respects it did not vary. At least six names are connected with those two forms, amongst whom I may mention Sir Charles Wheatstone, Mr. Greene, and Isoard. Here for the first time is used a reed in connection with a string, and that was used thirty years ago by a man named Pape. It consists of a free reed as used in harmoniums, and a string apart from it, the connection being effected by means of silk thread, so that the reed acts only upon the string in the backward motion. Then comes another form in which the string could be played upon by hand. That was made by a man called Julian twenty years ago. Here again is a string flattened at one point to a tongue, which lay between flanges in a frame, and that could be played upon as in a violin. The method upon which I founded all my investigations is here shown. It consists of a reed-tongue, instead of a flattened portion of a string attached to the end of a string. That was first invented by Mr. Farmer, our organist at Harrow. Here is the reed and string brought into direct and rigid connection, and here is the mode which I suggested, in which the string is set apart from the reed, so that it can be used in its own register as in a harmonium. When we get as far as this, and the reed-tongue and string are used, there is no longer any room for originality as regards either wind and string, or reed and string, but there is plenty of room for improvement. When the reed-tongue is put to the end of a string, you may make the string act, but the farther you get from the reed the intervals get wider and wider, because the amount of control exercised upon it gets less and less; and if you go as far as the sixth interval, it would be widened out immensely, but you seldom even get as far as that, because the string would bring up into new combinations of nodes and segments, or refuse to speak altogether. After trying a whole year to make a wind violin on this principle, I gave it up in despair, because I could not get enough intervals, and because they were so irregular; also because the tone varied. When you are close upon the reed, the reed is constrained in its motion and the tone is pure, but as you get away it is less controlled, and the tone becomes more loose, coarse, and reedy. It was only within the last four months that it occurred to me to use a conical form of string. The small end is applied to the reed, and as it departed from it it got larger and larger in bulk, and accordingly the intervals remained the same, and the tone remains the same, because you encounter the firm resistance of the larger string. There is another curious thing also, that whenever you tuned the string the intervals pulled out, because the relative intervals in the string did not remain the same; the tongue entered into its composition to such an extent; but when you tuned from the small end of this conical string, instead of the diminishing bulk of the string, you are able to maintain the same bulk, and accordingly this new form of string has recovered to us an instrument which I once abandoned in despair. But although I thought it impossible to do anything in the way of a wind violin, there was still a chance of doing something with the organ. If you use merely one string and one reed, you can take the best note they afford, and these difficulties of the intervals do not arise. Accordingly, I tried to make an organ with a separate reed and string to every note, and I found that whenever the string broke up into three segments, it gave a beauty of tone which I could not gain by the most deliberate means. The reasons for this superiority of tone, when a string is broken up, may be understood when one considers that any body, whether it is a plate, or a reed, or a string, or glass, when broken up into segments, and so made to sound, gives an intense harmonic tone. Each of those segments which compose that body are of course vibrating, with smaller amplitudes than the whole body would, if vibrating in the fundamental note. This is especially desirable when the string has to encounter the resistance of wind, and when the vibrator meets wind, the great thing to aim at is that the vibrations should be close and small. Therefore, the smaller the motion you obtain for the vibrator, the more intense and clean the tone will be. That was the reason why I always endeavoured to gain the presence of the node to ensure this harmonic tone. Nothing better could be desired than the tone when you got it, so long as time was no object; but you had to wait for it, because there was this resolving of the string, and it was also desirable that space should be no object, because if you got a structure about 8 feet high, it was a mild and convenient form of making a string organ. However, these difficulties of space have now been overcome, and the largest note required with a string corresponding to the 20 feet pipe of an organ is contained in this small box. It is the section of the register of the pedal stop of a string organ, and instead of the string being exposed, and liable to be deranged, it is stowed away inside, the wind escaping through a channel and pallet controlled by the ordinary harmonium treatment. The string is in that little coil, only a foot long, and instead of going out of tune, which it did when it depended on the string, it now no longer depends on the elasticity of the string, but upon a sort of bow or crook which is placed at the end, on which the tension depends, so that the tension now depends on a spring, instead of on the string. These improvements have only taken place during the last few months; for about this time last year, when I spoke at the Royal Institution, matters were in such an unsatisfactory state that I felt the only thing to do was to retire home, and never emerge until something satisfactory was done. Mr. Hermann Smith came with me, who entertained a singular theory on organ pipes and their functions, which has received a very remarkable verification. He regards an organ pipe as containing reciprocating action of two forces—a reed and a column of air which acts upon it. The column of air restrains, steadies, and purifies the vibrator, which would otherwise be coarse, and not worth hearing. The theory may be summed up in a few words, that an organ pipe is a spring cushion re-acting upon a lamina or plate. When we have a reed pipe, that is easily understood. Here is the reed, and here is the spring cushion of air, its elasticity depending on its bulk and proportion, but when you come to a flue pipe, in which the reed is not apparent, it is not so obvious; but Mr. Smith regards the reed as still existing, and considers that the sheet of air which passes from the mouth of the pipe acts in fact as a reed. Bearing this fully in mind, its application to the reed is easily seen. It one day occurred to me to make a double ring of wire, which I did; and fixed the two ends into a board, bored a hole in it, and applied an open reed, and I found there was the same result as with the string. We then determined we would analyze what were the real functions of the string and of the reed, and a long series of experiments took place, and at length it was found that in order to gain the æolian tone there were three things necessary: there must be constraint; there must be sympathetic resistance; and there must be transmission, in order to gain power; and those were all contained in the first object I had tried, namely, the ring. The ring first adopted was that of the simple form of a watch-spring, namely, a double ring; but one day on looking at Lissajous's diagrams, showing the different forms of vibration, an idea occurred to me that a different tone might be gained by a different form of spring. At first I had only a ring, and this afforded perfect constraining effect, but there was still the danger that to get perfect purity of tone you had to go on increasing the power of the ring, until there was so much resistance that the reed would not sound. The question was, how to get fulness of tone by getting that ring of a certain size and a certain elasticity, and yet that you should have the power of changing the quality of tone, without altering those favourable conditions? Accordingly, I have here eight different type forms of constraints upon the vibrator. Supposing I have a piece of wire, I begin with the frame of the reed, turn it over, come round, and attach the reed to it, and continue its course until it comes round on the other side of the frame, so that it forms a circle. That gave perfect resistance, and a perfectly pure tone, and was sufficiently elastic to allow it to speak at once, and afforded sympathetic resistance. Then, if you wished to change the tone, how could it be done? The next deviation was to flatten out the two circles, and a further development was to bring them to almost the shape of an almond. In passing from the simple circle to that form, you pass through all varieties of tone from the flute to the horn. The simple ring makes the simplest tone, and as you depart from that more harmonics are allowed to be introduced, because more play is allowed to the vibrator. If a rectangular figure is adopted, it allows more freedom than the circle, and that gives the tone of an open diapason pipe; and when that is changed into a more acute form, you get all the phases between the open diapason and the trumpet. I will sound two or three of these notes just to show you that there has been something absolutely accomplished, and that you may hear two or three qualities of tone. I will not detain you longer, but I have one thing further to say which is to show you what is the upshot of all this. It is not merely that one means is substituted for another. It does not concern a scientific body that there should now be success in place of a forlorn hope; but it means that that which has very naturally encountered ridicule has come true: that the natural division of a string being broken up, and made to sound by a reed, is a means of affording a more intense and more pure tone, and a tone is at once afforded by this natural coincidence better than any scale could arrange; and that a vibrator can receive from solid objects that reinforcement which has hitherto been considered peculiar to columns of air. When you once come to use solid surfaces, we know that a square yard of vibrating surface can afford to any amount of notes an equally good reinforcement. The whole object of these experiments is to show that we may use solid bodies, instead of columns of air, in re-acting upon a vibrator. This is the result which has been reached so far. The greatest progress has been made during the last year, but I think what I have said is enough to enable you to judge whether that result is worth obtaining, and whether it has been honestly and patiently worked for. *

---

* In the interval between this lecture, and the lecture delivered at South Kensington Aug 26th, so many improvements have been made, that this lecture is chiefly interesting as indicating the stage of progress then attained.

The CHAIRMAN: I now ask you to pass a very hearty vote of thanks to Mr. Hamilton for his very successful research, and for the extremely lucid way in which he has brought it before us.

The Conference then adjourned until two o'clock.

On reassembling, Mr. DE LA RUE, D.C.L., F.R.S., took the chair.

He said: There has been a slight change in the programme this afternoon in consequence of the modesty of M. Tresca, which prevents him speaking twice on the same subject. He has already brought before the Conference his wonderful researches "on the Fluidity of Solids," and he desires, being present this afternoon, and being connected with the "Conservatoire des Arts et Métiers" in Paris, to speak of the historical monuments of science and on the institutions which ought to preserve them.

UPON OBJECTS ILLUSTRATING THE HISTORY OF SCIENCE, AND THE MEANS OF ENSURING THEIR CONSERVATION.

M. TRESCA: Gentlemen, Mr. Spottiswoode informed me yesterday that he had put my name down for a conference upon the Fluidity of Solids. It would have been impossible for me to comply with his wishes, since I have already touched upon that subject on Wednesday last; and, consequently, I can only show you my good will by offering you another subject for discussion. I will accordingly replace a question, which, on repetition, might be considered too personal, by a few observations on the most advisable means of preserving the historical apparatus collected in your exhibition, but which are yet far from sufficient in number to satisfy, as fully as might be wished, our scientific curiosity. In England, Newton's telescope, Newcomen's and Watt's steam-engines; in Italy the apparatus of Galileo, of Torricelli, of Volta, of the Academy del Cimento; in Holland, the instruments by which Huyghens made his discoveries, and the apparatus of s'Gravesande and of Otto von Guericke, certainly form precious collections, but how many other important discoveries would not have been likewise represented, if as much care had been bestowed upon preserving the instruments, as in publishing their results. Without going far back into the History of Science, how many instruments have disappeared? After having been kept for some generations in families, with the respect they deserve, they are often passed over unnoticed and are lost for science, because they, in a very short time, cease to present any definite interest. If preserved in the laboratory of the philosopher, they are often disfigured for other reasons; professors, who have them at their disposal, are often carried away by the love of science, and cannot always resist altering an ancient instrument, in order to adapt it to the purposes of some experiment at that moment under consideration When once the modifications have been made, the instrument is never brought back again to its former construction, and often the new arrangement does not entirely answer the purposes of the fresh researches. And many a time has the very author of the first discovery mutilated his own instrument in order to follow up the investigation of some matter of secondary importance, and the original apparatus is for ever lost.

These observations are suggested by the difficulties which we have met with in collecting, as far as we could have wished, all the historical instruments connected with French discoveries in science, and if we have succeeded in bringing a certain number together at this exhibition it is owing to the readiness shown by the directors of our scientific establishments in responding to the call made by England.

To speak here only of the apparatus included in the Physical Section, the "Ecole polytechnique," the "Observatoire," the "Collége de France," the "Faculté des Sciences," have united with the "Conservatoire des Arts et Métiers" in order to call to mind, by a few original instruments, the discoveries of most of the celebrated French scientific men. This last institution, founded at the beginning of this century, and endowed with all the valuable objects found after the great political events of the preceding years, has been able to save from destruction many instruments interesting on more than one account. Those of the 18th century, however, are few in number; and we can mention but one or two that date as far back as the 17th century.

The immortal Pascal, whom we might place by the side of Torricelli, is only represented by his calculating machine; and if it affords but slight scientific interest it is, at least, undoubtedly authentic. As you may see, Pascal wrote on it with his own hand this inscription :—

"Esto probati symbolum hoc.
"Blasius Pascal Arvernus, inventor. 20 Mai, 1652."

How much it is to be regretted that the original instruments employed to prove and to measure the effects of atmospheric pressure should not likewise have been handed down to us. Let me also draw your attention to these two globes with clock movement by Just Burg (1586) and Jean Reinhold (1588), and which are, moreover, so remarkable for their chasing, attributed to Jean Goujon. The names of the makers are, no doubt, of less importance, but the inscriptions are contemporary with the change in the calendar, and the mechanism within them cannot but offer us curious problems which demand no little investigation to prove certain much controverted points in the history of clock-making

On the other hand, our collection, from the last years of the 18th century, is extremely rich, in spite of frequent gaps, and you may see the proof of it in the number of instruments which I have placed before you, and which—should you find the subject interesting—we will examine in the order which has been so suitably chosen for the classification of the objects sent to this exhibition.

But before entering into these divisions, let me call to your notice this small cathetometer by Dulong, the first that was ever constructed, and which was made under the direction of that celebrated man. It has been used as a model for all similar instruments, by means of which the numerical values of the differences observed in the principal phenomena have been able to be computed with exactness.

In the Section of Sound we should have wished to renew Savart's beautiful experiments on vibrations, but the plates on which he studied them have not been able to be arranged as they were originally, and so we have been forced to be satisfied with his musical instruments, near which you see the Register of Duhamel, who was the first to succeed in inscribing with sufficient precision the vibrations of sound. The determination of the velocity of sound is brought before you by M. Regnault's original apparatus and also by the one of M. Le Roux, who arrived, at about the same period, at a nearly identical conclusion.

To judge by the number and the perfection of the modern French optical instruments sent to the exhibition, there would be reason to believe that this branch of science is more attended to in France than anywhere else.

It is true that we have been fortunate enough to secure, in the various sub-divisions, histórical instruments bearing well-known names.

Fresnel's first lens—a true landmark in the History of Science—is accompanied by a large collection of models of the French lighthouses, from the first to those most recently constructed. And thus all the improvements which have been effected in this powerful means of averting loss of life at sea, can be seen at a glance. And England is, indeed, the most appropriate place where we could have exhibited such a collection which has been brought together, thanks to the engineer who carries out, with so much skill, the administration of our lighthouses.

Electric light has lately been applied to industrial purposes to an extent that cannot but go on increasing. The machines made by the "Compagnie l'Alliance," and the Gramme instrument, can light up vast workshops or timber-yards. And now M. Carré has succeeded in making carbon artificially, which, used instead of crayons of coke, gives a much greater regularity, although it burns faster. There is every reason to hope that this manufacture will allow of the use of sufficient compression, so as soon to get rid of the defect entirely.

The regulators used are those of M. Foucault, M. Dobosq, M. Serrin, and of M. Carré himself.

With regard to the determination of the velocity of light, here are the two instruments made use of by M. Foucault and M. Fizeau; it is known that it was by means of the latter one that M. Cornu succeeded last year, in his experiments between the Observatory and the Tower of Montlhery, a distance of twenty kilometres. We have placed next to them Wheatstone's revolving mirror, which belongs to the Paris Observatory.

To this establishment we likewise owe Arago's interferential apparatus, and the complete collection of his historical prisms; I can merely mention those of Biot, and Senarmont, and Jamin, and so many others who have made themselves celebrated by their researches in polarization, interference, and double refraction. M. Descloizeau must be named on account of the polarizing microscopes which he uses for the investigation of crystals, and which complete, in another branch, the series of goniometers of Charles, and Babinet, and Senarmont.

The spectroscope is now largely employed for industrial purposes. Here is the whole collection of instruments of M. Dubosq and Laurent, who have with such signal results followed in the steps of M. Soleil.

You know M. E. Becquerel's phosphoroscope, but I advise you to examine the small tablet upon which one of our most skilful glass-blowers has written in large letters this one word: *Phosphorescence.* If this little plate be exposed for a few instants to the rays of the sun, it is affected in such a way that all the letters appear luminous and assume different colours. The phosphates, with which the tubes are filled, then vibrate, each in its own fashion, conformably to the principles laid down by M. Edmond Becquerel, and the light which it gives out is of such brilliancy that it makes this phenomenon of phosphorescence one of the most beautiful of optical experiments.

I will point out presently the heliostats of s'Gravesande, of Silbermann, of Gambey, and of Foucault, which are in this exhibition. They form an almost complete collection of these instruments so precious for observations.

We have decided to entrust to you Daguerre's second proof. It is with great satisfaction that we have noticed the precautions you have taken—by placing it in a red glass frame—to preserve in a fitting manner this first specimen of a great art. And we are most happy to have been able to place by the side of the historical picture of the Photographic Society of France, in addition to Daguerre's attempts, those lent to us by M. Fizeau, and which represent the fixture by chlorate of gold, and his process of photo-engraving which marks a first and most important epoch in the annals of photographic reproductions. The stone and the proof by Poitevin call to mind undoubtedly the most important step which has been taken since then.

Daguerre's attempt carried out in red glass, without mercury, by M. E. Becquerel, and especially his coloured spectrum, are also very striking curiosities of photography. M. A. Girard's microscope is a more practical object: it gives directly magnified impressions of microscopic objects placed before the object glass.

The French cases exhibit, through the original apparatus of our principal inventors, the most complete history of the studies and discoveries in heat. The measurement of heat is represented by the instruments of Lavoisier, of Dulong, of Regnault, of Favre and Silbermann; observations on sideral heat by the actinometer and the pyrheliometer of Pouillet, and by the actinometer with thermo-electric pile of M. Desains. The metallic bars of Despretz are those which he used for his first experiments on the conductibility of solids; Gay-Lussac, Despretz, M. Dumas, and M. Regnault remind one of the uninterrupted series of studies on the density of gases and vapours; the collection of M. Regnault's apparatus is a proof of the gigantic labours which he has successfully undertaken to place upon a firm basis the actual facts connected with the science of gases and vapours.

With regard to the measurement of the expansion of bodies according to the method of Newton's rings it has become, in the hands of M. Fizeau, perhaps the most exact of all those connected with the science of heat.

For the very reason that discoveries relative to magnetism are comparatively of recent date, we have been able to procure some historical objects of great value.

M. Jamin has given us his great artificial magnet, made of thin plates of steel, which can lift 200 kilogrammes, and which this skilful experimentalist used as a starting point for his studies on the laws of magnetic distribution, which had been, up to that time, so little understood.

We have placed next to it the model of Gambey's deflecting compass, one of the finest instruments from the hands of that celebrated man.

In the collection of the Ecole Polytechnique, the small natural magnet belonging to M. Obelliane is remarkable as being, with regard to its weight, by far the most powerful magnet known; it can lift forty times its own weight.

With regard to electricity we are necessarily less rich in historical models, but nevertheless the series of discoveries made by French scientific men is tolerably complete.

Among the electrical piles may be noticed the fine collection of the different models reproduced by M. Ruhmkorff and the considerable effects of the secondary pile of M. Planté; among the thermo-electrical apparatus, the piles of Pouillet and of M. Becquerel, as well as his thermometer and his pyrometer. Thermo-electrical needles have, for some years, been put to a great number of different uses.

This support and this "solenoïde" are particularly deserving of your respect: They are those of Ampère, and the College de France wished that this table which forms part of the apparatus, and by means of which the celebrated philosopher varied his experiments and proved the laws of the action of currents upon currents, should be exhibited here. And thus, for the first time, the means of the discoveries of Volta, of Faraday, and of Ampère are to be seen under one roof.

Ruhmkorff's great induction bobbin (bobine d' induction) represents in itself alone a new conquest by science; but, in the Telegraph Division, in spite of the very interesting telegraphic apparatus of M. Mayer, of M. D'Arlincourt, of M Deschiens, of the military system of M. Trouvé, and of our telegraphic administration, we cannot show a collection at all comparable to fine English one, which commences with Wheatstone, and which forms a complete museum of this portion of the History of Science in its relation to the every-day wants of modern society.

On the other hand, you will no doubt have observed the models of the first optical telegraphs by Chappe, Bétancourt, and the last optical telegraph of Colonel Laussedat. Galvano-plasticism is only represented by Jacobi's first work, already perfect, and which the inventor presented to the Conservatoire in 1868.

With regard to modern astronomy, a few instruments only are to be seen at this Exhibition, but they have been selected with the view of interesting persons engaged in this science. Here is the siderostat, of which the mirror has been improved by Foucault himself, and the photographic telescope which was used for observing the transit of Venus from Campbell Island, one of the French stations in this astronomical campaign in which all the learned nations took part. The proofs given by these telescopes have allowed of tolerably numerous data being obtained, which measured, at leisure, under the direction of M. Fizeau, by micrometric means, already permit us to state the exactness of the figures which may be deduced from them for the chief results of the phenomenon.

Nor have we failed to exhibit at Kensington the pendulum, by means of which Foucault succeeded at the Panthéon in his first attempt to give a direct demonstration of the rotation of the earth; it is accompanied by the plaster sphere on which he drew his preliminary figures, and by the "entreteneur" which he had constructed afterwards in order to make the demonstration continuous. The "entreteneur" is the one which he sent to the Exhibition of 1855.

I will not speak at greater length on this subject, especially as M. Le Verrier intends soon coming among you to praise the English astronomers, and particularly Bradley, for whom I know him to entertain the utmost veneration, in consequence of the methods he has followed, and of the marvellously accurate results they have always produced.

In calling your attention, Gentlemen, to a few of the historical apparatus of French science, we have certainly not had the intention of exciting competition. If it gives us pleasure to show our instruments, we are also most happy to see yours; and judging from what we have already seen of your exhibition, there can be no doubt that if, under the direction of my excellent friend Mr. Owen, whom I notice among you, the idea was seriously entertained of devoting the great resources, which you have at hand, to the foundation of an institution similar to our Conservatoire, you would very soon obtain the most satisfactory results.

If such is your intention, we would seize the opportunity offered us of assuring you of our hearty co-operation and assistance.

We are, however, not generous enough to be careless of any return for our good wishes, and we are already thinking of asking your leave to have copies made of some of your most precious documents: it is good that it should be known everywhere with what a small instrument great discoveries could be made in Newton's time, and it is also necessary that our scientific men should have in their laboratories perfectly accurate copies of the apparatus of Wheatstone and of Joule.

When we shall possess authentic specimens of all your historical models, and when you have copies of all ours, the yet unsettled points in the history of science will decide themselves; the same discovery will bear everywhere the same name; indisputable dates will put an end to all competition, and science will henceforth have acquired, in a firmer manner, both with regard to the past and to the present time, that cosmopolitan character which it is so important for it to assume.

The CHAIRMAN: Ladies and Gentlemen, you have anticipated, as I am sure you would do, the proposal I intended to make of inviting you to thank M. Tresca for his very lucid review of the instruments contributed by France, and connected with some of the grandest discoveries of the French *savans*, but for the few words he has been so good as to utter, they might pass unnoticed by a great number of persons. The names of Biot, Arago, Becquerel, Ampère, Daguerre, Fizeau, Pascal, Savart, Fresnel, Lavoisier, Dulong and Regnault are all historical names in science, and we are very happy to have even a small portion of the apparatus which they used. I need only refer to this wonderfully simple piece of apparatus devised by Ampère for the foundation of the electro-magnetic discoveries with which science has been enriched by so many *savans*, not the least amongst them our own Faraday. The ingenious instruments, which M. Tresca has passed so rapidly in review, recall one's recollection to one's reading and experience respecting the progress of science during the last fifty years, and if the next fifty years are only as prolific in discoveries as the last it will indeed be a privilege to live in the days in which they occur. The last outcome of the work is the production of magnets of great power by M. Jamin ; I possess one which is so powerful that it will allow itself to be held out horizontally by its armature. Many years ago the Dutch were considered to construct the strongest magnets, but by this beautiful arrangement of magnetic plates attached to each other far more power is obtained. M. Tresca has also alluded to a photograph produced by the red rays by Becquerel. We all know that the red rays and the ultra red rays possess very little actinic power. Their study has been taken up recently by Captain Abney, and we may hope some day to get the whole spectrum, not only photographed but also fixed ; it was photographed some time back by M. Becquerel, and this case, which is very properly locked, so that it cannot be opened except under proper precautions, for exposure to strong daylight would spoil it, contains his photograph of the whole spectrum in its natural colors : And the day may come when we may not only get portraits and delineations merely in monochrome, as it were, but get photographs of natural objects in their natural colors. The review, which M. Tresca has been so good as to give of the progress of French science, illustrated by a selection of the apparatus used by philosophers of the great names he has enumerated, is highly important. He has alluded also to the difficulty he has experienced in collecting such objects, simply because they are very seldom taken care of when once they have served the purpose of their authors; and has also mentioned the grand institution in Paris, the Conservatoire des Arts et Metiers, with which he is intimately connected, in which are contained some of the most valuable records of the progress not only of science but of the mechanical arts. He has also alluded to a rumour he has heard that something of the kind is contemplated in England. I only hope that this may have effect, and that in this country we shall not in future have to regret the loss of apparatus which he has deplored on the part of the French; and that, eventually, all those pieces of apparatus which have served and will serve for scientific discovereries may be preserved for future generations. I point again to this very beautiful and simple apparatus of Ampère's to show what very small means are required to effect the most grand discoveries; and I ask you, again, to repeat your thanks to M. Tresca.

Mr. Ranyard had promised to give us a description of the astronomical instruments contributed by the Astronomical Society, but I am informed that they are not yet sufficiently well arranged for him to give you his intended discourse, and we will therefore defer that to next week; and I will now call on Lord Rosse.

ON THERMOPILES.

The EARL OF ROSSE, D.C.L., F.R.S.: I have been asked to describe in as few words as I can the thermopiles which I use with the moon, as I believe they are more successful than any other apparatus which have been tried. I am happy to do my best to make my method of observation intelligible to the public. There are very few points I need refer to. Of course if I went into the whole subject, and into the results, it would take much more time than we have at our disposal. This is the actual apparatus which was used in most of my experiments. I may mention that the reflector of my telescope is three feet diameter. The larger telescope, which is six feet, was not used for the purpose, because it had only a very limited range of motion, only about twenty minutes to half an hour on each side of the meridian, and therefore there was only that length of time on any night on which to make observations, and clearly, anywhere but especially in an uncertain climate like ours, it is desirable to seize on every moment when the moon may be sufficiently high in the sky. The focal length is about twenty-seven feet, and this apparatus was presented to the mirror, being fixed in the mouth of the tube, the small condensing mirrors being fixed in the focus of the large mirror. These mirrors are of about three inches focal length, and the thermopiles are placed in their foci, so that the whole light that enters the tube of the telescope, or falls on to a space of three feet in diameter is first concentrated by the large mirror on to the surface of each reflector, and then by them concentrated further on to the faces of the thermopiles $\frac{3}{8}$ inch diameter. The reason why I used two reflectors was to destroy the disturbing effects of the various strata of the atmosphere, the clouds which might be over it and those arising from heat and cold acting unequal on the thermopiles. The effect of using two mirrors is that they neutralize in one another the differences produced by heat and cold falling on the apparatus, as they act in opposite directions. If the heat of the moon be turned alternately on this reflector and on that, you simply get the effect of the moon doubled, but by throwing it on one and then off it, you get the single effect. Many of the details of the apparatus are hardly of general importance, because it is adapted to the instrument for which it is used. The mounting of the instrument was at that time an altazimuth which, as all astronomers know, is an inconvenient one to use for this purpose, as it is necessary constantly to work both the altitude and azimuth motions in following the motions of the heavenly bodies. The image being seen by the attendant on each mirror alternately, he swings the tube of the telescope which is suspended by a rope until he sees the moon's image on the reflector where the slight dust or tarnish allows it to be seen. There is only one point I think I need allude to further, and that is the construction of the thermopiles. I am not aware that anyone has brought this principle to bear either before or since. In the ordinary thermopiles the number of elements is very great; a plate of bismuth is soldered to one of antimony, and you have as many as you please, and the solderings all side by side form the surface on which the heat acts. If you applied this to the moon the moon's light would shine on each of the solderings and you would have some of the effects on all the separate solderings. I found some difficulty in getting thermopiles of sufficiently equal parts to use in the apparatus, and therefore I thought of a simpler form which was within my own powers of manufacture. I tried to make thermopiles with several alternations, but the matter is one of great delicacy; the bars are so thin and have to be handled with such a care that a maker told me that after making them for some years his sense of touch became so delicate, and he handled everything so lightly, that he frequently, without intending it, let fall heavier things. I adopted the plan of having one bar of bismuth and one of antimony soldered to a copper disk. I made the bars finer and finer every time, and the finer I made them the stronger was the effect, because the cross section to carry away the heat and the mass to be heated were smaller, and the limit of the thickness really was the difficulty of manufacture. By having a thin disk of copper I was able to work with much finer bars. I think this construction of a thermopile is worthy the attention of instrument makers, because it so much simpler to make. I do not see any reason *a priori* why as great an effect should not be obtained from that construction as from fifty or sixty pairs. The heat falls on the face of the copper which is a good conductor, and if it is perfect I see no reason why there should not be nearly as much power obtained from a single pair as from a great many. I do not think there is anything else I need say on this subject, unless it is to state the amount of heat obtained which I compared with that obtained from a vessel filled with boiling water. The heat derived from the moon would lead one to infer that the radiation was equal to what it would be if the moon's surface at full moon was at about the temperature of boiling water, or something like 200° Fahrenheit. Of course the assumption must be made that the moon has the same radiant power as a lamp-blacked surface of a tin vessel of boiling water, and that we have no right to assume at all. In fact I ought simply to say that that assumption is made in order to enable another hereafter to calculate what is the probable radiating power of the moon. I placed a sheet of glass so as to intercept the rays of the moon, and then I found the heat was all intercepted except ten or twenty per cent, showing that the moon's heat was of a different quality and lower refrangibility than the sun's heat, because the sun's heat on an average gives eighty per cent. of the power through a glass against ten to twenty per cent. of the moon's rays. I might also add that the concentration obtained by this apparatus is very large. The moon's light spread on a three feet diameter is concentrated on a third of an inch, which would represent a concentration of the moon's light of about 11000 times, and allowing for loss by reflection by the large mirror it would be about 4500 times.

The CHAIRMAN: I believe Lord Rosse will also describe to us a photometer, which is also on the table.

ZÖLLNER'S ASTRO-PHOTOMETER.

LORD ROSSE: This is not a piece of apparatus with which I am particularly connected, but I have brought it for description, because it appears to be very little known in the British Isles. It is a German instrument, devised by Professor Zöllner. Our climate is not at all well suited for photometric observations, because of the variability of the transparency of the atmosphere, and the number of clouds which perpetually interfere with the observations. But this instrument has been used in the German climate very generally, and appears to rank very highly. Its primary object was to compare the light of various stars. There is one considerable difficulty in comparing the light of stars, which does not occur in some other photometrical experiments, viz., that you have to compare them as accurately as possible when they are of *different* colours, and it is a problem, therefore, to which only an approximate solution can be given. In comparing two stars, you have simply to equalize the colour as far as you can of the lamp which is used, and then, by trying backwards and forwards, to seek to get the mean result as accurately as you can. There is a small pin-hole which admits the light of the lamp. It passes first through a double concave lens, then through a Nicol prism, then through a quartz plate, by means of which the colour is varied, and then through two more Nicol prisms. The pin-hole forms a minute point of light, and there is also a double concave lens which brings the light to a focus. The light from the natural star comes through an object glass, and is seen by the side of the artificial star, reflected by a diagonal clear plate of glass, whilst that of the star passes through the glass with very little diminished brilliancy. The size of the pin-hole may be varied, there being six different sizes attached to the instrument. There are various telescopes used with it; this is the smallest size, a very small one being used for very brilliant objects. For observing Venus you require a very small aperture, or else you never get an image faint enough to compare. The concave lens is to make the light smaller and more like the light of a star. Passing through the first Nicol, by turning it, in combination with the quartz, you are able to vary the colour. It then passes through the other prisms, which can be turned round, and you can thus vary the intensity according to the well-known law. It is usual to take four readings, and then find the average. Professor Seidell, of Munich, who worked at the stars with this instrument, obtained complete observations by four readings within the limits of probable error of about 8¾ per cent. I may mention that the error of one night's observation of the moon's heat by the thermopile was about 10 per cent. on an average. It is rather less accurate than this instrument was in Professor Seidell's hands, but I do not think, with the practice I have had with this, I could get quite as good a result as with the thermopile. Here is another piece of apparatus attached to it, which is used for measuring the moon's light. It is similar to that described above for forming the artificial star, except that there is no quartz plate. In using it you see in the focus of the eye-piece three stars, two from the lamp—one coming from the back surface of the lamp, and the other from the front, and one of the moon; and a very practised eye would be able by changing the position of the Nicol to make the image of the moon intermediate between the brilliancy of the two, the difference in their light being due simply to the absorption in passing twice through the thickness of the glass.

The CHAIRMAN: Lord Rosse, who is a most worthy successor to his illustrious father, has been good enough to explain to us some of the improvements which he has been making in the great telescopes which he inherited from their constructor. It will be in the recollection, I daresay, of many of you, that in all instruments at that time you had to get two movements. and it is no easy task to mount them on an axis inclined parallel with the earth's axis. That, Lord Rosse has now effected, and is now improving the equatorial mounting, and also the means for the observer to reach the telescope, which is not very easy with either of those enormous instruments. He has nearly completed the mounting of the three feet, and if that is farther carried on to the six feet telescope, we may expect far greater results than it has hitherto yielded, inasmuch as the objects which have to be observed are, of course, carried round by the diurnal motion of the earth, and the large telescope being mounted only with an hour's range on each side of the meridian, you perhaps can only catch them very occasionally. If, however, he mounts the six feet telescope to reach any part of the heavens, undoubtedly new features will come out. It is a very beautiful investigation he has carried out with respect to the heat radiation of the moon. It was thought at one time that no heat reflected from the moon reached us, but he has shown in the most effectual manner, eliminating all radiation from the atmosphere near the moon, that there is a positive amount of heat reaching us from the moon.

I ought not to omit calling your attention again to this beautiful apparatus, the thermopile—the application, as he has said, of an ordinary thermopile. The difficulty of getting two exactly alike in their indications is extremely great, but by this beautiful contrivance of receiving the image of the moon on a large plate, and allowing it to conduct the whole effect to one joint, was a very happy thought, and it has been most admirably carried out. I beg you to offer your thanks to Lord Rosse.

Lord Rosse here took the chair, whilst Mr. De La Rue gave the following account of his new battery:—

Mr. De La Rue: I have ventured to take somewhat out of order the description of a piece of apparatus which may have some interest for you. It is a galvanic or voltaic battery, particularly applicable to those cases where a very large number of elements is required. It is extremely simple, and consists essentially of a flattened wire of silver and a rod of zinc. At one end of the silver D is a cylinder of chloride of silver cast upon it. In order to prevent contact between the rods of zinc and chloride of silver, the chloride of silver is surrounded with a cylinder A, open at the top and bottom, made of vegetable parchment. B shows the rod of chloride of silver surrounded by the cylinder.

Figure 1.

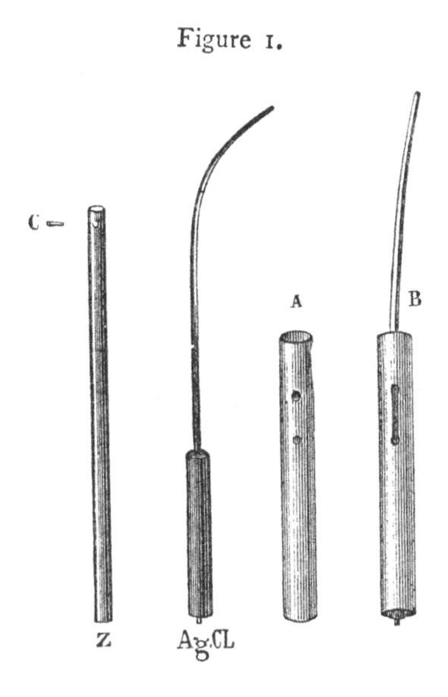

The cell is a glass tube; the stopper is a cork, made of paraffin, and a rod of non-amalgamated zinc is inserted through a hole in it; and then there is a second aperture, closed with a small paraffin plug, to enable one to fill the tube with a solution of twenty-three grammes of chloride of ammonium to one litre of water. I could not send the whole of my battery here; first of all, it is not easily conveyed; and, in the next place, it would be difficult to show its effects without special preparation; but I can show you one of this

Figure 2, showing ten cells of a rod—chloride of silver battery in their tray.

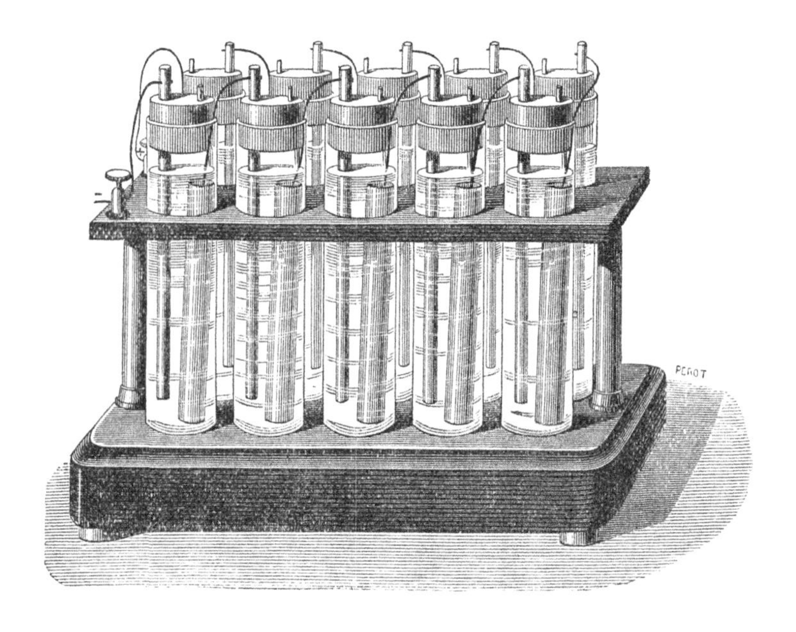

kind composed of forty cells. When I connect it with a Voltameter, which is a piece of apparatus for the decomposition of water and the collection of the products of decomposition, you will see that there is a rapid decomposition for the size of the battery, and that a certain accumulation of gases will take place. The object of this battery is to follow on the lines already laid down by our excellent and distinguished friend, Mr. Gassiot, who was the first, I believe, to construct batteries with any very large number of cells. I recollect he began with a water battery, in which there was a plate of zinc, one of copper, and distilled water as the electrolyte. With this he made some very interesting experiments, and showed that which was doubted at one time, that the voltaic spark will jump some distance before actual contact is made between the poles. We all know that when once contact is made, and the poles are afterwards parted, a beautiful arc of light, the voltaic arc, passes between the terminals, and the action is then continuous. He obtained some very beautiful results, with the discharge of the voltaic battery in residual vacua in tubes pretty nearly exhausted of the gases which they contained, and was one of the first, I beleive, to observe the stratification which takes place in these discharges. In a tube connected with the poles of a battery, or with the secondary wire of a Ruhmkorff coil, the discharge is sometimes continuous as a sheet of nebulous light, but mostly is beautifully stratified, and these stratifications we find to be different for different tubes. There has been no incontestable explanation yet given of the cause why we should have intervals of light and dark in the electric discharge in vacuum tubes. Mr. Gassiot carried on for many years a number of experiments, latterly with 3500 leclanche cells, with a view to elucidate the theory, and since he has in a manner relinquished them, I have taken up the subject in conjunction with my friend Dr. H. W. Müller. We have now 5,640 cells at work, and are adding 2,400 more, which will be at work in a few days.* If I could have had wires laid here from my laboratory, I could have repeated many instructive experiments. We have obtained some very curious results, which will form the subject of a communication to the Royal Society. The number and form of the strata varies, depending first of all on the individual tube, and the nature and tensions of the gases it contains, and also on the electro motive force of the battery and resistance interposed in the circuit. One may make the number of strata greater or less. Generally, if one introduces very great resistance their number augments, and as one takes out the resistance they diminish in number, and their width becomes greater, but this law does not always hold good, for there are some tubes which behave in exactly the contrary manner. But there is one point that I wish to bring particularly under your notice, and it

* 8,040 cells have been completed since this discourse, part of them, 3,240, have the chloride of silver in the form of powder, and are not so energetic as when it is in the form of a rod fused on to the silver wire. The striking distance of 8040 cells between a point (positive), and a plate one inch in diameter (negative), is 0.345 inch rather greater than one third of an inch.

is this :—formerly it was a point of doubt whether the spark would jump any appreciable distance before the circuit is closed. We have found the length of the discharge to be in the direct ratio of the square of the number of elements, and that leads to some very curious conclusions. For instance, in the paper, from the *Proceedings of the Royal Society*, I have before me the results given as follows :

Cells.	Rod chloride of silver.	Striking distance.
600	... ...	·0033 inch.
1200	... ...	·0130 „
1800	... ...	·0345 „
2400	... ...	·0535 „

which is very nearly the square of the number of cells. But as we have carried the number of experiments further we find that the striking distance increases in a far greater ratio. But even if the striking distance does not increase in a greater ratio than that and looking upon 1000 cells as unit—because with this high intensity we must have a larger unit than one cell—if we carry the battery up to 1000 units, or one million cells, the striking distance would be 764 feet, a true flash of lightning not only in distance but in quantity. I think it is not likely that a million cells will ever be made, but it is quite possible to carry up the number to 20,000 or 30,000 cells, and it is almost impossible to foresee the results one will obtain as the number increases. The advantage of this battery is that one can have it in one's laboratory always ready, walk in and make experiments at any time. Most other batteries from efflorescence or other causes continually get out of order, but with this one the first thousand was charged up in November, 1874, and to-day I was working with it and had the honour of showing some experiments to some of our foreign visitors. The acting electrolyte is the solid chloride of silver, which is insoluble in the weak solution of chloride of ammonium, with which the cells are charged, and yields up its chlorine only when the circuit is closed ; at other times no action takes place, and consequently no consumption of material occurs ; indeed, the battery remains always in action, and is very nearly constant. You notice figure 2 that the stoppers are made with paraffin, the feet are of ebonite, and the connectors are mounted on ebonite, so that the insulation is very perfect, and very little leakage takes place. It is most agreeable to be able to walk into one's laboratory and to find the battery always ready for work, and I am only sorry that the effects of 8,040 cells cannot be shown to you.

Lord ROSSE : I am sure you will all return thanks to Mr. De La Rue for his very interesting communication. The subject of batteries is one of very great importance, not only in experimental physics, in which a battery is a constant piece of apparatus, but in the telegraphic service and other applications to the arts. A constant battery which will not lose power by polarisation, or by working, or when it is not in work, which many batteries do, is a most invaluable thing.

Professor ANDREWS : I have one single observation to make. I must express my great admiration and delight at the fact that Mr. De La Rue is continuing those researches which Mr. Gassiot commenced, and I wish merely to correct one historical fact of which I happen to have an intimate personal knowledge. It was not Mr. Gassiot in this country but Sir Wm. Grove who observed these stratifications first. I should have been unwilling to make the statement but that the correction I think is better to be made at the time, and I should add that these stratifications were discovered about the same time in France. It has been a matter of some little doubt whether a Frenchman, whose name I forget, or Sir Wm. Grove was the first discoverer, but I believe Sir Wm. Grove was actually the first. I am quite sure if Mr. Gassiot were here, he would confirm what I say, that he does not claim the discovery.

Mr. DE LA RUE : I am much obliged to Professor Andrews for the correction, but as Mr. Justice Grove and Mr. Gassiot so frequently interchanged their thoughts the confusion was natural, and I qualified the statement by the saying I believed that was the fact.

Mr. DE LA RUE again took the chair, and called upon Cavaliere Professore de Eccher to read a communication on the instruments from Italy, especially those of Gallileo.

ITALIAN INSTRUMENTS AT THE EXHIBITION OF SCIENTIFIC APPARATUS.

Professor DE ECCHER: Your Queen was graciously pleased personally to thank the city of Florence and Italy for having sent the precious relics of Galileo and of the Accademia del Cimento to this first great International Exhibition of Scientific Instruments. Allow me, once more, to express here our most heartfelt and respectful thanks for such high condescension; which clearly shows how much importance England attaches to the historical instruments which you see before you. But with this accession to their value, I feel a proportional increase in the difficulty I have of speaking of them briefly, when thick volumes might easily be written upon them. I, therefore, humbly beg of you to excuse me if I fall short of your expectations. Having done my utmost in order that Florence should contribute her most precious relics to this solemn exhibition, I felt it my duty to accept the honourable, but at the same time onerous, task entrusted to me of explaining to you something of the principal instruments exhibited here. I am now present to discharge this duty.

I will naturally begin with those of Galileo.

This great man, born at Pisa on the 18th of February, 1564, was destined by his father for the medical profession; but in the year 1589, we find him mathematical lecturer at the university of his native town. His powerful intellect, adapting itself badly to the uncertainties of the art of medicine, had succeeded, almost unassisted, in mastering the exact sciences—the only ones which could satisfy his craving to scrutinize and become acquainted with Nature. When he was yet but a student at Pisa, he made the celebrated observation of the oscillations of the lamp, equable through different arcs, from which he afterwards deduced the theory of the pendulum and of the fall of weights. Here are two photographs of portions of Galileo's gallery. This third one shows the whole of the monument raised by grateful Tuscany to the memory of the creator of modern science. In the first one you see the student Galileo watching the oscillation of the lamp in the famous cathedral of Pisa; in the second, Galileo, as mathematical lecturer, repeating the experiments on the fall of bodies on inclined planes;

and in the distance the Tower of Pisa, from which, by letting fall bodies of different substances, he first showed how, but for the resistance of the medium through which they pass, they would all fall in the same time. Partly, perhaps, on account of being badly requited, Galileo left the University of Pisa, and in 1592 we find him mathematical lecturer at the University of Padua.

It was here that for the convenience of the youth whom he had to instruct in the arts of fortification and mechanics, he invented his proportional compass, also called military compass. This is one of the many which he himself made. On one side may be seen four sets of lines, they are :—

Arithmetical lines, which serve for the division of lines, the solution of the Rule of Three, the equalization of money, the calculation of interest.

Geometrical lines, for reducing proportionally superficial figures, extracting the square root, regulating the front and flank formations of armies, and finding the mean proportional.

Stereometrical lines, for the proportional reduction of similar solids, the extraction of the cube root, the finding of two mean proportionals, and for the transformation of a parallelopiped into a cube.

Metallic lines, for finding the proportional weights of metals, and other substances, for transforming a given body into one of another material and of a given weight.

On the other side of the instrument are :

Polygraphic lines, for describing regular polygons, and dividing the circumference into equal parts.

Tetragonical lines, for squaring the circle or any other regular figure, for reducing several regular figures to one figure, and for transforming an irregular rectilineal figure into a regular one.

Joined lines, used in the squaring of the various portions of the circle and of other figures contained by parts of the circumference, or by straight and curved lines together.

There is joined to the compass, as you see, a quadrant, which, besides the usual divisions of the astronomical compass, has engraved on it a squadron of bombardiers, and in addition, these transversal lines, used for taking the inclination of the scarp of a wall.

From Galileo's own correspondence we gather that in 1598 he had already presented his compass to the Prince of Holstein, and shortly afterwards he sent two others in silver, the one to the Archduke Ferdinand of Austria, the other to the Landgrave of Hesse; and he mentions besides that he had had 100 made in Padua by Antonio Mazzoleni, which he distributed among his patrons and admirers.

Here is the thermometer, or rather the thermoscope, as it first appeared from the hands of Galileo.

As far as my knowledge goes, the invention of the thermometer has been attributed to several philosophers; among the principal are Bacon, Robert Fludd, Cornelis van Drebbel, and Sanctorio.

In the first work of Bacon, "De Augmentis Scientiarum," the thermometer is not mentioned, nor is there any allusion to it in his second work, "De Sapientia Veterum;" but in the "Novum Organum," published in 1620, he describes a thermometer similar to the one you see before you without, however, declaring himself its inventor. With regard to Fludd, he published a description of a thermometer about the year 1603, after having visited Italy and spent some time at Padua. Sanctorio, without giving himself out as the inventor, describes a thermometer in his works published in 1625. In favour of Galileo, however, we have the testimony of Padre Benedetto Castelli, who in a letter written in 1638 to Dr. Ferdinando Cesarini, relates as follows: "I now call to mind an experiment shown me, more than thirty-five years ago, by Signor Galileo. It was this: a little glass bottle was taken of about the size of an egg, having a neck thin as a stalk of corn and about two spans long, and it was well warmed with the hands [you see that the instrument I am holding corresponds perfectly with this description, so that I am able to repeat the experiment before you]; then, the mouth of the bottle being turned upside down into a vessel placed underneath, in which was a little water, and the warmth of the hands being removed, the water immediately began to rise in the neck, and continued to do so until it was more than a span higher than the level of the water in the vessel. Signor Galileo had made use of this effect to construct an instrument for investigating the degrees of heat and cold."

With regard to this instrument we might well say that Galileo would thus have invented the thermometer even before the year 1603. We have, moreover, the evidence of Signor Giov. Francesco Sagredo, a noble Venetian, who, writing to Galileo on the 9th of May, 1613, says: "The instrument you invented for measuring heat, has been reduced by me into several very convenient and excellent shapes, so much so, that the difference of the temperature from one room to another can be seen up to 100 degrees. By the help of it I have observed several marvellous things, for example, that in winter, the air is colder than ice and snow; that at this season water appears colder than the air; that very little water is colder than a great quantity, etc.," adding that some peripatetics believe, "that the contrary effect ought to ensue, because heat having, as they say, an attractive power, the vessel in becoming warm ought to draw the water to itself. Then Vincenzo Viviani, in his life of Galileo, affirms that between 1593 and 1597 he invented the thermometer. Galileo's not having made any mention of his works ought not to create surprise; for the fact is that he undoubtedly communicated all his discoveries to his friends and disciples long before he made up his mind to write about them.

But in any case it is well known that the thermometer was for a long time called the Florentine Instrument; and this indeed is accounted for, since it was in Florence that it received its most important improvements, thanks especially to Ferdinand II. Sagredo, however, was the first to divide it into degrees (in 1612), and it was he also that closed it hermetically about the year 1615. The improvements effected by Ferdinand were described by Padre Urbano Daviso, who, after having mentioned that he gave it a special shape (corresponding to the one we now see), and that he filled it with coloured spirits of wine instead of with water, which, by freezing, breaks the glass, adds: "Hence it may be seen which of two liquids gives out the more or less heat or cold; and it will be possible to warm water, or a room, or a furnace to any degree; to keep it at that temperature, or raise it; to know when a thing has reached that state of heat necessary for cooking it properly; all operations from which it may be said that the chemical art has received its finishing touch; and in the same way it will be possible, with instruments made on the same principle, to find out the heat and coldness of any province, due observations having been taken beforehand, &c. And by this means it has been discovered that spring-water and also caves, cellars, grottoes, and other deep subterranean places, that in winter seem to our senses warmer than in summer, are at all times at the same temperature. And we are, therefore, compelled to say that the apparent difference comes from the way in which the air which circulates there affects our senses, and not from any variation in the degree of heat or of cold in the place.

Fra Paolo Sarpi, in a letter to Galileo Galilei, dated the 11th of September, 1602, says, that the same year Galileo had carefully examined the work of Guglielmo Gilberto, published in 1600: "De Magnete, &c.," and had repeated the experiments mentioned in it, and also made many new ones. The principal result to which he came was that of discovering the way to multiply the attractive power of magnets by arming them in a special manner. From a letter of Galileo himself, addressed to the Secretary of State, Curzio Pichena, we learn that he (Galileo) considered it probable that the same piece of loadstone did not preserve an equal power in all places on our globe. He states, moreover, that he was engaged in making the magnet bear three times its own weight; and that by dividing it into pieces, he could render it capable of raising thirty or forty times its own weight; finally he observes that the longer a magnet sustains a weight, the more it gains in strength. In another letter, to Marsili, Galileo announces that he had succeeded in making a magnet six ounces in weight bear 150 ounces. And the Abbot Castelli, in his lecture on the magnet, says: "I have seen a loadstone, only six ounces in weight, armed with iron by the untiring industry of Sig. Galileo and presented to his Serene Highness the Grand Duke Ferdinand, which lifts fifteen pounds of iron, worked into the shape of a sepulchre." Here is this historical magnet. As to the form of a sepulchre, this shape was probably given to illustrate the legend of Mahomet's tomb remaining suspended in the air.

And now, approaching the subject of Galileo's telescope, I cannot pass over in silence the fact that the invention of spectacles is likewise due to a Florentine. Ferdinando Leopoldo del Migliore, in his work printed in 1684, and entitled "Firenze città Nobilissima Illustrata," writes as follows: "There was a memorial (in Santa Maria del Fiore) which came to grief at the restoration of that church; it was, however, duly inscribed in our ancient register of burials, and is all the more dear to us, as it is, thanks to it, that we are made acquainted with the original inventor of spectacles. This was Messer Salvino degli Armati, &c. A figure of this man was to be seen, lying at full length on a stone slab, in civil costume, with an inscription around it, which ran thus—

> "Qui giace Salvino d'Armato degli Armati di Fir.
> Inventor degli occhiali.
> Dio gli perdoni le peccata.
> Anno D. MCCCXVII."

> [Here lies Salvino, etc., of Florence. The Inventor of Spectacles.
> May God forgive his sins. A.D. 1317.]

Redi, in a letter on the invention of spectacles, quotes Frate Alessandro Spina, a Dominican, as their inventor; speaking of whom, Frate Bartolommeo da San Concordio wrote in 1313: "Frater Alexander Spina vir modestus et bonus, quæcumque vidit, aut audivit facta, scivit et facere. Ocularia ab aliquo *primo facta, et comunicare nolente,* ipse fecit, et comunicavit corde ilari, et volente." From which it would appear that as Armati would not explain his method of making spectacles, Spina had found it out for himself. With regard to the date of their invention, Redi quotes a passage from the papers of the family of Sandro di Pippozzo, a Florentine, written about 1299, in which may be read: "I find myself oppressed with age and should not be able to read or write, without glasses called spectacles, invented lately for the comfort of old people when their eye-sight grows weak." These various quotations agree in establishing the fact that spectacles were invented shortly before the year 1299, by a native of Florence called Armati. Let us now return to Galileo.

In June, 1609, it was rumoured in Venice that an artificer in Flanders had presented to Count Maurice of Nassau an eyeglass so cunningly contrived that it made objects far off appear as if they were close at hand. When Galileo heard this he immediately returned to Padua, and after having thought over the matter for a day and a night, he set to work to make his telescope. When it was known in Venice that he had succeeded in constructing the enigmatical machine, he was invited by the Venetian Republic to present them with his telescope; he complied with this request on the 23rd of August, 1609, and dedicated it to the Doge. This solemn presentation you see represented in the photograph of the Gallery. Then the Senate, as a mark of appreciation, by a decree of the 25th of August, 1609, elected him a life lecturer of the Studio of Padua, at the same time granting him a provision of 1000 florins per annum.

As on several occasions Galileo was accused of passing himself off as the inventor of the telescope, and at the same time the great merit which is his due for having made it upon the very vague indications which he possessed, ought not to be ignored, it will, I hope, not be irrelevant if I quote a portion of what he wrote in answer to Padre Orazio Grassi, a Jesuit.

"What share of credit may be due to me in the invention of this instrument (the telescope), and whether I can reasonably claim it as my offspring, I expressed some time ago in my 'Avviso Sidereo,' which I wrote in Venice. I happened to be there when the news reached that a Dutchman had presented Count Maurice with a glass, by means of which things far away appeared just as clearly as if they were quite close at hand—nor was any detail whatever added. Upon hearing this I returned to Padua, where I was at that time living, and pondered over this problem; and the first night after my return I found it out. The following day I made the instrument. After that I immediately set to work to construct a more perfect one, which, when it was completed six days afterwards, I took to Venice; and there so great a marvel attracted the attention of almost all the principal gentlemen of that Republic. Finally, by the advice of one of my dearest patrons, I presented it to the Prince in full college. The gratitude with which it was received and the esteem in which it was held are proved by the ducal letters, which I have yet by me, since they contain the expression if his Serene Highness's generosity in confirming me for life in my lectureship in the Studio of Padua, with double the payment of that which I had previously received, which, in its turn, was more than three times what any of my predecessors had enjoyed. These facts, Signor Sarsi, did not take place in a forest or desert, they occurred at Venice; and if you had then been there, you would not have simply put me down as a foster-parent of the invention. But, perhaps, some one may tell me that it is no small help towards the discovery or solution of any problem to be first of all apprised, in one way or another, of the truth of its conclusion, and to know for certain, that it is not an impossibility that is being sought after; and that, therefore, the information and the certainty that the telescope had already been made, were of such use, that, without them, I should in all probability never have made the discovery. To this I answer, that the help given me by the information I received undoubtedly awoke in me the determination to apply my mind to this subject, and without it I should very likely never have turned my thoughts in that direction; but besides this, that I cannot believe that the notice I had had could in any way render the invention easier. I say, moreover, that to find the solution of a problem already thought out and expressed, requires far greater genius than to discover one not previously thought of; for in the latter, chance can play a great part, whilst the former is entirely the work of reasoning. We know that the Dutchman, the first inventor of telescopes, was simply a common spectacle-maker, who handling by chance glasses of various kinds, happened, at the same moment, to look through two, the one concave, the other convex, placed at different distances from his eyes; and in this wise observed the effect which followed, and thus invented the instrument; but I, warned by the aforesaid notice, came to the same conclusion by dint of reasoning; and since the reasoning is by no means difficult I should much like to lay it before you.

"This, then, was my reasoning: this instrument must either consist of one glass, or of more than one; it cannot be of one alone, because its figure must be either concave or convex or comprised within two parallel superficies, but neither of these shapes alter in the least the objects seen, although increasing or diminishing them; for it is true that the concave glass diminishes, and that the convex one increases them; but both show them very indistinctly, and hence one glass is not sufficient to produce the effect. Passing on to two glasses, and knowing that the glass of parallel superficies has no effect at all, I concluded that the desired result could not possibly follow by adding this one to the other two. I therefore restricted my experiments to combinations of the other two glasses; and I saw how this brought me to the result I desired. Such was the progress of my discovery, in which you see of how much avail was the knowledge of the truth of the conclusion. But Signor Sarsi, or others, believe that the certainty of the result affords great help in producing it and carrying it into effect. Let them read history, and they will find that Archites made a dove that could fly, and that Archimedes made a mirror that burned at great distances, and many other admirable machines. Now, by reasoning on these things, they will be able with very little trouble, and with very great honour and advantage, to discover their construction; but even if they do not succeed they will derive the benefit of being able to certify, for their own satisfaction, that that ease of fabrication which they had promised themselves from the pre-knowledge of the true result, is very much less than what they had imagined."

From the above-mentioned statements it is evident that if we have to thank Holland for having by chance discovered the principle of the telescope, we cannot but render due homage to the genius of the man who thought it out and constructed the instrument with full knowledge of what he was doing. And I may mention that very great was the number of telescopes made by Galileo, and that he himself gave away many, chiefly in compliance with demands for them, to distinguished persons, among whom were:—The Grand Duke of Tuscany, Prince D. Antonio de' Medici, the Elector of Bavaria, the Emperor Mathias, Cardinal Borghese, the Queen of France, the Landgrave of Hesse-Cassel, the King of Spain, the King of Poland, Giuliano de' Medici, Cardinal Dal Monte, the Dukes of Acerenza, Professor Wallis [Valseo] of London, and many others. And he, moreover, got reports from Holland, in which it was stated, on the authority of Signor Daniello Antonini, and of Spinola, that in that country, no one, not even the inventor, could make a telescope that would enlarge an object more than five times. And Constantine Huygens writes to Elia Diodati, in 1637, that in Holland magnifying glasses were not yet made with which Jupiter's satellites could be observed distinctly.

In the meantime, by dint of industry and perseverance, Galileo had succeeded in perfecting his telescopes, so that for some time he obstinately refused to impart to any one the manner in which he made them; and it was not until his eyesight began to fail him, that he consented to create a manufacturer in the person of Ippolito Mariani, commonly called Il Tordo; whose only authentic telescope now in existence, I am pleased to be able to place before you. It was only about the year 1637 that Francesco Fontana, a Neapolitan, began to make good telescopes. And since I am on the subject of telescopes, I must mention Torricelli, who, after the death of Galileo, having been made mathematician to the Grand Duke, set to work to construct telescopes of rare perfection. He had devised a more applicable method for whetting and cleaning the lens, of which he was the first to calculate the curve previously. Here is one of Torricelli's telescopes. This other one was made by Eustachio Divini di S. Severino, who, between 1646 and 1668, constructed telescopes of the extraordinary length of seventy-two "palmi romani." And Viviani, likewise, made telescopes in a masterly fashion. It is important to remember also that Cesare Marsilli, a member of the Accademia de' Lincei, devised a method of making telescopes by substituting a concave mirror, for the objective—an idea to which he alluded in a letter to Galileo in July, 1626; but which he never carried out, probably in consequence of the difficulty in obtaining the necessary mirrors. In his time, no man made more excellent telescopes than Giuseppe Campani, a Roman; there exist in many places in Europe beautiful ones made by him, some of which were even of the length of 210 palmi romani. This one of medium size is by Campani.

To return, once more, to Galileo. I am able to offer for your admiration all that remain of his venerable optical instruments, the two telescopes, and the broken objective, with which he made his most important astronomical discoveries; and these were described in a masterly way in his "Nuncius Sidereus," and in his letters and dialogues on the systems of the universe. Before all comes the discovery of the mountains in the moon; he showed the manner of calculating their heights, and proved that some of them are higher than some of our terrestrial ones. Then he explained that the lunar disc appears to us to shine but feebly in the various phases of the new moon, in consequence of the reflection of the earth, and how the movements of the moon influence the flux and reflux of the sea. He was the first to show that the milky way consists of a mass of innumerable stars; and, on the 7th of January, 1610, having his telescope fixed on Jupiter, he made the discovery that three small planets revolve round it; and on the 13th of the same month he observed the fourth satellite. Following up these researches in Rome, he determined the times of their conversions, described a figure of their movements, certified that they undergo eclipses, like the moon, and discovered that their progress is extremely fast, inasmuch as the slowest completes a revolution round Jupiter in little more than sixteen days; and that by their means, more than 1000 eclipses can be observed in the year, a matter of great importance for finding the longitude of any place. He dedicated these satellites to the Tuscan princes, calling them the "Stelle Medicee." In August, 1610, while examining Saturn, he was struck by its *tri-corporeal* appearance—as he himself expresses it. On the 30th of September, 1610, he discovered that Venus changes form, just as the moon does. In truth, whoever thinks over these most noble discoveries of Galileo, cannot but feel both heart and mind exalted and invaded by a sacred respect and veneration for the simple instruments that revealed to us so great a part of the heavens.

We are now approaching the moment when Galileo, urged by an ardent desire to return to his native city, and wishing to attend—unfettered by the trammels of public duties—to the publication of his works, is about to abandon the University of Padua, and establish himself in Florence, under the patronage of, and in the receipt of a stipend from the grand duke. Poor Galileo! Forgetting for a moment the uncompromising lealty to the Papacy, which thy future masters necessarily owed, since they had been re-installed in their dominions by Clement VII.—that head of Christianity and citizen of Florence, who did not hesitate to extinguish, by means of blood and treachery, the liberty of the Republic; dreaming of solitude, and tranquillity and study, thou foundest, on the contrary, agitation and persecution, and even torture! To which thou wast abandoned by thy princes, more through indolence and fear than from any ill-will against thee. How much more nobly would not the valiant lion of Saint Mark have protected thee! and alas! how prophetic was Sagredo in the affectionate, frank letter which he sent to thee as soon as he heard of thy departure! Towards September, 1610, Galileo had already settled down in Florence.

Viviani, in the inscriptions which he placed on his own house in the Via dell' Amore in Florence, states that the first microscope invented and made by Galileo was presented by him in the year 1612 to the King of Poland; nay, in the Life of Galileo, which he was commissioned to write for Prince Leopoldo de' Medici, Viviani attributes the invention of the microscope to Galileo. On the 26th of October, 1624, Prince Cesi, the founder of the Accademia de' Lincei at Rome, says in answer to Galileo, who had sent him one of his microscopes, that he had received the instrument which he (Galileo) had lately constructed for minute objects, but that as he had merely begun to taste it, so to speak, he would reserve an account of so admirable an instrument till he had had time to relish its merits more fully. And under similar circumstances (having likewise received the present of a microscope from Galileo), Imperiali writes on the 5th of September, 1624: "I have not words enough to thank you for the microscope which you have been so kind as to send me; it is very perfect in every respect, and is most admirable, as indeed are all your inventions." At the same period Bartolomeo Baldi, asking Galileo to give him one of these instruments, calls it *the little newly invented microscope* (occhialino). Hence we may conclude that Galileo invented the microscope, either about the year 1612, according to Viviani, or about 1624 according to the others. Christian Huygens in his work on dioptrics, written in 1678, remarks that he had heard say that about the year 1621 microscopes were seen in the hands of Drebbel. But there exists no document in proof of this. And besides, numerous were the inventions attributed to Drebbel, but the greater number of them belong more to witchcraft than to science.

Hence we Italians maintain, and I hope that others will agree with us, that Galileo is the inventor and first maker of the microscope. Unfortunately none of his instruments have been handed down to us. The one that I am now placing before you was found among the remains of the Accademia del Cimento; it is merely a simple tube—the lens is wanting. We may mention here that at a later date Galileo effected many improvements in his microscope.

To my great surprise, however, I found among the objects in this exhibition a microscope described as follows: Microscope invented and constructed by Zacharias Janssen, telescope maker, at Middleburg in the Netherlands. I cannot explain to myself how it is that Huygens, who was so bent on bringing under notice the works of his own country, should not have known of this microscope, which has been handed down even to us. But while acknowledging how difficult it is to obtain information regarding all these different discoveries, allow me to observe that I have never seen any document in which mention is made of a microscope before 1624.

And this is a suitable place to mention the binocular telescope, made by Galileo for observing with greater ease the satellites of Jupiter, and called by him the "Celatone" or "Testiera" (for the apparatus resembled a diving-helmet, having telescopes fixed in the apertures for the eyes). It was intended for use on board ship. And speaking of this same instrument, let me mention that towards the end of his life he offered it to Spain—at that time an important maritime power, in order that it should be of use for observing the eclipses of Jupiter's satellites, and hence, by knowing the ephemeris, to determine the longitude of any place even in a rough sea. The negotiations, however, with Spain having failed about the year 1636, Galileo offered the same instrument to the States-General of Holland, at the same time stating distinctly the four conditions on which the value of his instrument depended—namely, to know the theory and the tables of the "Stelle Medicee," to have perfect telescopes, to overcome the difficulties of the ship's motion, to have a perfect instrument for the measurement of time: all of which conditions he affirmed to be able completely to satisfy, for he had indeed invented a perfect measurer of time; and as to the motion of the ship, after having devised several other ingenious methods to protect the observer, the idea struck him of placing him in a boat floating in another boat filled with water, a spring being placed between them so that the two boats should not dash against one another. When the transaction which had been delayed through the death of the appointed agents, seemed, at last, to be coming to an agreement, Galileo himself died.

Allow me, after having spoken of these instruments of Galileo, the only ones which remain to us, and which I have the good fortune to be able to show you, to call to mind the principal discoveries which he made and afterwards published. And first of all the observation of the spots on the sun. Before he had given up his professorship at Padua, he had occasion while at Venice to point them out to Fra Paolo Sarpi, the celebrated Venetian theologian, and also to Fra Fulgenzio, his disciple and successor. Here is what the latter writes to Galileo, in a letter of the 27th of September, 1621, with reference to the Jesuit, Father Scheiner, who gave himself out as the discoverer of the solar spots: "It seems to me that that German Jesuit is a man of good judgment and deserves high praise; for as it is their peculiarity to make themselves a name by evil speaking, he could not, in his profession, have taken up a higher or more distinguished subject, or one that could bring his name in greater evidence; for even to be known as a slanderer is to have a certain reputation. But to return from this digression.

"I have the most distinct recollection that when you had constructed here the first telescope, among the first things which you discovered were the spots on the sun; and I should be able to point out the exact place, where you, by means of the telescope, showed them on a sheet of white paper to that father of glorious memory. I well remember the discussions which took place, first as to whether it were a deception of the telescope, or vapours in the interposed air, and then having repeated our observations, we concluded that the fact was such as it appeared and that it was deserving of serious thought. And that afterwards you left us. The recollection of all this is as fresh in my memory as if it were taking place at this very moment. But what beasts are to be met with! Truth conquers."

Thus it clearly appears that even in August, 1610, Galileo had discovered the spots on the sun. The following year he had occasion in Rome, to draw the attention of several dignitaries to them, as may be seen from what Angelo de Filiis wrote in 1613: "Besides this, he (Galileo) did not leave Rome until he . . . . not merely mentioned with words that he had found that the sun was spotted, but actually proved it. He pointed out the spots on several occasions, and once in particular in the garden of the Quirinal, in the presence of the most Illustrious Cardinal Bandini, and of the most Reverend Monsignori Corsini, Dini, Abbate Cavalcanti, Signor Giulio Strozzi, and many other gentlemen."

Nevertheless, the above-mentioned Father Scheiner did not hesitate to assume as his own this great discovery, when it was already known to most people. His brother monk, however, Padre Galdino wrote: "That he warned Father Scheiner that the spots which were to be seen on the sun, had been observed before any one else by Galileo; and Padre Adamo Tannero, who was not only a Jesuit, but also Scheiner's colleague in the college of Ingolstadt, in his "Astrologia Sacra," not only leaves him (Scheiner) unmentioned, but alluding to the solar spots expresses himself as follows: "Certe magnus astronomus Galileus horum Sydereorum ostentorum præcipuus inventor maculas. Solem incumbrantes aliud non vult esse, etc." Here it is beyond all doubt that Galileo was the first to discover the solar spots, upon which he afterwards wrote his "Lettere Solari," in 1613, in which he rebuffs, with solid arguments, all the objections raised against him; gives an account of the spots, teaches the way to draw them on paper, and, with happy intuition, compares them to terrestrial clouds; speaks of the rotation of the sun on its own axis, and admits that that luminary, to maintain its own heat, must be continually requiring fresh aliment (*pabulo*).

Another important work of Galileo is the one entitled, "Concerning the Things that Float on the Water, or that Move in it," in which, for the first time, light is thrown upon the principle of potential velocities. Most important, then, is his "Saggiatore," which was published in 1623, in answer to an astronomical work of the Jesuit Grassi, chiefly on the subject of comets. Among the many treatises which Galileo wrote, are some in which may be discerned the foundations of the sciences of hydrostatics and hydraulics; and it is certain that he conceived and made use of the theory of indivisibles before his disciple the celebrated Cavalieri. Coming to the "Dialogue on the Great Systems of the Universe," which brought such persecution on Galileo's head, it will not be out of place to recall how his incomparable knowledge and honesty and frankness brought down upon him the envy and hatred of the greater part of the clergy. That body knew full well that they could not for long battle against the light of truth, by means of the dark mist in which they had sought to envelop the minds of men, in order that every appearance of knowledge should belong to them alone, the absolute masters of the human conscience. At each new publication or discovery by Galileo there never failed a most bitter storm of reproof, incited by that caste which most of all hated and feared him, and which was often most disgracefully supported by some peripatetic, who, measuring Aristotle's vast mind by his own most limited intelligence in interpreting him, could not conceive that if that great philosopher were now to know of one of Galileo's sublime conjectures, very different would be his reasoning on the nature of things. It is enough to mention, that to such an extent was the hostility towards Galileo carried, mainly through the agency of the Dominicans, that one of them, Fra Tommaso Caccini.

lowered himself so far as to inveigh against him, in low and insolent language, in the Duomo of Florence, and did not blush to conclude with his opinion that "mathematicians, as the authors of all heresies, ought to be driven from every State." The narrow-minded monk ignoring the while, that of all the heresies which he so deeply lamented, not one was originated by a mathematician, while many had monks for their founders. Even the Bishop of Fiesole thought it his duty to allude to Galileo in his sermons.

So that when Galileo, who, it was well known, had, when yet but a student at Pisa, embraced the system of Copernicus, and both in that celebrated university and afterwards at Padua, had explained it with the greatest clearness, and with the addition of new arguments in its support to his scholars—when Galileo, I say, went to Florence, and began there, beyond the reach of the dreaded protection of the Lion of St. Mark, to publish the arguments which he was in the habit of communicating by letter to his friends, there was let loose against him a torrent of abuse and insinuations and subterfuges, for the purpose of obtaining his condemnation by the Inquisition. In fact, it was unavoidable necessity that compelled him to go to Rome to defend his views, and attempt to persuade the most obstinate that, as a matter of fact, the Bible was not opposed to the opinions of Copernicus, who, indeed, more than seventy years ago, had published his works, with a dedication to Paul III. Nor had there ever been any thought of prohibiting them. For all that, he could count it as a great piece of good fortune that he was able to return to Florence without any serious inconvenience, but with the sad annoyance, however, of having seen the book of Copernicus condemned by a band of ignorant and fanatical monks, by a decree of the 5th of March, 1616.

And for the moment everything seemed quiet; the fire, however, was but smouldering. In September, 1623, there succeeded to the pontificate, under the name of Urban VIII., Cardinal Maffeo Barberini, the friend of Galileo; and the latter went to Rome to do him homage, and received such honours and recompenses, that he was convinced that the time was come when he should be able to publish his "Dialogues on the Great Systems," a masterpiece both of language and science, in which he represents Signor Sagredo and Salviati, of Venice, as speaking in favour of the opinions of Copernicus, as well as Signor

Simplicio, a learned and sincere peripatetic. After many and great difficulties, the Dialogues, having been revised and approved, were finally published at Florence in 1632. But no sooner had they appeared, than his enemies decided to bring into play every evil art in their power in order to effect his ruin; and knowing that the Pope favoured him, and had even corrected, with his own hand, the introduction to the Dialogues, they quickly understood that, to carry out their purpose, they must turn Urban's friendship for Galileo into enmity. They therefore insinuated themselves hypocritically into the councils of the Pontiff, whom they made believe that Galileo, in his Dialogues, had meant to represent in the person of Simplicio no other than his own sacred person. At this suggestion anger and persecution straightway took the place of friendship in the mind of the Father of the Faithful. Of no avail was the protection, which indeed was but timidly proffered by the Grand Duke; nor yet the intercession of persons in high authority, nor even that of some ecclesiastics, in whom the feelings of religion were undoubtedly sincere, being inseparably bound up with those of truth. All was in vain; and although a pestilence was raging at the time, poor Galileo, old, in bad health, and sorely afflicted, was forced to leave Florence in the very middle of winter and proceed to Rome. At the Bridge of Centino, on the frontier, he was compelled to undergo twenty days' quarantine in a miserable old house, totally denuded of comfort. And in the meantime the Pope was revenging himself on all those who had had compassion on the unfortunate old man.

On the 13th of February, 1633, Galileo reached Rome, and was at first lodged at the Villa Medici, from which he was afterwards taken to the dungeons of the Holy Office, where he remained seventeen days, and was only released on account of his increasing ill health. Finally, after six months' distress and suffering, having once more passed four days in the dungeons of the Holy Office, he was conducted to the Church della Minerva, where, in the presence of the Inquisitors, he was obliged, on the 20th of June, 1633, to pronounce the famous recantation, which did not prevent him from uttering the now historical "Eppur s muove," which, while summing up in a word the most severe martyrdom to which a man outraged in his convictions could be subjected, expressed at the same time the boldest challenge that knowledge conscious of itself could possibly throw down to Ignorance. Endeavours have been made to ascertain whether Galileo underwent torture; in truth, there is reason to believe that he did, although not a word ever escaped from him on the subject of his treatment. He was held back, perhaps, by the threats and oaths, by means of which the Inquisition knew how to secure the silence of its victims. He was allowed, it is true, as a high favour, to have the Villa di Arcetri for a prison; but the state of his health after his condemnation, and the indignity with which he was treated, up to the time of his death, by the Court of Rome, lead one to believe that those barbarians must have vented their insolent wrath even on the body of so great a martyr for progress. But who can picture the heart-rending anguish inflicted on him when he was obliged to recant that which was the fruit of the great labours of his whole life? and he, who had risen higher than any one in the comprehension of Truth and of God, was forced to bow his head to a power—the direct denial of the Divine Essence—which, in order to rule more absolutely over the ignorant, has been compelled to have recourse to the monstrous Proclamation of Infallibility!

Pardon me if I have needlessly digressed from the duty entrusted to me of speaking only of what appertains to the instruments exhibited here; but in speaking of Galileo my inclination to discuss his merits in enthusiastic language is very strong indeed. Hence I will omit mentioning the principal thoughts contained in the Dialogues on the great system and the new sciences, and shall briefly relate what regards the application of the pendulum to clocks.

Galileo's observation on the oscillations of the lamp in the Duomo of Pisa is known; from that moment he thought of availing himself of the pendulum to determine the number of beatings in the pulse of a sick person; being then obliged, against his inclination, to occupy himself with medicine. Some time afterwards, reflecting on the means of obtaining in astronomy a more perfect measurer of time in order to determine longitudes with certainty, he returned with greater diligence to the subject. On the occasion of his offering his method for finding longitudes to the States-General of Holland, he wrote to Lorenzo Realio, under date of the 5th of June, 1637, as follows: "I come now to the second contrivance to increase to a vast extent the accuracy of astronomical observations. I am alluding to my time-measurer of which the precision is so great, that not only will it register the exact amount of hours, but minutes, primes, and seconds, and even thirds if their frequency could be numerated by us; and its punctuality is such, that if two, four, or six of these instruments be taken, they will keep so well together, that there will be no variation between them—not even as much as the beat of a pulse—at the end, not merely of an hour, but of a day, or even of a month. I derived the fundamental principle of this machine from an admirable proposition which I demonstrate in my book 'De Motu.'" And then, unfolding the theory of the oscillations of the pendulum, he continues: "The tediousness, however, of being obliged incessantly to count the vibrations can very conveniently be provided against in this wise—viz., by arranging that there should project from the middle of the circumference of the sector, a small, fine, thin pin, that in passing hits on a boar's bristle fixed at one of its extremities, which bristle rests upon the teeth of a wheel, as light as paper, which must be placed on a horizontal plane near the pendulum and having around it teeth like those of a saw, that is, with one of the sides placed at right angles on the plane of the wheel, and the other inclined obliquely; thus, it will serve this purpose, that when the bristle hits against the perpendicular side of the tooth, it will move it; but on the return of the same bristle on the oblique side of the tooth it does not move it, but bending over it, slides past and falls at the foot of the following tooth. In this manner, in the passage of the pendulum, the wheel will move for the space of one of its teeth, but at the return of the pendulum, the wheel will not move in the least. Hence its movement will be circular, and always in the same direction; and having marked the teeth with numbers, it will be easy to know the quantity that have passed and consequently the number of vibrations, and the particles of time run. Around the centre of the first wheel another wheel can be adjusted having a smaller number of teeth; which in its turn touches a third larger one; from the motion of which we shall be able to know the number of complete revolutions of the first wheel, by so disposing the teeth that, for example, when the second wheel shall have made one turn, the first shall have accomplished twenty, thirty, or forty, or as many as you like; but to explain this to you, who have men most exact and ingenious in the making of clocks and other admirable machines, is quite superfluous, since you yourselves, starting from this new principle that the pendulum, moving through greater or lesser spaces, always makes its reciprocating motions perfectly equal; you, I say, will deduce much more ingenious and sublime consequences than I can possibly imagine."

From this we gather that the first clock with a pendulum made by Galileo differed essentially from ours, inasmuch as the wheelworks were moved by the pendulum, which made it necessary for some one to impart a new impulse to it as soon as it was about to stop. If no other document existed in favour of Galileo, it would seem to me to be more than enough to secure the first place for him, with regard likewise to the application of the pendulum to clocks. The essence of the invention lies, of course, in the first idea, and we have here, moreover, an instrument that works very well.

But in Galileo's lofty mind it was impossible but that the thought should flash of making the motion itself of the clock maintain that of the pendulum, which would thus be reduced to a simple regulator. Not to draw out this lecture to too great a length, I shall only quote what Viviani wrote upon this subject for Prince Leopoldo de' Medici. In that account, after having described Galileo's experiments on the pendulum and the manner in which he applied it to the measurement of time, he proceeds thus :

" But as Galileo was most liberal in communicating his inexhaustible speculations, it frequently happened that the uses and newly discovered properties of his pendulum, spreading little by little, fell into the hands of persons who adopted them for their own purposes, or inserted them into publications, and by artfully passing in silence over the name of their true author, made such use of them that it was believed—at least by those who knew nothing of the origin of the discoveries—that the writers were the real authors of them." He next speaks of the observations of the "Stelle Medicee," of the tables relating to them prepared by the Padre Renieri, of the offering made by Galileo to the States-General of Holland of his method for determining longitudes by means of the eclipses of Jupiter's satellites, and of Galileo's determination to send his son Vincenzo and the aforesaid Padre to Holland, since he himself, being old and blind, was unable to travel thither. He then continues: "While, therefore, Padre

Renieri was employed on the composition of the tables, Galileo gave himself up to meditations on his time-measurer, and I remember one day in the year 1641, when I lived near him in the Villa d'Arcetri, that the idea struck him that it would be possible to adapt the pendulum to clocks with weights or springs, and avail himself of it instead of the usual regulator, hoping that the perfectly equable and natural motion of the pendulum would correct all the defects in the mechanism of the clocks. But as his blindness deprived him of the power of making designs and models which would answer to the designs which he had formed in his brain, his son Vincenzo having arrived one day at Arcetri from Florence, Galileo confided his idea to him, and many times afterwards did they reason over the matter, and at last they settled upon the method which is shown in the accompanying drawing, and they set at once to work, in order practically to overcome those difficulties which it is for the most part impossible to foresee. But Signor Vincenzo intended to construct the instrument with his own hand, in order that by this means the secret of the invention should not be reported by the artificers, before it had been presented to his Serene Highness the Grand Duke, his master, and the States-General (to be used for observing the longitude), but he put off the execution of his work so frequently, that a few months later, Galileo, the author of all these admirable inventions, fell sick, and on the 8th of January, 1641, 'ab Incarnazione,' according to the Roman style, he died. And consequently Signor Vincenzo's energies so cooled down that it was not until the month of April, 1649, that he actually began to make the present clock upon the idea explained to him by his father, Galileo. He then managed to obtain the services of a young man—who is yet living—named Domenico Balestri, a locksmith who had had some experience in making large wall clocks, and he made him construct the iron frame, the wheels with their axes and pinions, without cutting (*intagliare*), and the remainder he made with his own hand, constructing on the highest wheel, called the wheel of notches (*tacche*), No. 12 teeth, with as many cogs, divided between each tooth, and with the pinion in the axis of No. 6; and another wheel which moves the above-mentioned of No. 90. He then fixed on one side of the support which is at right angles to the frame, the key or trigger (*scatto*), which rests on the above-mentioned higher wheel; and on the other side he fixed the pendulum, which was made of an iron wire, in which was threaded a ball of lead, which could be loosened by a screw, so that it could be lengthened or shortened according as it was necessary to regulate it with the weight. When this much had been done Signor Vincenzo wished me (as one who was in the secret of this invention, and who indeed had urged him on to complete it), to see, by way of trial, the combined working of the weight and the pendulum. I observed the mechanism in operation more than once, and his workman was likewise present. When the pendulum was at rest, it prevented the descent of the weight. But when it was raised and then let go, in passing beyond its perpendicular with the longer of the two cords attached to the pivot of the pendulum, it raised the key, which fits into the wheel of the notches; which wheel drawn by the weight in turning round with its higher parts towards the pendulum, pressed with one of its levers on the other shorter cord, and gave it at the beginning of its return such an impulse that it served as a kind of accompaniment to the pendulum, which lifted it to the height from which it had started; so that when it fell back naturally and had passed the perpendicular, it returned once more to lift the key and immediately the wheel of the notches was set in motion and continued to revolve and push with the following lever the pendulum, and thus in a certain way, the swinging of the pendulum was rendered continual until the weight had reached the ground. We examined together the operation, connected with which, however, many difficulties arose; but Signor Vincenzo did not doubt but that he would be able to overcome all of them; indeed he fancied that he would be able to apply the pendulum to clocks, in a different manner and by means of other inventions; but since he had got so far, he wished to finish it on this plan, as the drawing shows it, with the addition of hands to show the hours and even minutes: for this purpose he set to work to notch (*intagliare*) another cog-wheel. But whilst engaged in this work, to which he was unaccustomed, he was overtaken by a very acute attack of fever, and he was obliged to leave it unfinished at this point; and on the twenty-second day of his illness, on the 16th of May, 1649, all his thoughts and aspirations, together with this most exact measurer of time, were for ever lost to him. He their author passed away to measure, in the enjoyment, let us hope, of the Divine Essence, the incomprehensible moments of eternity."

I think it is useless to occupy your time in further comments; the facts and evidence are clear, and you have moreover before your eyes the design for the application of the pendulum to clocks. You must not be surprised therefore if, without wishing to quarrel with the Dutch, fully recognising as I do Huygens's merits, I maintain and hope that all impartial thinkers will agree with me that Galileo must be considered the inventor of the system for the application of the pendulum for regulating the time of clocks, as he was likewise the author of many other useful discoveries.

And who knows how many others might not have sprung from that energetic and fruitful mind had his more mature years been less oppressed by bodily and mental cares, had the war carried on against him by the ignorant been less implacable?

He expired in the arms of his friends, Torricelli and Viviani, and returned immortalized in God that gigantic intelligence that had animated him. Let us venerate in him the father of modern philosophy, the leader of true progress, the redeemer of science.

In the meantime the number of the followers of his doctrines went on increasing every day. In Pisa, Castelli—who had been called to Rome by Urban VIII., while Galileo was yet living there—was succeeded by Niccolò Aggiunti del Borgo San Sepolcro, a man distinguished in geometry and physical science. To him are due many important observations on the resistance encountered by the pendulum moving in various liquids or in the air; and the discovery of the ascent of liquids in capillary tubes, which was applied by him to explain various phenomena, such as the ascent of chyle in the milky veins, the nutriment of plants, the preservation of flowers in moisture, the method of feeding peculiar to certain small animals, the rapid rise of water in sponges and wood, and the sperical shape of drops of water, attributed by him to what he called the *occult motion of water*, and which, from the general character of his works, may be set down as merely molecular affinity.

At Rome, Castelli, following in the footsteps of his great master, established the laws of hydrodynamics, explained the phenomena depending on the duration of luminous impressions on the retina; the apparently larger appearance of stars on the horizon; and found the ingenious manner of making sight more distinct without employing lenses, by making use of a little hole, and thus clearly demonstrating the utility of diaphragms in optical instruments. Then, forestalling the modern theories of heat, he clearly pointed out how different the warming of bodies is according to their nature and to the state of their surfaces; and in his publication on the magnet, he was the first to explain the arrangement which iron filings make on paper spread over the poles of a magnet according to lines, which were afterwards named "lines of strength." Castelli, who had made Torricelli's acquaintance in Rome, and who was convinced of his genius, introduced him to Galileo, who was at once eager to know him; but it did not fall to Torricelli's lot to spend in the society of that great man more than the last three months of his laborious existence.

Torricelli extended the mechanical discoveries of Galileo, ingeniously applied the method of indivisibles, conceived by Galileo, to the squaring of the circloide, which he was the first to demonstrate, and to the measurement of hyperbolic solids. Being appointed mathematician to the Grand Duke, he established himself in Tuscany, occupying himself, among many other things, some of which we have mentioned above, in the perfecting of telescopes. We also owe him a microscope simpler than Galileo's, being made of only one lens, or rather of a little glass ball, which he fashioned by the flame of a lamp. But the invention which contributed the most to his celebrity was, as every one knows, that of the barometer.

Galileo, by condensing air, had demonstrated its weight, and in his Dialogue on the resistance of solids, he says of water, that in suction-pumps, it does not rise higher than about eighteen "braccia," leaving the space above empty. Torricelli, pondering over this fact, was led to think of what would happen if in the place of water, mercury, which is so much heavier, were used; for he argued that by its means there would be much greater ease in obtaining a vacuum, in a much shorter space, than that necessary for water. And he then made a long tube of glass of the length of about two braccia which terminated at one end in a ball, likewise of glass, and remained open at the other; through this aperture he proposed to fill the tube and the ball completely with mercury; and then holding it with his finger and turning it upside-down, submerge the orifice of the tube below the level of more mercury in a large vessel; and that being done, take away his finger and open the tube, thinking that the quicksilver would detach itself from the ball, and having glided down, and remained suspended, according to the various calculations, at about the height of one and a quarter braccia, would, in all probability, leave a vacuum in the ball above and in part of the tube. He communicated this thought to his great friend, Viviani, who, most anxious to see the result, agreed to the experiment, which he himself carried out, and was, hence, the first, about a year after Galileo's death, to see Torricelli's ingenious idea confirmed by the fact. He hastened to his friend, who, most joyful at the news of this evidence, was all the more persuaded that the weight of the air was really that which was in equilibrium with the column of water or mercury. Indeed, being asked by Viviani what would have happened if the experiment had been made in closed space, Torricelli, after having reflected for a short time, answered: The same thing; since the air is already compressed in it. This most important discovery was communicated by the author himself to Ricci in Rome, and by Ricci to Signor de Verdus, who, in his turn, made Padre Mersenne acquainted with it, from whom Pascal learnt it and made it famous, as every one knows, in his celebrated Puy-de-Dôme experiment. It is very probable, however, that, without detracting in the least from the great merit due to Pascal, the experiment of the sinking of the mercury in the barometrical tube in proportion to the increase of the height to which it is carried, was made for the first time in Italy. Carlo Beriguardi, in his "Circolo Pisano," published in 1643, while endeavouring to clothe in an Aristotelian dress, both this and other experiments, says:—That the tube of quicksilver leaves more space empty when placed at the top of a tower or of a mountain, than at the foot. And in a letter to Ricci, Torricelli himself observes:—That it would be possible, by means of his instrument, to get to know when the air was lighter or heavier; and that it might be the case that this air, which is most heavy upon the surface of the earth, becomes more and more light and pure as we rise higher and higher to the tops of the loftiest mountains. There is, accordingly, reason to believe that he himself, or others following up his indications, really made the experiment and saw the mercury sink as the height increased. Here are two tubes which Torricelli used in his first experiments.

Before speaking of the Accademia del Cimento, it will not be out of place if I allude here to certain measuring instruments, which were made use of in its researches, and which are to be found described in the "Libro de Saggi" as belonging to the academy; but some of which, however, were made before its foundation, according to the directions chiefly of the Grand Duke Ferdinand II. And beginning with thermometers, it is useful to notice how even Galileo had substituted wine for water, which by freezing easily broke the thermometrical bulb; while the Grand Duke had replaced wine, first of all by coloured alcohol; and afterwards, in order that the deposit thrown down by the colouring matter on the inside of the bulb should not render the reading less clear, he substituted natural alcohol; which they call "acquarzente." Here is a series of thermometers of that period. But I must tell you that I have only brought to London a very small part of the ancient instruments existing in Florence. This is the thermometer, which we owe to the Grand Duke, it is called the "cinquantigrado," because it was divided into fifty parts in a masterly manner by means of these little grains of enamel or glass; and it is exceedingly interesting to observe how the thermometrical bulb was constructed, not merely in a spherical shape, but even in the cylindrical one which has now come into general use; as you may see in this second thermometer, which is also a "cinquantigrado." As to the way of dividing them into degrees, this is the method they employed:—They immerged the thermometer, just as we do, into melting ice or snow, and they marked the point to which the alcohol descended—which in thermometers of 50° was about 13°·5; while they remarked that the greatest cold of Florence could reduce it on some occasions even to 7°. Then they exposed it to the rays of the sun at midsummer, in the open air, without any object of reflection whatever; in this case the "cinquantigrado" rose, in Florence, to 43° at the highest, and in the shade, at the same season, to 34°; so that the difference between melting ice and the extreme heat was divided into 30°; and these were the points of comparison for their thermometers. Other thermometers were then constructed with arbitrary scales of 60, 70, 100, 200, or even

400°, chiefly by Giuseppe Moriani, called on account of his art, "il Gonfia," who was famous for making thermometers of 50° perfectly similar to one another. And such do some of those existing in the Royal Museum of Florence still remain. Here is a thermometer divided into 470°; and in this other one the bulb is exquisitely worked so as to form a hollow branched stand full of alcohol. To render these delicate thermometers less fragile here is another shape that they sometimes gave to the tube, by twisting it, as you see, spirally. The tube of this thermometer is 230 centimetres long, and yet the total height of the instrument is but 32 centimetres. Here is a thermometer on which is marked the exact temperature which the water should reach in order to be fit for bathing in. This one is rather a curiously shaped thermometer, or thermoscope; we owe it, also, to the Grand Duke Ferdinand II.; it is described in the "Saggi dell' Accademia del Cimento" as a "*thermometer lazier, or more slothful than the others.*" The tube contains alcohol, besides six little coloured glass balls. At the temperature of melting ice these little balls float. At 10° of their 70-grade thermometer one of the balls falls to the bottom, at 20° a second one sinks, and so on. At 70° they are all at the bottom. This other shape, like a small frog, used to be given to similar thermoscopes, in which cases the little balls were regulated so as to mark certain degrees of temperature nearer to one another. The thermoscope was tied as you see to the pulse of a sick person, and served to show the intensity of the fever. In certain experiments upon the amount of heat transmitted by different substances which were first warmed and then plunged into water, it is mentioned that of two thermometers used, the one of alcohol, the other of mercury, the first to show any sign was the one with mercury. It is nevertheless difficult to establish the precise date at which mercury thermometers were first made. And before finishing what I have to say on the subject of thermometers, I shall call to mind a most useful application that was made of them even at that time by the Grand Duke, that is to say, for meteorological observations. He thought of availing himself for this purpose of the monks, who, scattered as they were in every direction, could on account of their manner of life, apply themselves better than any one else to such observations. It is said that such a duty was entrusted to Padre

Luigi Antinori. The observations used to be taken at Florence, at the Palazzo Pitti and at the Boboli Gardens, and in the Convento degli Angeli; and as far back as the year 1654 they were regularly established at Vallombrosa, Cutigliano, Bologna, Parma, Milan, Warsaw, and Innsbruck. Observations were usually made at different hours of the day, of the state of the thermometers exposed to the north and to the south, the state of the sky, and the direction of the wind. From a manuscript of Viviani, however, we learn that in some places were noted the date, the hour, the temperature, the state of the barometer, the wind, the sky, and the humidity of the air. In this volume, entitled "Archivio Metereologico Centrale Italiano," you will see registered the observations of those days, and besides others made in this century. Now if the observations taken so long ago be compared to recent ones, it will be seen that after the due corrections have been made, or if observations be taken now with some of the best instruments of the Accademia del Cimento, the meteorological conditions of Tuscany have not changed.

Passing on to the subject of hygrometers, besides the one imagined by the great Leonardo da Vinci, and which is founded on the increase in the weight of certain substances through the action of moisture, others were used in which lengthening out or contraction of a given substance served to determine the humidity of the atmosphere in which it happened to be situated. Torricelli had already used oats. In 1664, Dr. Folli da Poppi constructed a new hygrometer founded on the expansion of paper with the variation of moisture, which, having been perfected by the Accademia del Cimento, was reduced to the shape in which you see it now. The movement of the hand to the right or to the left, according as the paper expanded or contracted, pointed out on the quadrant the relative quantity of moisture. Most interesting likewise is this other one, which has been called the condensing hygrometer. We owe it to the Grand Duke Ferdinand II. It consists of a truncated cone made of a tinned sheet of iron, covered on the inside with a layer of cork, and is supported on a tripod. Below the smaller aperture, turned downwards, there is suspended a hollow glass cone ending in a closed point, likewise turned downwards, and provided towards the upper portion with an escape-pipe. The superior cone is then filled with snow or ice, which in melting, runs into the glass cone which remains at the temperature of 0°, whilst the excess of water runs through the pipe into an appointed recipient. The moisture of the air in contact with the cold side condenses itself and covers it with dew, which collecting by degrees into drops, runs towards the point, whence it falls into a graduated vessel. If the time the experiment has lasted be taken into consideration, the quantity of water collected will be found to be in proportion to the humidity of the air. Experimenting in this wise, the members of the Accademia del Cimento found that the south winds are so charged with moisture that in one minute the hygrometer has given as much as thirty-five, fifty, and even eighty drops of water, whilst the north wind leaves the glass perfectly dry.

Here are various examples of the so-called Hydrostamms for liquids, used by the Accademia del Cimento, and some of which we owe to the Grand Duke Ferdinand II. This one with the ballast of mercury is employed for liquids lighter than water; this other one with the ballast composed of little balls of lead, is for those heavier than water. Just as is usually the case with thermometers, the divisions are marked with little balls of glass or enamel soldered to the pipe. Does not this other hydrostamm, provided with a kind of balancing plate, remind one of Nicholson's aerometer? It was used for determining the specific gravity of precious stones, by observing to what degree it sank without the precious stone, and with the precious stone placed upon the little metallic disc suspended by the three little chains. This is also a fitting place to draw attention to the so-called "Palla d'Oncia" (ounce-ball) of the Grand Duke Ferdinand, a ball of glass which displaces very nearly an ounce of water; into its pipe several rings were strung in order to make it sink, and then by the number of these rings it was known what the specific gravity of the liquid was in which it had been immersed. Other hydrostamms were afterwards constructed on the principle of the thermometer with the little balls, which has already been described; they were called "a gabietta" (like a small cage); because the little balls were placed in a sort of cage made of fine brass wirework. When it was immerged into the liquid which was to be experimented on, the number and colour of the balls that sank determined the specific gravity required. And it being known, even at that time, that the specific gravity varied with the temperature, it was thought that one experiment would suffice to give both the data necessary in order to determine it with precision. In this photograph you can see one of the methods employed—viz., by fixing a thermometer in a hydrostamm; but other ones of the Accademia del Cimento exist, which are true thermometers, having, in addition, the graduation of the hydrostamm.

The Grand Duke also occupied himself in perfecting a certain pendulum, in order to obtain equable intervals of time of greater or less duration; and from the figure which you see here in the "Libro de' Saggi di Naturali Esperienze," you can understand how for the oscillating ball always to remain in the same plane, it was suspended by two threads, which thus formed the two sides of an isosceles triangle, and of which it was possible to alter the length, at pleasure, by means of a convenient pincers. This pendulum was also used in the experiments made upon the velocity of propagation of light and sound. Indeed concerning the latter, which were afterwards repeated by the Accademia del Cimento, I cannot but say a few words, as in the usual works on physical science mention is only made of the experiments carried out by the French academicians, while those made so many years before, upon the same plan and with the same results, are omitted.

Here is an extract from one of Viviani's letters, in which he gives an account of these experiments. After having related certain discussions upon sound which had taken place with Signor Rinaldini and Signor Borelli, and had spoken of some bomb-shots that had been fired from the Petraja, he having been with the Grand Duke when they were determining its distance from Florence, he continues:—

"His Highness asked me the following questions—Which of the two sounds, the greater or the less, reached the ear in the shortest time? To this I answered that both would reach exactly in the same time. Secondly, of what impediment the wind could be to the propagation of sound? I answered: None. He then proceeded with his inquiries, and asked me what difference of time I thought there would be in the rate of the sound, between the discharge of the piece with the mouth turned towards the ear of the observer, or turned up perpendicularly, or turned the other way? To which I answered immediately, although the question was perfectly new to me, that I should have thought these periods perfectly equal to one another. His Highness did not tell me whether I had answered these questions rightly or not; but in the evening, when he came up to see the experiments, he assured me that in those which had already been made, and had been repeated two evenings before with a large arquebuse from the Petraja, it had been found to be actually the case that the rate of the lesser sound was equal to that of the greater; that the wind, which on the second evening was blowing from the south-east, did not affect it in any way whatever; and that the difference in the direction of the discharge made no variation in the rate of progress of the said sounds. Nor did his Highness's demands end here, for before I had left him, in order to ascend the terrace to make the observations, he finally asked me what I should think would be the rate of two sounds, the one made at a distance of two miles and the other at double that distance? I answered that I had also had a great curiosity to satisfy myself whether the motion of sound was in itself of a continually slackening velocity, or whether it were equable, because if it were found to be such, it seemed to me that more curious consequences might be drawn from it, and which might prove to be of great use. Upon this he urged me to say what I thought, as he wished afterwards to make the experiment. I answered—and indeed too boldly—that at double the distance the time would be exactly double, for I held that the progress of sound was in itself uniform—that is to say, that in any given equal spaces of time it will traverse equal distances. As I had reasoned on this particular point the day before, and I seemed to have greater reason to be convinced of this than of the contrary, I therefore threw no doubt in my answer, and for the time being our conversation stopped there."

Then after having related how the following day he had determined the required distance between the Petraja and Florence, about 9500 braccia, he continues: "Whilst conversing with his Highness and Prince Leopold about the experiments which had been made, and others which were to be made upon the subject of sound, I took occasion to let their Highnesses hear the contents of the enclosed writing, in which I had, the previous evening, noted down, more for my own remembrance than for any other purpose, all that had once suddenly struck me, and which would be able to be obtained, in case the progress of sound were always equable, or at least if the proportional rate at which it advanced were known.

"Serenissimo Granduca,—If the velocity of the motions of all sounds, both loud and soft, powerful and weak, are perfectly equal, as I am convinced they are, and are likewise unaffected by any change of the air or gust of wind, either favourable or unfavourable; and if moreover the progress of any sound is equable, as I believe to be the case—that is, that in any given equal spaces of time the distances traversed are the same (as the Discussion has persuaded me must follow), most useful and most curious consequences can be drawn.

"Let a most exact experiment be made of the time which the sound of a bomb (exploded, say, at the precise distance of three miles) takes to reach our ears, and having accurately found by measurement the time which it takes, we can say a third of a minute prime of an hour; or without caring to know what part of an hour the time may be, we can say that it took forty vibrations of such a pendulum; and from this observation alone we shall obtain the following, and many other results. And first of all, we shall be able expeditiously to know how far from us anything making a sound is, provided we can see the blow given and that the sound reaches our ear.

"It will be possible, without using any instruments (which for the most part prove incorrect), without moving from one spot, to know the distances of towns, villages, castles, &c., provided they be seen, and that the sound of a blow or a shot be heard at such a distance. Plans of great countries can thus be made, by merely knowing the angles, and with very great saving of positions; it will be equally effective in plains as on mountains, where great difficulty is experienced on account of the imperfections of the instruments, &c.

"It will be possible to ascertain the distance of an army from any place, to know how far off batteries are stationed, and obtain many other similar advantages in time of war.

"Moreover, we shall be able to know how far distant from any place on land or at sea an island may be, or a ship, or a fleet, or how far off a naval battle is taking place; we shall be able to measure the distance between two ships, or two islands, or two rocks, when they are in the open sea and there is no place to stop in order to take up the two stations, a thing which it is impossible to do by any other means, or at least has never yet been done, as the sailors tell me.

"Finally, we shall know with perfect accuracy the distance from us of the clouds from which issues the thunder, and this by measuring with the same pendulum the time which escapes from the moment when the flame of the discharge or the beginning of the lightning is first seen to the instant when the sound reaches the ear; for, by the Rule of Three, we shall say that if in forty vibrations exactly three miles (that is, 9000 braccia) are traversed by sound, how many miles or braccia will be traversed in ten, twelve, or thirty vibrations; for the fourth number given will show the distance required.

"It is well to notice that the measurements of very great distances will prove more accurate than those of the nearer ones, and this because the interposition of time between the flash and the arrival of the sound is so short that there is but little time to measure the vibrations even of the shortest and therefore quickest pendulum, which would be the one it would be necessary to use in similar operations, &c.

"VINCENZO VIVIANI.

"October 10, 1656."

"It so happened that their Highnesses thoroughly approved of these ideas and became all the more anxious to endeavour to find by means of experiments the proportion of these velocities. Shortly afterwards Signor Borelli appeared, and his Highness commanded that the next evening we should go to make the trial on the high road of the wood of St. Moro dalle Mulina, and appointed Ricci and Monsu Filippo to aid us. We went on the appointed day at 20 o'clock, having with us the same bombardier, who was the lame De Neri who had fired the shots at the Petraja, carrying with him powder, rockets, and a cannon (*maschio*). We employed the hours of the day in accurately measuring the length of the road, and we found that from the old wood to the Arno it was a little more than a mile and one-fifth; but in order that the experiment should be accurate we measured out exactly one and one-fifth mile, and at that spot we made the cart stop with the maschio on it. We then placed two pendulums, one at the very end of the above-mentioned road, the other at the precise middle; they were instruments the vibrations of which were perfectly equal. Signor Borelli and Signor Ricci stood at the end, with the youth De' Galilei, who seemed to behave well; and I was placed in the middle, together with Monsù Filippo d'Augusta, clockmaker to his Highness. When night had come on Signor Borelli made the first sign to the bombardier with the rocket, and then the firing began, but no shot was fired without a sign being given from another rocket. Each one of us was consequently previously warned on each occasion to watch the movement of the flame of the maschio, and we immediately set to work to count the vibrations of our pendulum; and the others further off did likewise. As many as fifteen shots were fired, and we always found, to my very great satisfaction, the same number of vibrations from the appearance of the light to the arrival of the sound, and ours were always less than eight, and we settled among ourselves that it might be called seven and a half. We finished the experiment at about two and a half o'clock in the night; and we, who were nearer Florence, remained waiting with the greatest anxiety for the others with the carriage, in order to hear their number of vibrations; and finally, not to keep you any longer in suspense, they, without knowing what ours had been, told us that they had always counted fifteen and a half; which was exactly double our time, just as one and one-fifth of a mile is double three-fifths of a mile; and so we all got into the carriage with the utmost satisfaction and entered Florence about four o'clock at night, and we immediately informed his Highness, who was awake, and expecting the news of the result of our experiment; and you can imagine with what pleasure he heard our communications. The following evening we repeated the firing from the Petraja, when there was a strong north wind blowing, and yet we found the same result as the previous evenings, which had been forty-one vibrations of the same pendulum. And to put a seal to this story, I must add that I have made the following calculation: If sound traverses 3600 braccia in fifteen-and-a-half vibrations, how many braccia will it have traversed in forty-one vibrations—the number between the Palace and the Petraja? and the fourth number is 9522 braccia, which is about the figure given by the greater number of instruments."

When the Accademia del Cimento afterwards repeated these same experiments, it was found that in five seconds sound traverses one

Tuscan mile=to kil. 1·65666, that is to say, it travels at the rate of 332 metres per second.

And at this place, I cannot but call your attention to the fact that it was Prince Leopold who was the life and soul of the Accademia del Cimento. This Macænas of science facilitated the publication of the most useful and distinguished works, he gave his advice and assistance towards the reprinting of the old works on geometry; he arranged and watched over the collection of Galileo's works and of the scientific essays of Padre Castelli; he urged Torricelli to make public the mathematical definitions of inertia; he encouraged Rinieri to bring to a conclusion the laborious charge, which he had undertaken, of finding the constitution of the Stelle Medicee; but in 1647 when the latter was giving daily information regarding Jupiter's satellites, and was on the point of publishing the tables, he suddenly died, and his valuable papers were, alas! very quickly scattered. It was, indeed, a year of ill-omen, for in it Rinieri, Torricelli, and Cavalieri descended, one after another, into the tomb. But their works, the germs of future disciples, outlived them. In fact, ten years afterwards, we find ourselves face to face with a great event in the annals of science, and one most auspicious for Italy, and particularly for Florence—namely, the foundation of the first scientific Academy. We are chiefly indebted to Prince Leopold for the great idea of establishing an academy which should be destined expressly to the study of experimental philosophy. That distinguished man, who was accustomed to gather round him for useful conversation the most illustrious persons of his time, thought that researches would be more systematically pursued, and the gatherings of many men would benefit to a much greater extent the progress of science, if meetings were held regularly and some rules and regulations laid down. Ferdinand joyfully agreed to his brother's proposal, and showed the greatest generosity towards the new institution; he presented all his own valuable instruments to it, and even endowed it with the results of his former experiments, several of which have been regarded as the work of the Academy, which was certainly not the case. On the 18th of June, 1657, there was held in the Pitti Palace, the first sitting of the first scientific Academy; it justly chose to name itself the "Accademia del Cimento" (Attempt, trial, essay), and it selected as its device the now celebrated motto: "Provando e Riprovando" (By trying, and trying again). The members present at this celebrated meeting were: Vincenzo Viviani, Alfonso Borelli, Carlo Rinaldini, Alessandro Marsili, the brothers Paolo and Candido del Buono, Antonio Oliva, Lorenzo Magalotti, Francesco Redi, and Carlo Dati. Among the Italian correspondents, Ricci, Cassini, Montanari, Rossetti, and Falconieri must be mentioned; and among foreigners, Stenone, Tevenot, and Fabbri.

It is not my intention to speak individually of each of these academicians; the limited time at my disposal would not permit of it; nevertheless, before speaking of the labours of the Academy, I cannot but add a few words more respecting one or two of them. And above all I must mention Viviani, Galileo's only disciple, a subtle and industrious academician, to whom we are indebted for many experiments which led the way to the discovery of the theory of undulations, the general idea of which, he even at that time, suspected. We owe him likewise a barometer, without a small well (*pozzetto*); experiments on capillary phenomena, independent of pressure; the proposition for finding the weight of ice compared with that of water; the experiments on the swim-bladder of fishes; but, above all, the machine for making a great vacuum, to which primitive shape of a pneumatic instrument we have at present returned, giving it the name of mercurial; and many persons dispute the invention of such a machine, forgetting all the while that the first instrument of the kind was made by Viviani.

From the explanatory figures of this machine, which are in the "Libro de Saggi," published by the Academy, and which you see before you, you will be able to understand how it was in reality a true mercurial machine, with which the academicians made a great number of experiments, to which I shall allude further on.

A no less powerful intellect was that of Borelli, a Neapolitan. A mathematician, a physician, an astronomer, he occupied himself with all subjects, and thus supplied himself with ample materials for future discoveries. He studied the reciprocal attraction of floating bodies (*galleggianti*), and discovered its theory; he was the first to observe the variations of the barometer with the changes in the atmosphere; he considered the question of the freezing of water; he planned the experiments which were decisive against the idea of the positive lightness (*leggerezza positiva*); he measured the greatest expansion of air, freed from the surrounding pressure; he determined the weight of air compared to that of water; he suggested the famous experiment with the silver ball, to test the compressibility of water; and another experiment on the propagation of sound in vacuo; he studied the contraction of various liquids in cooling; and in the researches on the velocity of propagation of light, he was the first to construct the Heliostat (*Eliostata*); he published, whilst lecturer in Pisa, the "Euclides restitutus," in which he reduces to 230 propositions the elements of ancient geometry; and afterwards the treatise on the force of the blow, which together with that on the natural motions depending on will, serve as an introduction to his great work "Del moto degli Animali;" he studied the optic nerve, and the organs of respiration in fishes, and was the first to occupy himself with the anatomy of the torpedo; he observed the comet of 1664, declaring that it was not an accidental meteor, or vapour, but a solid body moving, not round the earth, but round the sun, in a line resembling a parabola, and thus laid the foundations of the theory of the comets; he published the theory of the "Stelle Medicee," in which he clearly proved that the orbit of the satellites is not in the same plane as that of Jupiter, and he compared these satellites to the moon, and thus alluded to the principle of universal attraction; he was the first to make known that Venus can be seen two days running, as a morning and as an evening star; he wrote an account of the eruption of Mount Etna which took place in 1669; he studied the constitution of liquids, and was the first to point out the phenomenon of the contraction of the liquid vein; and he illustrated by new experiments Galileo's idea of the fall of weights in vacuo.

But we must now turn, for a short time, to the principal works of the Academy, although very much must necessarily remain unsaid of the individual academicians, and especially of Redi and Stenone and Cassini.

And coming now to the experiments, the first series refers to the natural pressure of the air. In a discussion on Torricelli's experiment, and on the reason why solid bodies do not lend themselves to it, and among liquids, mercury should adapt itself the best, there is to befound noted down,with regard to liquids, a most important hypothesis on their constitution : "Either on account of their slippery smoothness, or on account of the rotundity of their extremely small bodies or for some other figure which they may take, particularly inclined to motion, being just equilibrated, as soon as they are pressed they immediately give way on all sides and scatter themselves . . ." so that it would not be possible in our days to discuss their fundamental property any better. And another important law is added further on : "In all liquids this force of the pressure of the air is admirably shown, particularly when they are caught in some place where they have on one portion of their superficies an empty, or nearly empty, space into which they can withdraw. Since, in this case, being pressed on one side by contiguous air, which in its turn is pressed down by so many miles of amassed atmosphere, and on the other side, where there is no obstacle, are confined by the void which has no weight at all, they rise until the weight of the liquid raised up is equal to the weight of the air pressing on the other side." It is afterwards observed that by vacuum or void is meant that the air alone should be excluded, and not light, or heat, or ether.

We are indebted to the Accademia del Cimento for the usual experiment which is made under the bell of the pneumatic machine, with the bladder that swells by exhausting the air, and also swells when it is forced back into it ; the second of those two barometers, the one with the little well (*pozzetto*) outside, the other shut up in a bell in which the vacuum is obtained, and which was invented to prove that it is really the pressure of the air that holds up the mercury, and which was afterwards altered by immerging the barometer in a vessel full of water, so that the mercury was found to rise above the usual braccio and $\frac{1}{4}$ by $\frac{1}{14}$ of the height of the water in the vessel, the level being taken from that of the mercury in the little well (*pozzetto*).

Then availing themselves of Viviani's mercurial pneumatic machine, they instituted a number of experiments in vacuo, to see whether their operations prove contrary to, or in any way different from, those which manifest themselves when surrounded by air. They accordingly experimented on the spherical shape of the drops independent of the pressure of the air ; on the heat and the cold which cause the rise or fall of the barometrical column ; on the reflection of images in lenses, attributed by Kepler to the air ; on the attraction of amber in vacuo ; on the propagation of heat; on the propagation of sound in vacuo, when it was wisely remarked that the sound generated could be communicated to the exterior air, by the very partitions of the recipient in which the sonorous body was closed; on the attraction of the loadstone; on capillary phenomena, independent of atmospherical pressure; on the boiling of water in vacuo; on the bursting of air-bubbles by cold; and, finally, on the various ways in which many animals are affected when placed in vacuo, or in very rarefied air, such as leeches, snails, grasshoppers, butterflies, lizards, flies, little birds, crabs, frogs, and different kinds of fish, and they paid particular attention to the effects produced on the swim-bladders of the latter. With regard to these last-mentioned experiments, and especially to those which refer to little birds that die immediately, and even when succoured in time do not fly away, it is well known that in one of Boyle's experiments a lark lived ten minutes in vacuo, and a goose seven minutes. But, it is added, that whoever reflects upon the different ways of producing the void in these two cases will perceive that the experiments, far from proving contradictory to one another, agree most admirably; inasmuch as where in the one case (Boyle's) the air is thinned by succeeding attractions with very slow and little less than insensible acquisitions; in the other, by the extremely rapid descent of the quicksilver, it is immediately reduced to that last degree of rarity and thinness, which cannot be of any avail for respiration. And from this we can gather that even at that time the superiority of the mercurial pneumatic machine over all others was perfectly well known.

As an introduction to the experiments on the compression of water, Magalotti writes—"Although the truth is not always arrived at by the first experiment, that is not the case because the first idea of the experiment is not very often quite adequate to obtain the truth; but it may sometimes happen because the materials and means which are used to carry it out practically are not adapted to that purpose; and although these experiments cannot contaminate the purity of the theoretical speculations, they are nevertheless unfitted to second them, on account of the materials employed. But not for this reason must these experimental inquiries into natural phenomena be deemed to have failed; because, although at times we do not succeed, by means of them, in coming to the bottom of the truth, which was first of all sought after, it is indeed rare, if some glimmering of the fact be not derived from them, or the falsity of some other contrary supposition discovered by their means."

Of the three ingenious experiments that were made to compress water—viz., by the force of rarefaction, by the pressure of mercury, and by the force of a blow—the last has remained the most universally known. It was made with a subtle ball of silver containing the water, which, when the ball was beaten with a hammer, sweated through all the pores of the metal, and looked like quicksilver coming out drop by drop through some skin in which it was squeezed. Here is the ball which was used for this celebrated experiment, and it still contains some water. This other one, which likewise has water in it, was prepared by the academicians for a repetition of the experiment, which did not, however, take place. Other most ingenious and conclusive experiments were made by the academicians on the question of positive lightness—namely, "whether those things which are commonly called light are so by their own nature, and whether they float upwards of their own accord, or whether their rising be merely the effect of heavier bodies driving them away; these weightier bodies having more vigour and greater power to descend and place themselves lower, squeeze the lighter ones away (so to speak), and compel them to rise." Nor did they forget the marvellous operations of the magnet, a vast field, in which though much be already discovered, there undoubtedly remains a great deal more to be found out. They clearly showed that iron and steel are essentially magnetic; that attraction and repulsion can take place, even through solids and liquids; that the poles of the magnetic needle are much enfeebled when held in the position of the magnetic equator. With regard to the electrical action, the academicians merely explained which substances possessed such a property, and which not, and concluded that it was common to all transparent stones and not to opaque ones, but that it was more especially a property of amber, which even attracts smoke.

With regard to the experiments on sound, we have already spoken of them at sufficient length; those on projectiles are but experimental proofs of three propositions of Galileo. The first, that a ball discharged horizontally from a tower always takes precisely the same time in reaching the earth as it would take had it been let fall freely from the same height; and that would always be the result whatever the force of the discharge. The second, that, on account of the resistance of the air, there is a maximum limit of the velocity with which a body falling from any height can reach the earth. The third, that a motion cannot be destroyed by another motion coming upon it afterwards; and this was illustrated by the well-known experiment of a cannon mounted on a waggon drawn by six horses; when the ball was discharged upwards, it always fell close to the mouth of the cannon, whether the waggon was at a standstill, or whether it was going at full speed. Other experiments refer to the relative weight of air and water; others to the impermeability of glass to moisture and smells; some are directed, in accordance with one of Galileo's ideas, in determining whether light takes any time in propagating itself; others again clearly demonstrate that even white substances can be set on fire by means of a powerful burning-glass; nor did they fail to observe the phenomena called by us phosphorescence in sugar, in rock salt, rock-crystal, in agates, in jasper, &c.

Chemistry, likewise, was included in the field of their researches: they observed that waters distilled in lead, render turbid all natural waters, which are afterwards made clear by means of strong vinegar (*acetoforte*); that oil of tartar and oil of aniseed infused in water occasion the formation of a little cloud, which disappears on the addition of spirits of sulphur; and they insist strongly on the fact that the thickness is less with purer and lighter waters, and hence point out the above-mentioned liquids for the analysis of waters; that the tincture of roses, extracted with oil of vitriol, becomes green when oil of tartar is poured upon it, and returns to its original colour by means of spirits of sulphur; that citric acid, spirits of vitriol, and spirits of sulphur change lac into violet, and the tincture of violets into vermilion, which can be made violet by the addition of oil of tartar, &c. But there remains to tell of more interesting experiments—namely, of those made on artificial freezing, and on the dilation of bodies by heat. It was Galileo's opinion that ice was rather rarefied water than condensed water; since condensation brings about diminution of bulk and increase of weight, whilst rarefaction, on the other hand, causes greater lightness and enlarges the bulk. But water, in freezing, augments in bulk, and ice already made, is lighter than the water on which it floats. Now the academicians by means of intelligent and varied experiments, clearly set forth, that as a matter of fact water by freezing increases in volume, so as to rend in pieces the vessels in which it is contained, and they proved, at the same time, that the bursting of the vessels is not to be attributed to the formation of a vacuum, but solely to the dilation of the contents. In this way they caused the bursting of balls of the thickest glass, and of brass, and of copper, and of silver, and of gold. They also determined the relative expansion of ice and water, which they established as 9 : 8. Then the experiments made to investigate even in its most minute details the act of freezing are marvellously interesting, clear, and precise.

Here are these subtle observations and their classification; and I leave you to judge whether at that time, or even in our days, I was about to add, more could be done. They used for these experiments balls of glass ending in long tubes; they were filled with water and then immersed into a freezing mixture of salt and ice, to which they at times added a sprinkling of alcohol (*aquarzente*). The natural state means the degree of temperature of the water or other liquid in the tube of the vessel before it was placed in ice. "Salto dell' immersione" is that first rebound which the water is seen to make the moment the ball touches the ice. This does not come from any intrinsic alteration of the water, but from extrinsic reasons of the vessel. Hence it is that it sometimes varies slightly, and thus brings about some change in the other conditions through which the liquid passes before freezing. But as that leap altogether is but very little, so ar its variations but very slight, and infinitesimal are the effects which it produces in the subsequent changes. "Abbassamento" denotes the degree to which the water, after the above-mentioned *salto dell' immersione*, descends when the cold is beginning to affect it.

"Quiete" is the state in which the water remains for some time after the *abbassamento*, without showing any apparent signs of motion.

"Sollevamento" is similarly the state which the water from the lowest point of *abbassamento* attains by means of rarefaction, with a very slow and apparently equable motion (in all respects like the first), with which it contracts itself.

"Salto dell' agghiacciamento" means the state into which the water is thrown with the utmost velocity, at the moment of freezing. So that in quoting the terms by which they signified the different conditions which they observed, and in telling you what they meant to define by each one of them I have already informed you of the important observations which they succeeded in making. They proved the *salto dell' immersione*, not only by thrusting the ball into ice, but into hot water. The *quiete* corresponds to our maximum of density of water, which was indicated in their thermometers by 119°. *Sollevamento* is the expansion which the water undergoes from 4° down to the point at which it freezes. Then the word *salto* (leap, jump) shows us how the phenomenon of freezing takes place rapidly; in fact the academicians well observe the almost instantaneous freezing of water in certain cases—namely, when left to itself, its temperature descends to below 0°, and being then shaken (the ball having been taken from the snow, as they were accustomed to do, to see the *salto dell' agghiacciamento*) all of a sudden the whole of its mass became solid. And conditions similar to these they observed in spring water, and in water distilled from various substances, in wines, in vinegar, in spirits of vitriol, in oil, in which, however, they found no rarefaction, so that frozen oil sinks to the bottom of fluid oil, and in brandy, which contracts itself regularly but does not freeze. As to natural freezing it was so minutely observed and described that it fills one with astonishment and admiration. They proved that ice produced in vacuo is more compact and weighs more than in the air; and that when water freezes slowly in a glass it makes first of all a crust, which, being pressed, in some place or other, by the expansion of the remainder of the water affected little by little by the cold, breaks open, so that from the aperture there issues a certain amount of water, which, freezing in its turn, renders the surface of the ice convex; and sometimes if the cold be intense, wrinkled, or rather forms an excrescence of ice, which can even reach the height of a finger. And this observation recalls to my mind the experiments of the illustrious Gorini, and his theory on the origin of volcanoes and mountains. But if the crust be too thick, and the expansion underneath be not able to break it, it generally happens that the vessel bursts. Then they pointed out the difference between the ice of ordinary water, of distilled water, of sea water, and the congealing action of common salt, and of the more efficacious salts of ammonia; and even the vapour which rises from ice, and which they with happy intuition compared to clouds; and finally all that you see illustrated here in the photograph of a part of Galileo's Tribune—namely, the reflection of cold by means of a burning-glass, which thus outstrips the modern theories of radiating caloric and its tendency towards equilibrium. To clear up the phenomenon of the *salto dell' immersione*, they were led to make other and no less important experiments on the expansion of bodies by heat, and their contraction by cold, using glass and many metals; and they were the first to make the well-known experiment with a bronze armlet in which when heated moves sensibly a cylinder made expressly, but when it has cooled down this last does not fit in to it.

And this Academy, the first, and one so rich in most important discoveries in every branch of science, after a duration of only ten years, ceased to exist. Its founder was made a cardinal; reasons of State required. it; and truly for a cardinal to be the head of an Academy which was named Del Cimento, would have been too much for those times. And indeed the disestablishment of the Academy was but a corollary to Galileo's trial.

Instituted in the year 1657, it was the first scientific academy; the one of Vienna, begun by Doctor Bausch was not fully established till the year 1670. The Royal Society of London had its true origin in 1663, and in the first volumes of its reports it relates the experiments of the Accademia del Cimento. The Academy of Sciences of Paris dates from 1666.

The CHAIRMAN: Ladies and Gentlemen, I have no doubt a great number of the audience were able to follow the Cavaliere de Eccher better than I have been able to do in the interesting history of the instruments he has described, but we cannot too much thank the Italian Government for permitting them to leave Italy. As I might miss much in giving you a summary, from my imperfect knowledge of Italian, I have asked my distinguished friend the Rev. S. J. Perry, whom you have all heard of as being at the head of the expedition to Kerguelen Island to observe the transit of Venus, to run briefly over the principal topics of the paper.

The Rev. S. J. PERRY having given a *resumé* of the paper above printed—

The CHAIRMAN said: I must thank Mr. Perry for having helped me out of the difficulty in giving you this very interesting review of science developed in the Florentine Academy, and I hope we shall have the full document preserved and translated into English, for it will be a great pity if such an excellent *exposé* should be lost to the English public. I imagine that in the review which Father Perry has been so good as to give, he omitted to state that a claim was made also for the determination of the maximum density of water, which I fancy was among the matters which the Florentine Academicians undertook. The experiments upon the supposed non-compressibility of fluids were very remarkable. Several attempts were made to compress different liquids, but in all cases they escaped and no appreciable compression was observed. I must say again that we are deeply indebted, not only to the learned Professor who has given us this excellent discourse, but also to the Italian Government, and I believe also to the municipality of Florence, to whom many of these invaluable records of the past belong. I ask you, therefore, to return by acclamation your thanks, not only to Professore de Eccher, but also to the Italian Government, and to the municipality.

The Conference then adjourned.

SECTION—PHYSICS (including Astronomy).

May 24th, 1876.

The PRESIDENT : The first paper on our list to-day is that by Professor Clerk-Maxwell, on the Equilibrium of Heterogeneous Bodies. He is here now to speak for himself, and I beg to call upon him to give us his communication.

ON THE EQUILIBRIUM OF HETEROGENEOUS SUBSTANCES.

Professor J. CLERK-MAXWELL, M.A., F.R.S.: The warning which Comte addressed to his disciples, not to apply dynamical or physical ideas to chemical phenomena, may be taken, like several other warnings of his, as an indication of the direction in which science was threatening to advance.

We can already distinguish two lines along which dynamical science is working its way to undermine at least the outworks of Chemistry, and the chemists of the present day, instead of upholding the mystery of their craft, are doing all they can to open their gates to the enemy.

Of these two lines of advance one is conducted by the help of the hypothesis that bodies consist of molecules in motion, and it seeks to determine the structure of the molecules and the nature of their motion from the phenomena of portions of matter of sensible size.

The other line of advance, that of Thermodynamics, makes no hypothesis about the ultimate structure of bodies, but deduces relations among observed phenomena by means of two general principles—the conservation of energy and its tendency towards diffusion. The thermodynamical problem of the equilibrium of heterogeneous substances was attacked by Kirchhoff in 1855, when the science was yet in its infancy, and his method has been lately followed by C. Neumann. But the methods introduced by Professor J. Willard Gibbs, of Yale College, Connecticut,* seem to me to be more likely than any others to enable us, without any lengthy calculations, to comprehend the relations between the different physical and chemical states of bodies, and it is to these that I now wish to direct your attention.

In studying the properties of a homogeneous mass of fluid, consisting of n component substances, Professor Gibbs takes as his principal function the energy of the fluid, as depending on its volume and entropy together with the masses, $m_1, m_2 \ldots \ldots m_n$, of its n components, these $n+2$ variables being regarded as independent. Each of these variables is such that its value for any material system is the sum of its values for the different parts of the system.

By differentiating the energy with respect to each of these variables we obtain $n+2$ other quantities, each of which has a physical significance which is related to that of the variable to which it corresponds.

Thus, by differentiating with respect to the volume, we obtain the pressure of the fluid with its sign reversed; by differentiating with respect to the entropy, we obtain the temperature on the thermodynamic scale; and by differentiating with respect to the mass of any one of the component substances, we obtain what Professor Gibbs calls the potential of that substance in the mass considered.

As this conception of the potential of a substance in a given homogeneous mass is a new one, and likely to become very important in the theory of chemistry, I shall give Professor Gibbs's definition of it.

"If to any homogeneous mass we suppose an infinitesimal quantity of any substance added, the mass remaining homogeneous and its entropy and volume remaining unchanged, the increase of the energy of the mass, divided by the mass of the substance added, is the potential of that substance in the mass considered."

These $n+2$ new quantities, the pressure, the temperature, and the n potentials of the component substances, form a class differing in kind from the first set of variables. They are not quantities capable of combination by addition, but denote the intensity of certain physical properties of the substance. Thus the pressure is the intensity of the

* Transactions of the Academy of Sciences of Connecticut, vol. iii.

tendency of the body to expand, the temperature is the intensity of its tendency to part with heat; and the potential of any component substance is the intensity with which it tends to expel that substance from its mass.

We may therefore distinguish these two classes of variables by calling the volume the entropy, and the component masses the *magnitudes*, and the pressure, the temperature, and the potentials the *intensities* of the system.

The problem before us may be stated thus:—Given a homogeneous mass in a certain phase, will it remain in that phase, or will the whole or part of it pass into some other phase?

The criterion of stability may be expressed thus in Professor Gibbs's words—"For the equilibrium of any isolated system it is necessary and sufficient that in all possible variations of the state of the system which do not alter its energy, the variation of its entropy shall either vanish or be negative.

"The condition may also be expressed by saying that for all possible variations of the state of the system which do not alter its entropy, the variation of its energy shall either vanish or be negative."

Professor Gibbs has made a most important contribution to science by giving us a mathematical expression for the stability of any given phase (A) of matter with respect to any other phase (B).

If this expression for the stability (which we may denote by the letter K) is positive, the phase A will not of itself pass into the phase B, but if it is negative the phase A will of itself pass into the phase B, unless prevented by passive resistances.

The stability (K) of any given phase (A) with respect to any other phase (B), is expressed in the following form:—

$$K = \epsilon - vp + \eta t - m_1 \mu_1 - \&c. - m_n \mu_n$$

where ϵ is the energy, v the volume, η the entropy, and m_1, m_2, &c. the components corresponding to the second phase (B), while p is the pressure, t the temperature, and μ_1, μ_2, &c. the potentials corresponding to the given phase (A). The intensities therefore are those belonging to the given phase (A), while the magnitudes are those corresponding to the other phase (B).

We may interpret this expression for the stability by saying that it is measured by the excess of the energy in the phase (B), above what it would have been if the magnitudes had increased from zero to the values corresponding to the phase B, while the values of the intensities were those belonging to the phase (A).

If the phase (B) is in all respects except that of absolute quantity of matter the same as the phase (A), K is zero; but when the phase (B) differs from the phase (A), a portion of the matter in the phase (A) will tend to pass into the phase (B) if K is negative, but not if it is zero or positive.

If the given phase (A) of the mass is such that the value of K is positive or zero with respect to every other phase (B), then the phase (A) is absolutely stable, and will not of itself pass into any other phase.

If, however, K is positive with respect to all phases which differ from the phase (A) only by infinitesimal variations of the magnitudes, while for a certain other phase, B, in which the magnitudes differ by finite quantities from those of the phase (A), K is negative, then the question whether the mass will pass from the phase (A) to the phase (B) will depend on whether it can do so without any transportation of matter through a finite distance, or, in other words, on whether matter in the phase B is or is not in contact with the mass.

In this case the phase (A) is stable in itself, but is liable to have its stability destroyed by contact with the smallest portion of matter in certain other phases.

Finally, if K can be made negative by any infinitesimal variations of the magnitudes of the system (A), the mass will be in unstable equilibrium, and will of itself pass into some other phase.

As no such unstable phase can continue in any finite mass for any finite time, it can never become the subject of experiment; but it is of great importance in the theory of chemistry to know how these unstable phases are related to those which are relatively or absolutely stable.

The absolutely stable phases are divided from the relatively stable phases by a series of pairs of coexistent phases, for which the intensities p, t, μ, &c. are equal and K is zero. Thus water and steam at the same temperature and pressure are coexistent phases.

As one of the two coexistent phases is made to vary in a continuous manner, the other may approach it and ultimately coincide with it. The phase in which this coincidence takes place is called the Critical Phase.

The region of absolutely unstable phases is in contact with that of absolutely stable phases at the critical point. Hence, though it may be possible by preventing the body from coming in contact with certain substances to bring it into a phase far beyond the limits of absolute stability, this process cannot be indefinitely continued, for before the substance can enter a new region of stability it must pass out of the region of relative stability into one of absolute instability, when it will at once break up into a system of stable phases.

Thus in water for any given pressure there is a corresponding temperature at which it is in equilibrium with its vapour, and beyond which it cannot be raised when in contact with any gas. But if, as in the experiment of Dufour, a drop of water is carefully freed from air and entirely surrounded by liquid which has a high boiling point, it may remain in the liquid state at a temperature far above the boiling point corresponding to the pressure, though if it comes in contact with the smallest portion of any gas it instantly explodes.

But it is certain that if the temperature were raised high enough the water would enter a phase of absolutely unstable equilibrium, and that it would then explode without requiring the contact of any other substance.

Water may also be cooled below the temperature at which it generally freezes, and if the water is surrounded by another liquid of the same density the pressure may also be reduced below that of the vapour of water at that temperature. If the water when in this phase is brought in contact with ice it will freeze, but if brought in contact with a gas it will evaporate.

Professor Guthrie has recently discovered a very remarkable case of equilibrium of a liquid which may be solidified in three different ways by contact with three different substances. This is a solution of chloride of calcium in water containing 37 per cent. of the salt. This solution is capable of solidification at -37° C., when it forms the solid cryohydrate having the same composition as itself. But it may be cooled somewhat below this temperature, and then if it is touched with a bit of ice it throws up ice, if it is touched with the anhydrous salt it throws down anhydrous salt, and if it is touched with the cryohydrate it solidifies into cryohydrate.

The PRESIDENT: We must return our best thanks to Professor Clerk-Maxwell for his very curious communication illustrating the unexpected condition in which matter may be found and its unexpected tendency to change in various ways. To myself it has been particularly interesting to see how all these conditions of matter and their tendency may be represented by a model of some tolerable simplicity and how the geometrical properties of that model are capable of illustrating and expressing the laws of these peculiar conditions of matter.

I believe Dr. Andrews, who has paid great attention to the molecular condition of matter, is present, and we shall be glad to hear any communication which he has to make.

ON THE LIQUID AND GASEOUS STATES OF BODIES.

Professor ANDREWS, M.D., F.R.S.: It is with some difficulty, not being accustomed to address meetings of this kind, that I attempt to speak on the subject of which I have given notice, more particularly as it is a very wide subject and represents the work of nearly ten years. The facts are so numerous that I am afraid in attempting to give even a sketch of the entire subject I may end in being partially unintelligible. At the commencement I wish to say that researches of this kind are not researches on conditions of matter which cannot be realized in Nature. It is highly probable that in the stellar regions, and also in many of the larger planets of our own system, all the conditions of matter to which I shall refer may actually exist. Certain it is that we could produce them without any artificial means, if it were possible to descend to the depths of our existing ocean, and there to establish at certain intervals experimental laboratories. Many years ago, as some of you will remember, Professor Leslie, of Edinburgh, imagined that at the bottom of the ocean there was a layer of condensed air heavier than water, and that there might be inhabitants in that condensed atmosphere. Certain it is we can reduce the gases to very nearly the same density as water—to a density at which, as I have seen myself, they barely rise in that liquid. I may perhaps just mention, to give a popular idea of the question—I am speaking roughly—that if you descend to 1200 feet below the level of the sea, and there examine the properties of carbonic acid, it would exist, at the pressure there produced and at the temperature of 0° C. (the freezing point of water), in the state of a liquid. If you descend 1700 feet, not a great depth in the sea, you would find that this body would remain a liquid at the temperature of 15° C., and if you heated it above that point, it would boil; it would resemble in this respect the chemical compound hydrochloric ether which boils at the ordinary temperature of the atmosphere. If you descend to a depth of about 2500 feet, and there make your observations, you will find that the carbonic acid would cease to show the properties of an ordinary liquid at all temperatures above 31° C. At the temperature of 15° C. and at a depth of 2500 feet, carbonic acid would be a liquid. If you heated it, instead of boiling, it would change in a most remarkable manner, the surface would lose its curvature, the concave surface with which every one is familiar in an ordinary liquid would gradually become flattened and disappear, and at last the liquid would change, not into the gaseous state, but into one of those intermediate conditions which I have described as connecting the liquid and gaseous states together. At this depth of 2500 feet and at 31° C., which I have designated the "critical temperature of carbonic acid," new conditions of matter supervene, which cannot be referred either to the liquid or gaseous states, but connect those states by an unbroken continuity with one another. If the temperature or pressure be varied by 1° or 2° when the carbonic acid is at the critical point, you will have flickering movements of the same kind as those which every one has seen on a fine summer's day, or still better in a telescope, produced by the changing densities of the air, but here those conditions are so intense and extraordinary that you would not be able, I believe, to see one foot before you. You would be in a transparent atmosphere, but at the same time it would be rendered useless for all ordinary purposes by the intensity of these conditions. I may just further mention that the experiments of which you see a representation on the diagram, are experiments which have been carried on with exact measurements to pressures corresponding to the depth of 9000 feet below the level of the sea; and in the same apparatus I have without measurement gone as far as pressures of 500 atmospheres, which correspond to a depth of 16,500 feet below the sea. I mention the subject in this popular way so as to give an idea of the general conditions corresponding to those which can be seen in a few moments with the apparatus before you. With regard to that apparatus itself I will very shortly describe it (Professor Andrews then explained the apparatus by means of a diagram). The lower ends of the glass tubes containing the gases dip into small mercurial reservoirs formed of thin glass tubes, which rest on ledges within the apparatus. This arrangement has prevented many failures in screwing up the apparatus, and has given more precision to the measurements. A great improvement has also been made in the method of preparing the leather-washers used in the packing for the fine screws, by means of which the pressure is obtained. It consists in saturating the leather with grease by heating it *in vacuo* under melted lard. In this way the air enclosed within the pores of the leather is removed without the use of water, and a packing is obtained so perfect that it appears, as far as my experience goes, never to fail, provided it is used in a vessel filled with water. It is remarkable, however, that the same packing, when an apparatus specially constructed for the purpose of forged iron was filled with mercury, always yielded, even at a pressure of 40 atmospheres, in the course of a few days. The carbonic acid gas, which is now under a pressure of about 40 atmospheres, can be liquefied by a few turns of the screw, and then you can go on visibly compressing the liquid, because liquid carbonic acid is much more compressible than water. Here is a single apparatus, but the diagram shows a double one—a communication being made between the two tubes, so that although we have two screws we can operate with either or with both. One tube is filled with air and serves as a manometer, and the other is filled with the gas to be examined. In making observations of this kind it is necessary not merely to make them at ordinary temperatures, but to examine the properties of the gases under different temperatures, and for that purpose there are two modes. You may work with ordinary glass cylinders, but experiments of great accuracy can never be made in this way, because the inequalities in the glass produce an error of sometimes half a millimetre, and therefore it is necessary to make observations through plate glass. Accordingly this arrangement has been made which I described the other day. I should mention that some of these instruments are in the hands of different Professors, amongst others of Professor Rijke of Leyden. I have sometimes found difficulty in operating at the temperature of 100° from the tendency of the tube to become greasy, and drops of water collecting upon it. This for a long time gave me a great deal of trouble, but at last I found a remedy, which was to pour boiling water over the tube,—an operation which even under a pressure of 300 atmospheres is unattended by any danger provided the apparatus be at exactly the same heat as the steam—namely, 100° C. I should have further mentioned that in order to make experiments above 100°, you must bend the tube, and introduce the end of it into a suitable medium, such as sulphuric acid or melted spermaceti, or heated air; the only limit to the temperature being the softening of the glass. In all these cases so perfect is the apparatus as regards pressure that all our difficulty is with the temperature. I should mention another matter of importance. If you run up the apparatus suddenly to a pressure of 200 atmospheres or more, it will appear to leak; this is caused by the compression of the leathers. It requires some little time, in short, before everything becomes steady. Therefore, in some experiments, it is desirable to run up the pressure a little above what you want for a few minutes, and then bring it down to the proper point.

Now, having given a general account of the apparatus, I will very briefly refer to the results, and in order to make the subject as intelligible as possible, I shall refer for a moment to a series of effects which I think establish pretty clearly the remarkable fact, that between water or any other liquid—for it is not confined to one liquid more than to another—between any body in the liquid state and the same body in the gaseous or vaporous state, not only is continuity possible, but we have been able to establish it by experiment. This continuity has not been extended to the solid state, nor do I believe that the ordinary plastic condition of the solid is a passage between the solid and the liquid condition. That such a passage may be hereafter effected is possible; but as far as I can see my way at present, it is a question which will involve an apparatus capable of bearing a pressure of many thousands of atmospheres. I may be wrong on that point, but it is the view which I have taken for many years, and I believe it is the true view.

Here is a tube which illustrates some of these conditions. It is a sulphurous acid tube, and when you heat the liquid it first boils, and then changes into the intermediate conditions to which I have already referred. Here is another tube, which I am almost afraid to open, not because it contains carbonic acid, but because it is not so strong. It was made by Professor Dewar, and contains liquid carbonic acid.

I will very briefly refer to the recent investigations I have made on the properties of matter in the gaseous state. There are three important laws which are known with regard to the gaseous state: the law of Boyle, according to which gases vary inversely in volume according to pressure; the law of Gay-Lussac, according to which they vary in volume by a definite amount of their volume at zero, or any other fixed temperature; and the law of Dalton, according to which gases in mixture act as if the other gas were not present, as if each gas occupied the whole space. At high pressures these laws fail. They are true in the ordinary conditions of gaseous matter approximately, very closely indeed in the case of some of the gases, but when you operate under these great pressures all these laws fail. The failure of the law of Boyle has been known to some extent, but that of the law of Gay-Lussac, I believe, has not been observed till my last communication to the Royal Society.

I will conclude by writing down three formulæ which express ascertained facts. The expansion by heat of a body in the ordinary gaseous state, whether measured by its expansion at constant pressure, or by the increase of elastic force under constant volume, is not a simple function of the initial volume or initial elastic force but a complex function changing with the temperature. There are certain points to which I have given the name of homologues. The following are the formulæ $\mu = \frac{\mathrm{p}}{\mathrm{p}'}$. If you take any two isothermals, the value of μ is the same throughout the whole range of isothermals.

$$\rho = \epsilon p$$

$$p v = p' v'$$

$$(1 - p v) v = c.$$

The PRESIDENT: I am sure our thanks are due to Professor Andrews for the account he has given us, and for the very interesting and delicate experiments which he has performed. I am sure you will with me regret that we, like himself, are working under such extreme pressure that we were obliged to ask him to make his communication very short.

Before calling on M. Sarasin-Diodati I will venture to ask Mr. Dewar to give us a short communication on charcoal vacua. We are fortunate in having him here this morning, and his communication may be properly and naturally placed between the two you have already had and the lecture which is to follow. I have no doubt it will come in very well between the two other subjects best, and he will further imitate nature in allowing that state of transition to be one of extreme brevity in point of time.

Professor DEWAR: Especially since the recent experiments of extraordinary interest of Mr. Crookes on motion *in vacuo*, a method of readily getting a very perfect vacuum has been one of considerable importance; and my friend, Professor Clerk-Maxwell, has mentioned my name in his report in the department of physics, dealing with a method of effecting this result. All I have to say is that this plan was originally used by Professor Tait and myself, to improve on the mercurial vacuum. Of course, Sprengel's method may be carried to a very great extent, and will produce a very perfect vacuum as compared with an ordinary air-pump. The plan we adopted at that time was as follows:—Suppose this tube is to be made a very perfect vacuum. Charcoal is placed in the tube, and the whole exhausted by means of a mercurial air-pump at a red heat; it is then sealed up, and the last traces of gas, the vapour of mercury, sulphuric acid, and other traces of impurity are absorbed by the charcoal, and in a very short time we found that an electric spark could not pass. That has been shown very often at the Royal Institution. The advantage of this method was that by heating the charcoal again you could at once produce *striæ*; it goes backwards and forwards as often as you like. The improvement which I have made on this within the last two or three days, is one which, I daresay, may interest Mr. Spottiswoode, in connexion with his experiments on *striæ*. This vacuum is made by charcoal without the use of any mercurial pump whatever. The secret is a very simple one. Unless I wanted to use the charcoal tube it would be sealed off, and there is virtually no loss of matter nor of charcoal, because, if I seal this off, the same charcoal can be used for another tube, and then the matter which is left in the tube is so excessively small that the original matter I put in is virtually recovered. I will show you first that it is a tolerable vacuum, and then I will show you that the charcoal does contain a very large quantity of matter by heating it. I will attach a coil to it, and Mr. Spottiswoode will be the best judge if it is a tolerable vacuum, and you will see by the *striæ* when I heat it, that it contains a very large quantity of matter. The *striæ* are very wide at first, and by this simple method you can study them in all stages of their formation, from an innumerable number there is a gradual widening until they nearly disappear altogether. You see at once it is a tolerable vacuum from the enormous width of the *striæ*, and the condition of the negative pole, which is always the best evidence of a good vacuum, being isolated altogether from the other portion. I will now heat the charcoal in a spirit lamp, and you will see that it holds a large quantity of matter. As I heat it gradually, the gas comes out of the charcoal, and I may tell you at once the substance is bromine. I place a little fluid bromine in this tube, then place it in a water bath, the bromine distils over. The charcoal is heated during the process to drive out all gases, and it absorbs the bromine vapour and creates the vacuum. In this way you can produce a very good vacuum without any machine at all. You see at once the alteration in appearance, as the heat acts on the charcoal, and in a few moments the spark will not pass, the tube getting quite full of bromine vapour. If I had a Bunsen flame I could make it give up all the bromine. There is no special care required in this process; this is cocoa-nut charcoal, but any charcoal will do. The quantity of bromine used was about a cubic centimetre. You can now see quite plainly the colour of the bromine vapour, but on cooling it will be all again absorbed by the charcoal.

The PRESIDENT: Our thanks are due to Professor Dewar for his account of this remarkable instance of the simplification of processes which only those who have been engaged in these matters know how to appreciate. We sometimes get a vacuum by air-pumps and so on, and here we have seen a way of getting it without any pump at all.

I now beg to call upon M. Sarasin-Diodati to give his communication on "M. de la Rive's Experiments in Statical Electricity."

M. SARASIN-DIODATI'S ADDRESS UPON AUGUSTE DE LA RIVE'S LAST RESEARCHES IN ELECTRICITY.

M. SARASIN gave an account of the last works of Auguste de la Rive, conjointly with whom he had often worked, and he described some of his instruments, which were sent to the exhibition, from Geneva.

He stated that M. de la Rive and he himself had repeated—in order to study them more accurately—the observations previously made by Plücker, by De la Rive himself and by others, upon the modifications which the electric discharge undergoes, in rarefied gases, under the action of magnetism. They first of all considered the case in which the spark is produced perpendicularly to the line of the poles of the magnet, working for that at pressures included between 1^{mon} and 10^{mon}. In this case the discharge is deviated and tends to describe a circular curve around the axis of the magnet; moreover the luminous jet is more condensed and has all the characteristics which it shows at a higher pressure. MM. de la Rive and Sarasin have verified the fact that the resistance offered to the passage of the discharge then increases in a considerable degree, different for every gas and greater in proportion as the conductibility of the gas is weaker. They have besides, observed that the condensation of the luminous jet is accompanied by a real condensation of the gas in the part in which the magnet acts. This can explain, to a certain extent at least, the observations made by M. Chautard upon the modifications of the spectrum of gases under the action of magnetism. This fact is easily proved by means of a special Geissler's tube, divided into two chambers, both traversed by the same discharge, one of which is placed under the influence of the magnet, the other not. These two compartments being separated by the turning of a cock whilst the discharge is passing, it will be afterwards found that there is a greater pressure in the compartment placed between the two poles of the magnet than in the other.

A very brilliant experiment made about thirty years ago by A. de la

Rive shows the action of the magnet upon the electric discharge when the latter radiates round the pole of the magnet which then forms one of the electrodes, the other electrode being a concentric ring at the pole. The electric discharge revolves, in this case, round the magnet, like a needle on a dial, and with a velocity which varies considerably according to the nature of the gas, and which is nearly in an inverse ratio to its density. MM. de la Rive and Sarasin have, moreover, satisfied themselves with regard to the fact that the electric spark carries along with it, in its rotation, the gas itself which transmits it and any sufficiently light movable body, such as a little pendulum, or, still better, a small mill having its fulciment concentric to the annular electrode. In hydrogen at a pressure of 1^{mon} the discharge, revolving under the action of the magnet, has imparted to this mill a velocity of 120 and even 140 revolutions per minute. If the direction of the rotation be changed, the discharge very quickly stops the mill, and then causes it to revolve in the opposite direction at a speed which very soon equals the former.

Finally, M. Sarasin drew attention to the fact, that this experiment of De la Rive, upon the rotation of the electric spark around the pole of a magnet, has given him new and important evidence in support of the electric theory of the aurora borealis, of which he has been one of the most earnest defenders. Indeed this experiment can perfectly be compared to the well-known fact of the rotation of the arcs of the aurora round the poles of the earth. M. Sarasin gave a rapid sketch of the theory of the aurora borealis such as De la Rive had stated it, and explained the apparatus which he devised in order to reproduce the phenomenon, for the sake of experiment.

The PRESIDENT : We must now convey our thanks to M. Sarasin-Diodati for his interesting communication. He has explained very clearly the origin of the pieces of the apparatus which are happily now well-known to us ; thanks to Professor de la Rive and himself.

I will now call on M. Lemström to read his paper on "The Aurora Borealis."

Professor Lemström, of Helsingfors, Finland, then delivered the following Address on his Polar-light Apparatus, and on the Theory of the Polar Light.

Professor LEMSTRÖM : This apparatus serves to prove that the polar light or aurora borealis is an electric current flowing from the higher regions of the atmosphere down to the earth.

A sphere of brass, fixed on a bar of india-rubber or ebonite 0·6 metre long, is screwed in the board of the cross-shaped foot. A cylinder of india-rubber, 3 metres long, is fixed to the same board at about 0·7 metre from the sphere. From the cylinder comes out a branch with a bow, both of india-rubber. On the bow are fixed sixteen Geissler's tubes, wherein the air has a pressure of about 0·5 millimetre. The lower ends of the tubes are pierced by platinum wires, which are directed towards the sphere, whilst at the upper end the platinum wires are, by means of their copper wires, in a metallic union with a button, and also in metallic union with the earth. From underneath the sphere a copper wire, well insulated with india-rubber, leads to the negative pole of a Holtz's electric machine (a machine of Carré (Paris) was employed with great advantage), of which the positive pole is in metallic connexion with the earth. As soon as the machine is put in movement, the sphere being charged, becomes negative-electric, and at the same time there goes through all the tubes a current of reddish-lilac light, so that they altogether form a shining bow-shaped belt. With an ordinary machine this phenomenon may still be observed when the lower ends of the tubes are at a distance of *two metres* from the sphere. This proves evidently that the electricity flowing out from (or into) the sphere not only traverses the layer of air that is between, but goes also with such power through the tubes that the gas therein becomes glowing by the heat that the electric current produces, as is well known. In order that the electricity might more easily flow out in the air from the sphere, this latter is furnished with points. These points, as well as the metallic union between the upper end of the tubes and the earth, are of no absolute necessity, for the phenomenon may be produced without them, the distance between the sphere and the tube must, however, then be considerably reduced.

The described light-phenomenon produced by the apparatus proves clearly that a current of electricity may go through a layer of air of ordinary pressure 760^{mm} without producing the light-phenomenon, but if it meets in its way a space of rarefied air of low pressure (from 0 to 30^{mm} to 40^{mm}) there arises immediately a light-phenomenon, caused by the fact that the current makes the molecules of gas glow.

As regards the theory of polar light, the knowledge we have acquired of the electric state of the earth proves that it is a conducting body, charged with a small quantity of negative electricity, and surrounded by the atmosphere, in general charged with positive electricity. Though this latter might be produced by an influence from the earth, it is still very probable that it proceeds from the process of evaporation, either directly by this phenomenon itself, or by the friction of vapour against particles of air. The atmospheric air possesses a very small conducting power for electricity when dry and of ordinary pressure, but the conducting power increases considerably as soon as the air becomes moist and rarefied. It has been proved by experiments that the conducting power is highest at a pressure between 5^{mm} and 10^{mm}, and goes then 10,000 times beyond the conducting power at a pressure of 760^{mm}. If the rarefaction of the air is carried further than 5^{mm}, the conducting power diminishes again, but very slowly. It is known that in proportion to its elevation over the surface of the earth the air becomes more and more rarefied according to an irrefragable law, which finds its expression in the formula given by Laplace, and that consequently, at a certain elevation the earth is surrounded by a layer of air that has a pressure of only 5^{mm}; the conducting power for electricity in this layer is sufficiently great to allow of its being regarded as a conductor in comparison to the air in lower regions, and even in the highest. The negative electric earth is thus surrounded by a conductor for electricity concentric with it. All the positive electricity that attains the space of rarefied air of about 5^{mm}, or, as it might be called, this conductor of air, submits almost to the same laws as if it were in a real conductor, and must thus set in a restricted manner according to the influence of the electro-negative earth. Part of the electricity, conducted by the vapours, remains on the clouds in the atmosphere and discharges in form of lightning and thunder; another part attains the space of rarefied air or conductor of air, by the fact that the vapour itself, submitting to well-known physical laws, rises to this elevation, and also because electricity, according to its nature, endeavours always to set on something.

The manner in which the electricity divides itself on the two conductors depends on their reciprocal position to each other as well as on their form. The earth might, without a remarkable difference, be considered as a sphere, and likewise the conductor of air, but in their reciprocal position to each other it appears that the space of rarefied air of 5^{mm} approaches much nearer to the earth at the poles than at the equator, principally in consequence of the inequality of the temperature of the air in the two places. If we assume the mean temperature of the air round the equator to be 25°, at the poles −12°, and everywhere on the conductor of air −60°, and we suppose at the same time the air everywhere half saturated with moisture, and that the temperature is reduced in proportion to the elevation, we find, if the above-mentioned formula of Laplace* is applied, that the conductor of air, at the equator, must be at an elevation of 37·47 kilometres, and at the poles but 34·25.

In consequence of this relative position, and if the two conductors are regarded as conducting surfaces, the electric density on them both becomes about 9 per cent. greater at the poles than at the equator, and the power, by which the two electricities endeavour to join again, at least 20 per cent., but probably 30 or 40 per cent. greater, if all the circumstances are considered, at the poles than at the equator. It is in these facts we have to seek the principal cause of the accumulation of electricity at the poles of the earth and of the phenomenon that occurs there, and is called polar light or aurora borealis.

It is a remarkable fact that thunderstorms diminish as well in number as in intensity in proportion as we remove from the equator, and that at the 70th degree of latitude they cease completely, after having shown once more in the highest north vestiges of their primitive intensity. In Finnish Lapland, for instance, thunderstorms are very uncommon, but when they occur they are extremely severe, and are almost always accompanied by thunderbolts. This peculiarity has probably its cause in the fact that the region of thunderclouds lowers towards the earth in accordance with the same rule as the before-mentioned conductor of air. The reduced number of thunderstorms

---

* $X = 18{\cdot}393$ metres $(1 + 0{\cdot}002837 \text{ Cos. } 2\phi)\ (1 + 0{\cdot}004\frac{T+t}{2})\ l\frac{H}{h}$ where X signifies the elevation, ϕ the latitude, T the temperature at the surface of the earth, t at the upper point H, and h the stand of the barometer for the same points, but duly reduced.

is caused by the fact that the very source of electricity in the atmosphere, that is to say, the evaporation, is very much reduced; however, another important cause is here active—namely, the heightened conductive power that the air possesses in consequence of its greater quantity of moisture, whereby the electricity becomes unable to keep itself, beyond a certain latitude, upon the clouds, until it has attained a greater tension, but is conducted down to the earth in form of a slow current, visible in the polar light.

It results from experience, with a high degree of probability, that the polar light is an electric phenomenon, for its effects are of the same nature as those of the electric currents. Thus the polar light causes disturbances in terrestrial magnetism, induces currents in the telegraphic wires, and furnishes a spectrum of nine bands, which coincide, except one, with the spectral lines produced if an electric current goes through a rarefied space of air. Thus there is no doubt that the polar light is caused by an electric current going down from the upper rarefied layers of air to the earth; this current, during its passage through the rarefied air, produces light phenomena that cannot arise in denser layers of air.

The polar-light apparatus now exhibited shows that an electric current flowing out from an insulated body does not produce any light phenomena in air of normal pressure, but as soon as it rises to the rarefied air in the Geissler's tubes, there is directly produced a light phenomenon very like the real polar light. In the apparatus the upper end of the tubes is in union with the earth; this is by no means necessary, for the light phenomenon is also produced if this union be removed, provided that in such case the tubes be brought a little nearer to the insulated sphere. For the rest, the earth represents here the wide space of rarefied air that we find beyond the limits of the conductor of air, and which serves here as an electric reservoir.

Let us now consider how the polar light on a large scale is formed in nature. As before said, the earth, and the conductor of air, hold to each other the position above-mentioned, and the two electricities, the negative electricity of the earth and the positive electricity of the conductor of air, endeavour with a certain force to unite in a belt around the north pole. The insulating power of the denser air prevents this reunion; but if we assume that perfect equilibrium is attained, the reunion will take place as soon as this equilibrium is disturbed, either by the insulating power being diminished or the electricity of the conductor augmented. The first case, which probably is the most ordinary, happens if a southerly wind carrying a quantity of vapour attains the polar regions; for instance, the belt, where the vapour, in consequence of the cold, is condensed into a fluid form, reduces thereby considerably the insulating power of the air and enables the electric current to flow through it. The same thing would occur if a layer of clouds happened to enter into this belt; the upper end of the cloud would become negatively electric, the lower one positive, and thus the distance between the two conductors would in fact be diminished. The electric current would go from the conductor of air to the cloud, and through this latter to the earth. Similar phenomena are observed in the polar regions, or the upper edges of the clouds are not unfrequently seen shining with a yellowish light stretching considerably upwards, whilst no light is discernible under the cloud because of the air there having attained a density sufficient to prevent the current from producing light.

For special knowledge of the polar light and its theory, we refer to essays inserted in the Archives des Sciences Phys. et Natur. de Genève, 1875 (Sept. and Oct.), and in January, 1876, as well as to two essays published in the years 1869 and 1873, in the same scientific journal, all which articles are more or less the result of observations made in the arctic regions. Besides these we may refer also to the works upon polar light of the American natural philosopher Loomis, Rep. of Smithsonian Inst., 1865, &c.

The PRESIDENT: We must return our best thanks to Mr. Lemström for his elaborate paper, and express a hope that he may some day publish it, when we shall be better able to do justice to it than we can to-day.

In consequence of the lateness of the hour Baron Ferdinand de Wrangell and Il Commendatore Professore Blaserna will give their communications in the afternoon.

(The Conference then adjourned for luncheon.)

The PRESIDENT: The clock having struck two, I will call on Mr. De La Rue for his communication on astronomical photography.

Astronomical Photography.

Mr. De La Rue: In the Loan Exhibition there are, as you know several astronomical photographs, and some apparatus with which these photographs were taken. There is one telescope which is absent necessarily on account of its enormous size; but I have a small model of it, the Melbourne telescope, to which I shall presently allude. In speaking of astronomical photography, I wish my audience to understand that it is not merely the pictorial representation of celestial objects that astronomical photography concerns itself with mainly; it is the production of records which can afterwards be measured, and which afford data for astronomical investigations. I had better, in the first instance, just state what happens. Suppose, for example, we have a lens or a mirror directed on to a celestial object and that the image is received on a sensitive plate. I will imagine we have a fixed star in focus whose image is a very small point indeed, scarcely to be distinguished from specks in the collodion. But the lens or mirror not being provided with any movement to follow the star, it would happen that in consequence of the star's apparent path in the heavens, if the atmosphere were perfectly quiet, we should have a straight line impressed on the plate instead of a spot, the line being longer or shorter in proportion to the duration of the exposure; but what does really happen is that we get an irregular wavy line on account of atmospheric disturbances. Now I will suppose we use a telescope mounted equatorially, that is to say, on an axis parallel with the earth's axis, and that we drive it by means of clockwork machinery in order to follow the apparent path of the star perfectly. Then the star would, if undisturbed, stand still in the centre of the field, if it were placed in the centre to start with, but instead of its being depicted as a single spot, we get a series of dots, in consequence of the agitation of the image arising from the currents of air differently heated and differently refracting, if it were a double star we should get a conglomeration of each of the two pictures, for the same reason. The eye looking at a star sees it sometimes steady and well defined, at others blurred and moving about in different positions of the field, but the mind selects the best images, and in all astronomical drawings the disturbances are eliminated; the memory discards those, and the hand only draws the appearances at moments of finest definition. A photographic plate, however, retains all the impressions; it is a retina on which all the disturbances are permanently recorded, and consequently no photograph yet obtained in any way conveys the sharpness of outline of the finest definitions of the telescope, it is always more or less blurred. Nevertheless very valuable results, as you will presently see, are obtainable. For example, if we want to ascertain the position angle and the angular distance of two double stars, the disturbances do not prevent our finding accurately the centres, and we can get, by means of a micrometer, these data just the same as though there were no blur.

The size of the focal image of any celestial object, other than a fixed star, which is always a mere point, I need scarcely tell you, depends on the length of the telescope; in my own telescope the focal length is ten feet, and the image of the moon is about an inch; I say "about an inch," because it varies in consequence of the nearer approach of the moon to the earth at one time than at another. I have here a group of original negatives of the moon, showing the sizes they are obtained by my telescope when she is in different positions in her orbit. The time of the exposure of these photographs is marked on them,—it varies generally from about one second or less than a second, for a full moon, to about eight or ten seconds when at the crescent. The duration of exposure to produce an image depends, with equally sensitive chemicals, mainly on the relation of the aperture to focal length, and hence it is very desirable to get as large an aperture as possible with respect to the focal length—or, in other words, to make the focus of the telescope as short as is consistent with good definition. It is quite possible, by means of clockwork, to follow a star almost perfectly, that is to say, if a telescope is put on a star and adjusted so that its image falls on the cross wires in the field, one may leave the telescope and come back again after an hour, and find the star there still; so that that mechanical difficulty to obtaining good photographs is quite overcome. The great enemy, however, is always the atmospheric disturbance; but with regard to the moon there is another difficulty: the equatorial telescope moves simply in a circle parallel with the earth's equator; but you all know that the moon's orbit is inclined to the earth's equator; and, consequently, if we have placed the moon centrally in the telescope adjusted to follow her in right ascension, after a time she will have moved up or down (in declination) some distance, in consequence of her orbital motion. The motion of the moon in declination, as it is called, is greatest when she is near her nodes, and then in one minute she may move as much as 16 seconds of arc, consequently in 10 seconds she may have moved $2''\cdot7$ in arc, or nearly 3 seconds of arc; and as we can depict on the moon objects whose diameter is not greater than 1 second, which is equal to about a mile, it is very clear that when the moon has the greatest motion in declination, the lunar pictures cannot be as perfect as when she is at a greater distance from her nodes. The Astronomer Royal, Sir George B. Airy, whose name is not only connected with mathematical astronomy, but also with astronomy as a great mechanist, has proposed to add to the ordinary equatorial a second axis, which would be carried round so as to remain parallel with the axis of the moon's orbit in its diurnal path, and provided with an independent clockwork driver carrying the telescope in a direction inclined to the equator, so as to follow the moon in her orbit from west to east. I need scarcely say that this appliance would complicate a large telescope so much that it would be almost impossible to adopt it; still it is conceivable that we may, besides the motion in right ascension, give an adjustable motion to the declination axis, varying from nothing up to 16 seconds of arc either in north or south declination in a minute. But this has not yet been done, and consequently it has to be accomplished in order to make it possible to photograph the moon perfectly at all times. However, the first photograph I ever obtained I took with the piece of apparatus I hold in my hand, by which I followed the motion of the moon both in right ascension and declination. It is unfortunately now getting broken. It was put on to the eye end of the telescope which was allowed to remain at rest, and by means of this milled head, the sliding piece carrying the sensitive plate having been placed so as to move in the direction of the motion of the moon in her orbit and in right ascension, I was able for a short time, by making three quarters of a turn, to follow the moon by hand; and here is a picture which was obtained in 1853 in that way. But I was not the first to photograph the moon. I have here a daguerreotype, copied from one which was exhibited by the late Mr. Bond, of Cambridge, Massachusetts, in the International Exhibition of 1851.

You are aware that the moon does not always present precisely the same hemisphere towards us in consequence of her unequal angular motion in her orbit, while her axial motion, her rotation on her axis, is perfectly equable. Although they coincide for a whole revolution, and bring a crater of the moon exactly again to the same place—at other times we see a little on one side and a little on the other side of the hemisphere which is turned away from us. Moreover, the moon's axis is inclined to the earth's equator, and, consequently, sometimes we see a little more to the north, and sometimes a little more to the south, of the moon's equator as she moves round in her orbit. These effects are called libration in longitude and libration in latitude, and we also have the diurnal or parallactic libration dependent on the position of the observer. The moon may be in the horizon or overhead, at one time or the other, and in consequence of parallax, which is quite sensible, being rather more than a degree, we sometimes see a little on one side or the other of the moon in varying directions. We have here those conditions which enable us to obtain a stereoscopic view of the moon, by combining two pictures taken at different times, showing the moon as a globe. But besides these three librations, it has been suggested that the moon has a real or physical libration, in consequence, as it is supposed, of a protuberance of matter on that hemisphere which is turned towards the earth which would tend to follow the redundance of matter about the earth's equator, and to adjust itself towards it with a sort of balancing or wobbling motion. Whether there is or not a physical libration of the moon, is a problem which astronomical photography can solve, and I do not know that there is any other method by which it could be done so perfectly. For this object the original negatives may be placed on an instrument called a micrometer, and adjusted concentrically with it; then by bringing any object, a selected crater for example, under the microscope, which is effected by turning the divided circle, and then drawing out the slide, we can measure the angular position and the distance from the centre or periphery of the moon of that object, and obtain data for ascertaining whether there is a physical libration of the moon or not, after taking into account the effects due to libration in longitude and latitude, and to parallactic libration in shifting the crater in question from the position it occupies in reference to the lunar disc at the period of mean libration. The telescope with which these negatives were taken is now at Oxford, and, I believe, Professor Pritchard, the Director of the University Observatory at Oxford, intends to devote the telescope to determining this problem, and a very interesting one it is. I said that the size of the negative depends essentially on the length of the telescope and the position of the moon in her orbit. I have here a model of the very splendid instrument, constructed for Melbourne by Messrs. Grubb, of Dublin, under the superintendence of a Committee of the Royal Society. In that colony they are more public spirited than we are here, and they have found the money for constructing this very large and perfect equatorial telescope, the mirror of which is four feet in diameter, whereas the telescope, with which my photographs were obtained, is only thirteen inches. Here, through the kindness of Mr. Ellery, the Melbourne Astronomer, are some negatives of the moon three inches in diameter, made with the Melbourne telescope, and very beautiful productions they are. These especially would be particularly well adapted for measurement with the micrometer for determining the question of the libration, but even pictures so small as those obtained with a thirteen inch reflector may be magnified to a suitable size. There is one on the wall thirty-eight inches in diameter obtained in a negative taken at Cranford, and here is another, about eighteen inches from diameter, a very beautiful one, taken by a Mr. Rutherfurd, an American gentleman, in his private observatory. I do not know whether that is his best, but there is one he has taken under particularly fine conditions of the atmosphere, which is the finest in existence.

It may be interesting to you to know, as all the necessary apparatus is before you, in what way the enlarged photographs are obtained. One of these original negatives is placed in the focus of this long camera. Lenses, specially made by Mr. Dallmeyer, are used for projecting the image along the axis of the camera on to plates about one-foot square, and then the first enlargement takes place. Here are specimens taken from downstairs, this is a first enlargement to nine inches in diameter, and is a transparent positive on glass; by a second enlargement it furnishes the negative which is used to print the paper positives. There are a great number of such along the wall of different phases of the moon enlarged to about eighteen inches. The largest here is nearly one metre or rather over three feet two inches diameter; and at the time this was made, some years back, we did not possess cameras large enough to produce so large a photograph at one operation, so that it was done on four plates. You will notice that in the successive enlargements there is always a loss of definition, and that more details are to be seen in the original negative, to which recourse must be had for exact measurements.

I now come to solar photography. There are some early French photographs downstairs; but solar photography was first systematically followed up at the Kew Observatory, in consequence of a suggestion of the desirability of photographic sun observations having been made by the late eminent astronomer, Sir John Herschel, to the Royal Society. The Kew photo-heliograph is before you. This instrument was in use about 1854 at Kew, but was not systematically worked. It was taken to Spain in 1860, and there used for photographing the solar eclipse. It was afterwards worked from 1862 to 1872 systematically in observing the sun's spots. Eleven instruments on the Kew model, but improved, have since been constructed by Mr. Dallmeyer under my direction, and took part in the observations of the transit of Venus. The more recent photoheliographs have the object glass four inches in diameter; this is about three and a half. The image is not allowed, as in the case of lunar photography, to fall directly on the sensitive plate, but passes through a secondary magnifier, and is thus enlarged to four inches. In the common focus of the object glass and the magnifier are cross wires, most conveniently placed at the angle of 45° to a terrestrial meridian passing through the sun's centre, and a negative of the sun like that I show you has the cross wires depicted upon it. They are of great use in the subsequent measurements for obtaining the position angle from the north of any spots that may be depicted. The pictures obtained during ten years are now undergoing measurement by this micrometer, which I will describe. It has a divided circle, on which the photograph to be measured is placed and fixed; with the aid of the microscope it is then adjusted so that the centre of the sun exactly coincides with the centre of the divided circle whose axis is a hollow cylinder, and the north point made to correspond with the zero of the circle. By drawing out a slide we can bring the periphery of the sun under the cross wires of the microscope, and by reading off on a vernier can see the distance passed over, and obtain the value of the radius of the sun in inches and decimal divisions. The photograph is then crossed over to the other side so as to eliminate any error of centring, and then we have the measure of the whole diameter on the arbitrary scale. We then bring any sun spot underneath the microscope by turning the circle and drawing out the slide, and we read on the circle its angle of position and on the slide its distance from the centre. With these measurements after they are reduced we are enabled to ascertain the helioscopic position in latitude and longitude. But there is another disturbing cause to the application of photography to exact measurements in astronomy. There is always some optical distortion. In order to ascertain its amount, a scale, of which this is a model, was fixed on the Pagoda at Kew, distant 4398·24 feet from the Observatory, and photographed. Each of these plates of the scale, one foot wide, is depicted in the photograph, and occupies rather a large space in the picture the further it is from the optical centre. We were able to measure, by means of the micrometer, the width of the image at the centre, and the width at other parts of the field up to the edge. That gives us the means of ascertaining what allowance ought to be made for distortion, and we can apply this correction and obtain most accurate measurements of the sun.

I ought to say that the Kew photo-heliograph had the honour, by its observations in 1860, of first proving that the luminous prominences which are depicted in this photograph, and which were only to be seen at the period of a total eclipse, absolutely belong to the sun. We do not want that proof now, because we can see them at any time under favourable atmospheric conditions, whether there is an eclipse or not.

I have already alluded to the transit of Venus. The English, the Russians, and our colonies employed the eleven instruments of the Kew model, before spoken of, but the Americans made use of a long telescope, I think nearly forty feet in focal length placed horizontally in the direction of the meridian, and the image was thrown into it by the optical instrument which you will see in the lower gallery called a siderostat or heliostat, which was adjusted to reflect the image of the sun constantly in the axial direction of the telescope. This has some advantages and some disadvantages—the advantage is that distortion is reduced to a minimum. Our Chairman reminds me that I ought not to forget to call attention to a set of instruments for the optical and photographic observation of the transit of Venus, kindly contributed by the Astronomer Royal, which will be found on the balcony. It is a complete equipment, and it is well worth inspecting because it shows how very much had to be done for every fully equipped station in order to secure accurate results. In the Kew photo-heliograph the diameter of the sun on the occasion of the late transit of Venus would be nearly four inches;—the diameter of Venus is 1·26 of an inch, and the maximum displacement of Venus nearly one-tenth of an inch. Measurements can be made accurately within 0·2 of a second of arc, by means of the micrometer which I have described, but an error to that extent would only produce an error of less than 0·04 second of arc in the deduced solar parallax. It is to be hoped, therefore, with what has been secured in these observations that we shall obtain the solar parallax—in other words the earth's distance from the sun—to a very great degree of accuracy; moreover we have another eclipse coming off in 1882, and the same instruments or similar ones will be ample to obtain the observations requisite for a still closer approximation to the true parallax.

I now wish to call your attention to some photographs of the sun on a scale of four feet for the sun's diameter. They were obtained with the Cranford instrument, which is only thirteen inches in diameter, the image passing through a secondary magnifier, by which it was magnified, before it fell on the sensitive plate. Not only is it possible to photograph the sun directly on this large scale, but it is also possible, as you will see by this print, made in 1862, to obtain solely by the action of light and electro-metallurgic processes, printing blocks without the touch of a graver which will print the sun spots. This appeared in the monthly notes of the Astronomical Society in 1862, but unfortunately it was very badly printed.

You are all aware that many attempts have been made to connect solar phenomena with meteorological phenomena, and there can be no question whatever that as solar radiation varies so must vary also the whole atmospheric conditions of moisture and rainfall; hence it is most desirable that sun pictures should be obtained in a sufficient number of places, in order that we may have a solar picture at least every day, not only on a small scale for the position and area of the spots, but on a large scale, so that we may study special phenomena of the sun intimately on a large scale. Solar photography has this advantage, that it can be pursued in a town;—the sun has so much actinic power that it matters very little where we are placed so that the sun shines. I am happy to say that in Austria they think of making such observations. In Paris there seems to be a chance also, and I trust that through the action of the Royal Society and the Astronomical Society we shall get the Indian Government to take it up, and I certainly do hope that in England we shall have a solar physical observatory.

The PRESIDENT: I am sure you will all join with me in offering our best thanks to Mr. De La Rue for so complete an exposition in most respects and so lucid in all, of this important subject of astronomical photography. I say emphatically complete in most respects because there is one point in which it is decidedly incomplete. He has given us a very clear and fair idea of how these processes were brought successfully into action, and how it has been managed to bring out of them these great results of which he has given us specimens, but he has by no means conveyed to the minds of his audience how much these results are due to his own individual efforts and contributions. I think I shall not say too much if I speak of astronomical photography in this country at all events, and through this country to most parts of the civilized world, as being indebted more to him than to any one individual from its commencement to the present time.

[Mr. De La Rue here took the chair.]

On the Variable State of Electric Currents. By Professor Blaserna, of the Royal University, Rome.

Gentlemen,—

I regret much not being able to address you in English, and I must therefore ask your indulgence for speaking to you in a language which is neither yours nor mine.

The question of the Variable State of Currents was originated and has been treated by Ohm.

By following up the ideas which lead him to the discovery of one of the most important laws of physical science, he arrived at the conclusion, with regard to the origin of the current, that the fixed normal state, from which the current derives its permanent intensity, is preceded by a variable state, in which, from the moment of interruption, begins at zero, and reaches, in a very short space of time, its permanent intensity.

Ohm endeavoured also to establish the law of this movement, which may be represented by a curve, at first convex, and then concave towards the axel of the abscisses, and which has consequently a point of inflexion.

We can understand that the variable state should exist. There are no phenomena in Nature which do not require a certain time for their formation and development, and the question is merely to find out whether this time is sufficiently long to be indicated or measured by instruments of the most delicate make, and the most perfectly adapted for the object to which they are to be applied.

Faraday's great discoveries of inducted currents and extra currents gave to this question a new and wider aspect. The question may be asked, What are the nature and duration of the extra current? either of cessation or interruption; and as the extra current is nothing but the current inducted on itself, the general laws of inducted currents may be looked for in wires and in fixed helices.

These different questions have been discussed by a great number of scientific men. I will mention, among others, M. Helmholtz, who, by calculations and by experiments, has succeeded in representing the variable state of currents by an exponential curve, differing from that of Ohm, and without a point of inflexion; M. Rijke, who, by an ingenious, but, as he himself confesses, not sufficiently accurate process, tried to determine the duration of extra currents; M. Guillemin, who confirmed Ohm's theory with regard to long telegraphic lines; and finally Sir W. Thomson, who arrived at far different conclusions from the others.

This was the state of the question when I began to interest myself in it. After me, it occupied the attention of many scientific men, among them MM. Helmholtz, Lemström, Bernstein, Felici, Donati, Cazin, and others.

A certain number of instruments have been made, of which I shall mention the most important: The apparatus of M. Helmholtz upon a system of levers; it has also been employed by M. Lemström. My apparatus, with a revolving cylinder, and with metallic contacts, after the idea of MM. Wartmann and Dove. The second apparatus of M. Helmholtz, in the form of a pendulum, with metallic contacts. The apparatus of M. Bernstein, revolving, with contacts of mercury. The apparatus of M. Cazin, a weight, which in falling forms a current. Finally, the apparatus of M. Felici, revolving, with an inscribing diapason for the measurement of time.

The principal fact which resulted from my researches, and which has been confirmed and amplified by other scientific men, is that the variable state of currents is occasioned by a phenomenon far more complex than it was supposed to be. The intensity of the current, which is at first zero, reaches a degree nearly double that of its normal value; then descends nearly as far as zero—but without quite reaching it; then attains a second maximum, rather less than the first; then descends to a second minimum, less marked than the first. Thus the current makes a series of fluctuations around its final value. The duration of an oscillation does not appear to be always the same. The conditions, on which it depends, are not yet known; but it seems that it can vary from a half to four ten-thousandth part of a second. The number of these oscillations is very great. I have found traces of them to the hundredth parts of a second. The number which can be observed depends evidently upon the delicacy of the galvanometer used It depends also upon the shape of the circuit. When in the circuit there are helices, capable of giving strong extra currents, the oscillations are strong, and consequently more numerous. They may be expressed by a formula which contains an exponential and, at the same time, a periodical function. In rectilineal circuits they are very feeble, and perhaps they do not exist at all. So much for the variable state. If the extra current be separated from the principal current, and be considered as an independent thing, it may be said that the inverse extra current is formed of a series of alternate currents, which begin by being very strong and become gradually very weak. They are oscillations positive and negative.

As to the direct extra current, it is also formed of oscillations, which are even more energetic and rapid.

Sir W. Thomson had arrived, by analysis, at the conclusion, that, under certain circumstances, the current may become oscillatory. The oscillations, which he discovers, have a different character from those which experiment has, up to the present time, revealed. According to what has as yet been experienced the fluctuations of the current, in its variable period, never reach zero, whilst Sir W. Thomson proves that the oscillations are positive and negative—that is to say, there are contrary currents. It would be very difficult to say whether the conditions, under which Sir W. Thomson's beautiful analysis brings about this result, could ever be practically realized, for it becomes most difficult to turn these mathematical conditions into experimental ones. In any case, it is interesting to see how calculation had, to a certain extent, foreseen these phenomena.

With regard to inducted currents, judging by all the results obtained by experiment up to the present moment, it may be said that they are, like the extra currents, formed of very numerous oscillations. The exact conditions on which they depend are as yet unknown, and many questions relative to them must yet be studied. It is indeed a broad field that has still to be gone over. You will permit me not to enter into the details of this question, regarding which there exists still much difference of opinion, which will, however, disappear under the influence of time and research.

I prefer rather to call your attention to an idea which readily suggests itself to the mind, and which, according to my opinion, deserves to be considered. It has been somewhat difficult to prove the existence of the oscillations of the current, but they cannot be considered very rapid. They occur at the rate of about 5000 a second. In the vibrations of sound far higher figures than this are attained. And thus the molecules might yet undergo vibrations.

Now, all phenomena can be divided into two perfectly distinct classes: into phenomena which propagate quickly and into those that propagate slowly. To the first division belong light and radiating heat, which are formed by the oscillations of ether, and which have a propagating velocity of several hundred thousand kilometres a second. Sound, communicated heat, &c., belong to the second class. They attain a speed of some hundreds of metres, or, at the most, some kilometres a second, and this speed is produced by the motion of the molecules. The difference is immense between the first and the second category.

As to electricity, it propagates itself, in good conductors, at a rate which can be compared to that of light. But the oscillations of the current belong to the second division: they are slow. The conclusion might be drawn that the motion of electricity in metals is caused by ether or by another fluid of about the same elasticity and delicacy, whilst in the oscillations of the current the molecules themselves vibrate, or else they acquire in it a predominant influence. The oscillations of the current are the result of a reaction of the helices on themselves. They are phenomena of induction through bodies which are bad conductors. They are consequently slow, and must not be confounded with the vibrations, which the ether probably makes in vibrating electrically in bodies, which are good conductors. They stand in the same relation to these latter as the great waves of the sea do to the small calorific vibrations, which the molecules of the sea can, at the same time, produce.

The CHAIRMAN: You have anticipated me by your applause in asking you to pass a vote of thanks to Professor Blaserna for his most interesting and lucid account of these most remarkable experiments. In confirmation of what he has been stating, with respect to the induced currents and the principal currents in an induction coil, I may state that your President, in his way of experimenting, and I and my colleague, Dr. W. H. Müller, in ours, with a voltaic battery consisting of several thousand cells, have come very much to the same conclusion—that there is no such thing as a perfectly continuous uniform electric current. There is what appears to be a continuous flow, but at certain periods this flow is very much increased; and I think that the more we experiment the more we shall see that the electric current passing through any medium—metallic wires and other solids, fluids, or gases—does take up certain pulsations, and there is a maximum and a minimum of transmission—there are starts as it were in the current. Mr. Spottiswoode has been pursuing for a long time some most interesting experiments on electric discharges in vacuo. A portion of his researches was communicated the other day to the Royal Society, and I and Dr. Müller had the honour of communicating one with him, in which we believe we detected that when stratification took place in vacuum tubes there was a fluctuation of the current. We are still pursuing our experiments, and he is pursuing his, and they all go to confirm that which Professor Blaserna has stated. Professor Blaserna wishes me to say also, that Sir William Thomson mentioned to him that, on mathematical grounds, he considered it was quite possible—in fact probable—that there were oscillations in the electric current.

MAGNETIC REGISTRATION.

Mr. BROOKE, M.A., F.RS.: The subject to which I am desirous of directing your attention is that of the instruments connected with the automatic registration of the variations of the magnetometers, which are exhibited in this building. In order that it should be better understood by those who have not paid any particular attention to the subject of magnetism, I may state that the magnetometers are instruments which have now for many years past been subjected to continuous registration by the photographic method, and those I may briefly describe are three in number. The first is the declinometer, a magnet suspended by a single skein, which takes up its position according to the direction in which the magnetism of the earth acts upon it, and if the direction of the action of the earth's magnetism varies, it will vary the position of this magnet, just the same as it would that of a compass needle: the declinometer is intended, therefore, to record changes in the direction in which the earth's magnetism is acting. The second instrument is called a bifilar magnetometer: it is a magnet suspended by two parallel or nearly parallel threads, and the upper attachment of those threads is twisted, so as to bring the magnet round exactly into the position of magnetic east and west, that is, at right angles to the magnetic meridian. In that position the tendency of the suspended weight to untwist is balanced by the pull of the earth; and supposing its tendency were to untwist towards the south, the magnetism of the earth tends to pull the marked end of the magnet towards the north, and it is so arranged that the torsion force is exactly balanced by the magnetic force. Now if any increase of magnetic force takes place, it would tend to overcome the torsion force, and to pull the marked end of the magnet further towards the north; but if the magnetic force of the earth diminishes, the torsion force will preponderate and the magnet will untwist a little, and it will move slightly in the contrary direction. The third instrument is what is called the balanced magnetometer, which I hold in my hand. For that purpose a magnet is balanced on a knife-edge resting on an agate plane, and placed at right angles to the magnetic meridian, so that while the weight of the magnet tends to elevate the marked end, the vertical force of the earth draws the marked end downwards, and that force will deflect a dipping-needle about 70° to the horizon. The vertical force of the earth tends to draw the marked end down, and if it diminishes, the weight would cause that end to move in the other direction and rise a little upwards; consequently the movements of this magnet in a vertical plane indicate the changes which take place in the earth's vertical force.

For many years observations were made with these instruments solely by means of the eye with the aid of a telescope; and they were made by attaching a small mirror to the magnet, which was viewed by a fixed telescope with cross-wires in it, and by the telescope a fixed scale placed at a certain distance was seen by reflection from this mirror; and as the magnet moved, and, consequently, the mirror moved with it, it is clear that if the cross-wire in the telescope were adjusted so as to fall on the scale, as the mirror moved, the cross-wire would appear to rise or fall upon the scale, and by that means the position of the magnet at the time was recorded simply by eye observation. But in the Magnetic Conference, which was held at a meeting of the British Association at Cambridge in 1845, there was a general expression of opinion that it was a great drawback that no means had been hitherto found for making the magnetic instruments record their own changes of position. Attempts had been made to effect this by means of an attached needle-point which was periodically impressed upon a surface so as to mark it, so that by that alteration in the position of the mark of the needle, an indication of magnetic change would be obtained ; but it was found totally impracticable. In point of fact, the amount of actual force that is exerted is so very minute that it was quite clear that it could actuate no pencil but one which moves without friction or any mechanical resistance—viz., a pencil of light. The desirability of accomplishing this object attracted the attention of many persons to it—amongst others that of the late Sir Francis Ronalds and myself, and the instruments which were devised for this purpose by both of us are to be seen downstairs. In the apparatus of Mr. Ronalds the plan pursued was to attach a screen with a slit in it to the magnet. A light was burnt behind the screen, and as the magnet moved, of course the screen would move by minute quantities, and the light transmitted through the slit was allowed to fall upon a sensitive photographic surface. But for various reasons which I need not go into now, this system was found impracticable. The idea that suggested itself to my mind was that of attaching a concave elliptical mirror, as you see here, to the magnet. The same system is applied to all three instruments such as I have described. A concave elliptic speculum is attached which has its conjugate foci at about two and seven feet from the surface of the mirror. A light, either a lamp or jet of gas, is placed at a distance of two feet from this mirror. The light passing through the small slit in the opaque chimney of the lamp or gas burner, as the case may be, falls upon the mirror, and an image of that slit is formed at a distance of about seven feet. The reason for using a slit, and not a point, is that the image of the line of light is received upon cylindrical lenses which contract an image which is about one inch and a half long, into a narrow point not exceeding one-sixteenth of an inch in width, and consequently the whole of the light is accumulated into a very narrow space. That point of light falls upon a sheet of photographic paper which is placed round a cylinder, and the cylinder is carried round by clockwork once in twelve hours. For the balanced magnetometer the axis of the cylinder is placed vertically, but for the other two instruments, horizontally; so that the combination of the motion of the point of light on the paper, with the motion of the paper in a perpendicular direction presents a means of tracing out the magnetic curve. Here I have an actual photograph of a magnetic disturbance which took place in the observatory at Toronto, where, for many years, as at other places, these instruments had been in action. The cylinder goes twice round in the twenty-four hours, and therefore there will be two photographic traces on the paper. In one of these you will see the changes are very small, while in the other they are very large, and there is a constant vibratory disturbance of the magnet which has been maintained through nearly the whole period of twelve hours. It is quite impossible that any eye observations could ever have made us acquainted with the details of magnetic disturbance which are shown by photographs of this kind. With regard to the amount of movement, I would state to you that the distance is about seven feet one inch to seven feet six inches from the mirror to the point of light which falls on the photographic paper, and inasmuch as the angular motion of the magnet is doubled in the angular motion of the point of light upon the paper, this would represent the actual amount of change which would take place in a magnet of about thirty feet in length, that is, the end of it would move to the extent here indicated. That is the general system of photographic registration which has been adopted at Greenwich, at Lisbon, in America, in Canada, and many other places.

There is one further point to which I wish to draw your attention, and that is to the mode of correcting these instruments for changes of temperature. It is perfectly well known that as the temperature of a magnet rises, it loses its magnetic force, and where the changes are not beyond a certain amount, as it cools down again it regains its power when it arrives at the same temperature from which it started. It is therefore evident that the movements noticed would be due to two causes, to changes in the earth's magnetism and therefore to the induction upon the magnet, and also to alterations in the force of the bar itself, because the position which the bar takes is the joint action of the earth on the magnet, and of the magnet on the earth.

For this reason it became necessary either so to place the magnet that it is not liable to changes of temperature, or else to provide it with some means by which those changes can be compensated. Unquestionably the best course is to place the magnets as is now done in the Observatory at Greenwich, in an underground apartment, which is liable to very small changes of temperature; but if it be not convenient to do that, it is then desirable to employ some means by which the change of temperature can be compensated. In the bifilar magnetometer that is accomplished by attaching to the magnet a rod of glass to the ends of which two zinc tubes are clamped, and at the ends of these near each other two hooks are placed to which the double skein is attached. It is quite clear that owing to the greater expansion of zinc over glass, heat will have the effect of approximating the hooks by which the magnet is suspended by a minute quantity, and if they are made to approach each other the torsion force is diminished. If the interval between the hooks be diminished so as to diminish the torsion force proportionate to the diminution of magnetic force by change of temperature, it is quite clear that in that case the indications will be unaffected by temperature: that is shown in the bifilar instrument downstairs. In the balanced magnetometer the compensation is effected in a different way: a thermometer tube is clamped to the magnet, the bulb being on one side of the point of support and the end of the thread of mercury in the closed tube somewhere on the other side. It is quite clear that as the temperature rises and consequently the thread of mercury travels along the tube, a minute quantity will be transferred from one side of the balance to the other; as the energy of the bar diminishes a little additional mercury would be thrown over from one side to the other, and the weight of mercury in the tube is such that it just counteracts the diminishing force of the bar, so that in spite of change of temperature, supposing the earth's magnetism remained constant, the magnet will retain the same position. There is a further advantage in this mode of compensation, which is this: the amount of the temperature correction may be represented by a formula of this kind ($A\,t + B\,t^2$), supposing t to be the number of degrees of temperature at the time above the freezing point, and A and B are coefficients which have to be determined. Those I have determined in the instrument to which this correction has been applied by suspending the bar-magnet as a bifilar magnetometer without any correction, the zinc tubes being clamped at their proximate ends so as to prevent any alteration in distance. Then, while the registration is going on, the temperature on the magnet is gradually raised by a water envelope, the water in which is gradually heated by a jet of gas burnt outside the apparatus. This goes on for a period of about six hours, and the heat is so arranged that the temperature of the water should be raised gradually from that of the atmosphere up to about 90°, which is as high as is necessary. The jet is then extinguished, and the water allowed to cool down again, in about the same period of time. It is found that the line which would represent the actual magnetic variation during twelve hours is deflected in the direction of diminished force as the temperature rises, and goes back again as it cools down, and thus it is found that the register line will again return to its normal position. It is quite clear, therefore, that these changes are due entirely to temperature. The temperature having been recorded at intervals of a quarter of an hour or any convenient time, there are means of identifying these epochs with the register line when removed from the cylinder and developed, by shutting out the light for a very brief period, and then letting it in again, when you have a little interruption in the line which identifies the line with the particular time when you have produced that spot. By equating the formula $A\,t + B\,t^2$ with the differences of the ordinates of the normal and displaced register lines, at the times at which the temperatures have been recorded, and reducing these equations by the method of least squares, the values of the coefficients A and B may be obtained. With regard to the bifilar magnetometer the coefficient A only is taken account of. In the balanced magnetometer B is also given by the relation of the bore of the thermometer to the weight of the mercury in it, because it is evident that the finer the bore the further the minute quantity of mercury will travel for each degree of temperature, the more will it affect the balance, and that will determine the coefficient of t^2.

The set of instruments constructed by the authorities at Kew is not yet arranged, but it will be shortly; the principle is exactly the same, the registration being obtained by reflected light. There is, however, one deviation from the system which I adopted which I am free to confess is an improvement, that is, instead of having a concave mirror attached to the magnet, there is a plane glass mirror, and the light is collected into a focus by means of a fixed combination of lenses. The advantage is that the mirror is much lighter, the magnet is not so much loaded, and you then get a direct image in which the axis of the pencil of light falls centrally on the paper; whereas by the system I have adopted you get to a certain extent an oblique reflection, and the image so obtained is not so distinct as that obtained by direct reflection or refraction.

There is a little additional apparatus here for the purpose of testing the scale coefficient, or value of the displacement of the indication on the paper, by adding a minute weight of ·01 of a grain in this scale. You will observe how much that displaces the bar, and thence obtain the value of the ordinates of the curve. I have only further to say that the magnets in some cases are flat bars; but as set up by myself at the Paris Observatory, they are of this form—a hollow magnet, with a lens at one end and a collimater scale at the other, so that the photographic records may always be checked by observations with the telescope of the magnet itself. To each instrument a small plane mirror is also attached, so that the indicated amounts of variation may be checked by a fixed telescope and a scale.

The CHAIRMAN: I am sure you will all agree with me that we are much obliged to Mr. Brooke for his very able description of the photographic recording instruments, to which he has so largely contributed. Mr. Ronalds, formerly director at Kew, he has already alluded to, but since Mr. Ronalds' time the directors of Kew have not been idle; the late Mr. Welch contrived the above-mentioned modification of this very beautiful instrument of Mr. Brooke's, and a complete set will be shortly erected in the grounds of the observatory. The Kew instrument is now being worked under the direction of Mr. Whitwell; and beside that there are instruments at St. Petersburg, Lisbon, Coimbra, Florence, Toronto, and in many other places. It is extremely important that simultaneous and continuous records of magnetic changes should take place at a great number of points on the earth's surface. We have the curves recorded at Lisbon and at Kew, and the disturbances are found to take place at the same time, and prove that magnetic changes are cosmical and that they are synchronous—that is to say, the instrument at Kew was disturbed at the same time as the one at Lisbon. Moreover, two astronomers, both now dead—Mr. Carrington and Mr. Hodgson—were looking at the sun at the same time, and happened to observe a sudden outburst of light. Mr. Chambers, who was then at the Kew Observatory, observed a simultaneous disturbance of the magnet. You are aware that the period of sun spots and magnetism appear to be closely connected; and it is extremely important that these observations, which have been going on now with the same instrument for sixteen years, should be continued for a great number of years still, and then we may have some light thrown possibly on the cause of magnetism. I beg to return your thanks to Mr. Brooke, not only for the account he has given us, but for what he has done for science.

I now call on Professor Rijke, on the Historical Instruments from Leyden.

Professor RIJKE: Ladies and gentlemen,—The first thing I have to do is to implore your indulgence for the bad English which I am about to speak, but I hope that you will remember that, if you were obliged to lecture in Dutch, your Dutch would not perhaps be much better than my English. I hope also that you will forgive me if I use a foreign word when I cannot find the English one. I shall not lecture on the services which my countrymen have rendered to science, but I intend only to speak a few words on the most interesting Dutch instruments of the seventeenth and eighteenth centuries which have been here exhibited. The first instrument to which I wish to draw your attention is the first compound microscope which was made in Middleburg, and which you see here. It consists only of two convex lenses, and was made by Hans and Zacharias Janssen. There is a letter written by William Boreel, Dutch ambassador in France, stating that Hans and Zacharias Janssen, whom he knew perfectly well, having been their neighbour at Middleburg, and having often played with them when they were young,—he says in that letter that compound microscopes were made by the Janssens long before the year 1610. The art of making compound microscopes made hardly any progress at all during the 150 years which followed upon the invention of the Janssens; and indeed no progress could be made until their theory was further completed. It was for this reason that the great discoveries in the microscopical world were made by single lenses, and the man who was foremost in the field of science was an usher to the sage magistrates of a little town in Holland called Delft. His name was Van Leuwenhoek, a name which is now known throughout the whole world. Last year we had not only a festival in Holland, but also in different parts of Europe, to commemorate one of Van Leuwenhoek's greatest discoveries of which I am to speak, and all these discoveries were made with the little instrument I am showing you, of which every part was made by himself. It consists only of a little lens, which is not as large as the head of an ordinary pin, and the objects which had to be observed were placed on a pin underneath. When we remember how well and carefully arranged are the microscopes which we use now, we cannot too much wonder how it was possible to make with such an apparatus the discoveries which he made. The most important discovery was that of the *Infusoria*—those little microscopical animals which are to be found nearly everywhere, and which give so much trouble to modern science. In order to observe those infusoria he was obliged to change the form of his apparatus a little, and constructed a microscope of this form. His whole life was devoted to microscopical researches, and you will understand this when I tell you that at his death 247 microscopes with their frames were found, and also 172 lenses without frames. As often happens, his merits were much sooner acknowledged by foreigners than in his own country, and I am happy to say that it was Englishmen who were the first to declare how great a man he was. He was made a Fellow of the Royal Society, and that was, as he often declared, the greatest blessing he ever received. I will now speak of another scientific man, who lived quite another life—namely, Christian Huygens, who was not a man of humble extraction, for his grandfather had been Secretary to the first Prince of Orange, and his father held the same office under the following stadtholders, even under William the Third, who was as much yours as ours. He at once saw that if astronomers were to make greater discoveries it was necessary to have lenses of greater focal length, and as these lenses could not be found in any part of Europe, he, with his brother Constantine, made them himself. After some trials he succeeded very well, and two or three days after his first lens had been made, he was fortunate enough to discover a satellite of Saturn, and afterwards he was able to solve a problem which had not yet been solved—namely, to say what was the true nature of those little bodies which Galileo discovered near Saturn. He first discovered that they were not two, but one body only, and that that one body was a ring. He made that discovery, not with the telescope which is here exhibited, but with one nearly the same. He made more lenses, but not as many as his brother Constantine. The object lens of this old telescope has been made by Christian Huygens, and here is another one. Here also is a lens on which is written the name of Christian Huygens, but it is not genuine. The lenses of the greatest focal length are here in England, and belong to the Royal Society. There is one of 210, another of 170, and one of 120 feet, those lenses however were not made by Christian Huygens, but by Constantine; but I am happy to add that the lenses of Constantine were as good as those which Christian himself made. We have sent to the Exhibition this planetary, because it was invented by Huygens, and was in fact the first instrument of that kind ever made.

We have exhibited several instruments invented by 'sGravesande, not because great discoveries have been made by each of them, but because you find amongst them the first specimens of instruments which were devised to illustrate by experiment discoveries which had been made by mathematical research. It was his object in the first instance to make popular on the Continent the great discoveries of Newton, and he succeeded in fact very well. I think I do not say too much when I say that in his time he was the first lecturer on the Continent. Unfortunately he had to deliver lectures not only on natural philosophy, but on mathematics and on moral philosophy, and that is the reason why he did not make a great number of discoveries. Some of those instruments have been brought upstairs. For instance, this is an instrument to show that the velocities acquired by falling bodies are to one another as the square roots of the heights. Two little balls are made to roll down a curve, called cycloïde, side by side, and one being placed at the height marked 4 and the other at the height marked 16, it is seen that the velocity of the body which has fallen the height of 16 is twice as great as that which has fallen from the height of 4.

Here is an apparatus which was made to show the properties of centrifugal forces, and it was so good that it is still used every year in our Lectures. Here is also a little apparatus for demonstrating the properties of the wedge. If 'sGravesande did not make a great deal of important discoveries, he at all events invented an instrument which has rendered great services to science—viz., the Heliostat. This is the first which has been constructed. You all know that when a physicist is experimenting upon light he desires that the pencil of light with which he is experimenting should always keep the same direction. Now a pencil of light which comes directly from the sun changes its direction at each moment, and therefore 'sGravesande contrived an apparatus through which the pencils of light fall upon this mirror and are always reflected in the same direction. He also made some experiments which had at that time a very great importance. Scientific men were divided on the question in what manner the power of a body, which has received a certain quantity of velocity, should be calculated. They all agreed on one point, that the power was in the ratio of the masses of the two bodies, but did not agree on the question if the power was in the ratio of the velocity or whether it was in the ratio of the squares of the velocities. 'sGravesande thought that the powers were in proportion to the ratio of the velocities, but he thought it would be very useful to solve the question by direct experiment, and the experiments he made were the following. He took two pieces of wet clay and two bodies whose masses were different and those bodies he made fall from different heights on to the clay. Falling on the clay they made holes, and if the power of the two bodies were the same the holes must be the same too. He began by making such arrangements that the mass of each body multiplied by its velocity was the same, and that he could do by allowing them to fall from different heights. They made holes which were different. Then he made them fall from such heights that the mass of the bodies multiplied by the squares of the velocities were the same and then the holes were the same in each piece of clay. He found thus that he was wrong. But look what a man he was! His brother-in-law was in the same room in which he was experimenting, and after 'sGravesande had ascertained that in the last experiment the two holes were the same, a shout escaped him so that his brother-in-law came to ask what was the matter. "What is the matter, my dear fellow!" he said, "the matter is, I am quite in the wrong, and Leibnitz is quite in the right." He was as joyous, his relative relates, as if he had just discovered that he was in the right. It was by that experiment that the matter was settled for ever. You will also find downstairs a very ingenious apparatus invented by 'sGravesande, by which he has been able to prove by experiments that the same quantity of labour produces always the same quantity of *vis viva.* I should certainly have some more things to add, but I have already spoken, I fear, too long.

The CHAIRMAN: I do not think that Dr. Rijke need have made any apology for his want of fluency in speaking English, for he has given a most lucid description of these wonderful contributions to science of his countrymen. The compound microscope of the Janssens, but more remarkable still, the simple microscope of Leuwenhoek, whose name every microscopist is well acquainted with, certainly performed wonders. These simple lenses showed an amount of detail in the objects he examined which for a long time was unrevealed, and certainly was not revealed for a long time by the compound microscope. Of course it is not to be compared with the microscopes of the present day, but in its day it was a remarkable instrument. Dr. Rijke has also alluded to the two Huygens, and these lenses upon which I place my hand belong to the Royal Society, and as you are aware they were mounted with a ball and socket and a balance, and placed at the top of a high pole. The eyepiece was held in the hand, and the object glass was controlled by a long rod and a string. The observer had first to find his object glass with a lantern, and when he found it, turn it on to a star and seize the blaze of light in the object glass, and then he went on observing. In the lower room is a tower constructed on piles, which was intended by the Royal Society for mounting, at the suggestion of Mr. Struve, the celebrated Russian astronomer, these object glasses of Hugyens, with a view of ascertaining whether there was any change in Saturn's rings. A theory had been started that the rings were altering and collapsing. It was found to be rather expensive, and Mr. Struve on reconsideration did not press for the expenditure of 400*l.* or 500*l.* for mounting them. I have myself looked through these object glasses and can attest that they are very good ones. I did not look at a celestial object, but at a test object on the Pagoda at Kew. Now I believe Mr. Lockyer has employed them with a siderostat for making investigations on the solar phenomena. These experiments of Dr. 'sGravesande's are very remarkable, and show what a great philosopher he was. At all events, when he found himself in error, he was only rejoiced to have discovered the cause and to give credit to Leibnitz for the correctness of his theory. I have now only to ask you to return your thanks again to Dr. Rijke.

On a New Form of Voltameter.

Baron Ferdinand de Wrangell: The instrument to which I have the honour to draw your attention has been devised by my friend Professor Robert Lenz of St. Petersburg, and as he is unable to attend he has requested me to explain the principle on which it is founded. I will not go into the details now, but if any one is interested in the subject I shall be very glad to explain the construction of the instrument in its details at any time. The instrument is intended for volta-metric measurement of the strength of a current, and it is based upon the principle that the quantity of matter decomposed by the current which flows through it is proportional to the strength of the current, all other conditions remaining equal. This assumption that all other conditions remain equal is the most important point and the most difficult to attain in a voltameter. Besides this first condition of equality of conditions from one measurement to another, in order to compare the two currents, of course brevity of time and ease of management are very important conditions, and I think that this mercury voltameter fulfils all these conditions in a much higher degree than those which are generally used. The best mode of which I am aware consists in measuring the decomposition of a salt of silver solution by a current. The solution is contained in a platinum vessel, and after the decomposition has taken place for a certain time, the vessel is washed out, then dried perfectly, and the amount of silver which has been decomposed by the current in a certain period of time is ascertained by weighing. This process of washing out the silver, secondly, of drying it, and thirdly, of weighing it by a chemical balance, takes a great deal of time, is very troublesome, and is liable to some errors, of which the most important is that in washing out the silver contained in the vessel some parts of it are easily broken off; because it covers the black platinum vessel with a rough surface, and so some particles are lost. Then if it is not well washed, foreign matters remain in it, and thirdly, if it is not well dried, a great deal of water may adhere to it; for instance, upon a square centimetre of a plate covered with silver by this process, eight milligrammes of water will adhere under ordinary circumstances.

This instrument, devised by Professor Lenz, measures the decomposition of mercury not by weighing it but by volumetric measurement. There are two vessels connected by an inclined tube, and in the bottom of the upper one a little mercury is put; the lower vessel ends in a glass tube which dips into a cylindrical iron vessel, which forms the chief part of the measuring apparatus. Some mercury is poured into the upper vessel and drips into the lower one, and goes through the glass tube and into the measuring apparatus. Platinum wires, enclosed in glass tubes, are immersed into both vessels, and they form the electrodes. The platinum wire is quite immersed into the mercury, so that the mercury really forms the electrode. Over the surface of the mercury in both vessels a solution of mercury salt is poured. There is a micrometer screw which works an inner cylinder which fits into the iron vessel quite tightly, and which is carefully calibrated so that the value of one division of the micrometer screw is easily ascertained; and by this screw one can lift the mercury in the tube or let it fall. Before the commencement of the experiment, when the solution is poured in, the mercury contained in the tube is lowered to a standard point, which can be read off by means of a lens, and it is then lifted again so that it forms a bead at the bottom of the vessel, which is sufficient to cover the end of the platinum wire. The experiment then begins; the current flows through one wire to the other; the decomposition begins, and the mercury is evolved at the lower electrode, and just as much as is evolved there would enter into combination from the acid solution contained in the other vessel. By the decomposition of the mercury in the lower vessel the fluid becomes lighter, but that is prevented by the slanting tube, which allows it to flow from the upper vessel, and it can also be stirred. After a certain lapse of time the current is broken, the screw is lowered down, which brings the mercury again to the standard point; and then, by means of the micrometer, you can ascertain the exact volume of mercury which has been decomposed by a given current in a given lapse of time. That is easily converted into weights, and so you have an exact measure of the effect produced by the current in a certain time. You can perform the whole measurement in five minutes with great facility and great precision, for by a series of experiments it has been proved that the error is not more than 04 per cent. of the quantity measured. If the current is not strong enough crystals may be formed on the surface of the mercury, but that is prevented by heating it a little. The correction for temperature is very small indeed, because the bore being so very small it is scarcely noticeable; but still it can be taken into account, because all the dimensions are known.

The CHAIRMAN: This ingenious instrument for measuring electrical currents is, I think, likely to prove of great value; and it is interesting as having been contributed by Mr. Lenz of St. Petersburg, the son of the great physical philosopher of that name. We are very much indebted to the Baron de Wrangell for his description of it.

I will now call upon the Rev. Robert Main, the Radcliffe Observer.

ON A NEWTONIAN REFLECTING TELESCOPE OF SIR W. HERSCHEL.

The Rev. R. MAIN, M.A., F.R.S.: I am rather taken by surprise in being asked to say anything about this telescope, which I sent from the Radcliffe Observatory, but it may perhaps give me an opportunity of saying a few words to those not accustomed to the instruments in an observatory. The telescope in question is a ten-foot Newtonian reflecting telescope, and almost the only interest it can have here is that it was prepared and brought down to the Radcliffe Observatory by Sir William Herschel himself, and his correspondence is preserved there. It was made in the year 1812, and was received by the Observatory in 1813. I wish I could assure you that there was any series of observations made with it worthy of the telescope. I fancy there were a few casual observations, but nothing much was done with it.

It will be well perhaps to say a few words about the Observatory itself, and the way in which observatories at that period were furnished with instruments. The want of some such institution in Oxford had been much felt near the end of the last century, about 1770, and the University not being able to furnish anything, the Radcliffe Trustees undertook it and built that magnificent erection with which most of you are no doubt familiar. But the instruments were servilely copied from those at Greenwich, which was a very great misfortune for astronomy. Here was an opportunity for bettering a very bad class of instrument which had been used for the determination of the absolute places of bodies, by a due consideration of what could be done in this new observatory; but unfortunately Bird, who at that time was a great instrument maker, had a reputation for his quadrants, and two quadrants were ordered; and a transit instrument of the character usual in those days, with a small object glass, was also furnished, and that was the equipment of the Observatory. It was thought by Dr. Robertson that some instrument for observing casual phenomena would also be desirable, and Sir William Herschel gave a good deal of consideration to it, and recommended him to have this Newtonian reflecting telescope. It is very similar to the one you see here on the table. It has a small mirror of eight and a half inches, which was not considered small then, and the focal length is ten feet. The epoch at which this telescope was given is an important one and an interesting one. The first mural circle had just been established at Greenwich, and then began that series of observations which have only been improved upon very recently, and which totally superseded all observations of zenith-distances of bodies which before had been obtained by the quadrant. You may consider that as an epoch in the new astronomy. The Radcliffe Observer did not for a long time get any new instruments; he had not the power, in fact, of getting out of the groove in which things then were. The quadrants continued in use up to almost the termination of the Directorship of Professor Rigaud, and it was only just before the time when Mr. Johnson became Radcliffe Observer that a meridian circle, based on that of Dr. Robinson of Armagh, was established, and this you may consider another epoch in the astronomy of the age. From this time began an uninterrupted series of star-observations, rivalling those of Greenwich in the continuity and the value of the definite series of observations which were made. Similar observations have been kept up to the present time.

These things may not seem to have much connexion with the particular subject I have before me—namely, this telescope, but you may consider that the same kind of improvement has gone on in everything else; and yet if we were to observe one thing more than another, we are not so much astonished at the improvements which have been made in this long interval of time, but rather at the tenacity with which old methods utterly unsuited to the purpose for which they were intended have been kept to. Nothing could be more clumsy or ill devised for the object in view than the old quadrants, but they were kept up on the Continent of Europe long after they were given up in England, and long after Pond had superseded them by the mural circle at Greenwich. In the same way it was through an unfortunate mistake of Newton that the reflecting telescope, without those improvements and the mode of mounting which have made it a very apt and proper instrument at the present day, was kept up in contradistinction to the refracting telescope. It was supposed that the want of achromatism was hopelessly insuperable. Newton laid down the principle, and others servilely followed it, and thus was delayed the making of large object glasses for the greater part of a century. The tax on glass also, in England at least, was another reason why great object glasses could not be made. It is only within the last few years since the tax has been taken off and that glass has become an article of commerce which could be used freely, that we have been able to reap the full advantage of the scientific improvements which have been made in the construction of glasses, the shortness of focal length, and everything of that sort which renders telescopes of very considerable apertures as manageable as small ones used to be. I make these few rambling and cursory observations with respect to these things to show in some degree the way in which we have got to our present position. The wonder is not so much that, when the human mind is bent on any particular discovery, improvements are so rapid, but that in the preceding century they were so slow. It is to be hoped that, as time goes on, the rate of discovery, rapid as it is at present, will be still further increased. There is no want of genius, no want of scientific means for improvement in material things; it was want of opportunity and want of interest in the general public which stood in the way. That want of interest has now vanished;—all classes and both sexes, in fact the world at large, take interest now in what would have been formerly considered very recondite researches. All are eager and anxious to learn something, if but little, of various sciences, and to learn that little well. The fervour with which the public take up these things react on scientific men themselves. Each is anxious to do something in his vocation, and is only baffled by finding that, however early he may have been in the field, some one else, either at home or abroad, has forestalled him. At this hour it is not desirable to keep you any longer, but I would call your attention when you go below to those four telescopes of the class I have been mentioning as being so materially connected with the history of astronomy. There is first Newton's original telescope in a paste-board case, very likely the cover of an old copy-book; there is a telescope of Sir William Herschel's with which he began his researches; there is another of seven feet, although in rather an imperfect condition; and finally there is this ten-foot one, which is a very good representative of the telescopes which he made when they became with him an article of commerce. They will amply repay you for the little attention you may bestow upon them, though they form such a very small portion of even the astronomical instruments in this grand collection, which contains so vast a variety of interesting objects. There is the learning of a whole life here; in fact the ordinary span of human life would hardly suffice for the study of all one sees; but in this one particular I am glad to have been enabled to say a few words to show how improvement has gone on, and by what means, and what are our hopes for the future.

The President here again took the Chair.

Mr. De la Rue: The President wishes me to ask you to return your thanks to Mr. Main, as he has not been here during the whole of his address, which has dealt with some very important things. I may say that I am intimately acquainted with the work which the Radcliffe Observer is doing, and can attest that the observatory in his hands is doing as good work, and that that work is as rapidly brought before the public on the reduction of the observations as ever it was before—perhaps faster. He has alluded to Sir William Herschel taking down to the observatory part of its first equipment—namely, an eight or nine inch reflector, and has alluded to the very inadequate instruments which were for a long time employed when the observatory servilely copied what existed in another. But he has also shown that in recent times invention has proceeded so fast that there has been no necessity to continue the use of those old instruments. He himself has acquired a transit circle designed by the late Mr. Carrington, and has continued that very important series of star observations which his predecessor commenced. He has also alluded to the first reflecting telescope of Newton. I had recently to allude to the four-foot reflecting telescope made by Messrs. Grubb, of Dublin, mounted equatorially, and I may say that I recollect when first a four-foot equatorial was spoken of, it was thought to be impossible to mount it on a polar axis, and a contrivance to guide such telescope temporarily in the diurnal path was proposed. We are very much obliged to Mr. Main for having alluded to the older form of instruments in this interesting historical account which he has given.

The PRESIDENT: Professor Eccher now wishes to address a few words to the meeting in addition to what he said on a former occasion, with reference to the instruments from Italy.

Professor DE ECCHER: Your celebrated Faraday, in a letter to his mother, giving an account of his journey on the Continent, thus expresses himself with regard to Florence: "Florence, too, was not destitute of its attractions for me, and in the Academy del Cimento, and the Museum attached to it, is contained an inexhaustible fund of entertainment and improvement; indeed, during the whole journey, new and instructive things have been continually presented to me. Tell B. I have crossed the Alps and the Apennines; I have been at the Jardin des Plantes, at the Museum arranged by Buffon, at the Louvre, among the *chefs-d'œuvre* of sculpture and the masterpieces of painting, at the Luxembourg Palace, among Rubens' works; that I have seen a GLOW-WORM!!! waterspouts, a torpedo, the Museum of the Accademia del Cimento, as well as St. Peter's, and some of the antiquities here, and a vast variety of things far too numerous to enumerate."

And those among you who have visited Florence can bear witness to the great number of celebrated scientific instruments which that city is fortunate enough to possess. It was, I can assure you, no easy task to choose from the midst of such abundance, a few objects only, to bring to this Exhibition. All are worthy of such a distinction. But it having been decided that the selection should be of Italian instruments, there was chosen from among them a small number, that would give an adequate idea of the History of Science in Italy, and especially in Tuscany.

Two instruments, however, although not Italian ones, were chosen as a mark of homage to your noble country: the one was the invention of a countryman of yours; the other was made use of, in many celebrated experiments, by another Englishman. In the *Times* of the 20th of this month, there is this statement: "Besides this, there are other astrolabes of different countries—French, German, Arabic, and Persian, but unfortunately no English specimen." Now you have before you here the astrolabe of Lord Dudley, and in return for having placed before his eyes an English astrolabe, I would ask the *Times*' reporter to be good enough to notice, also, the Italian astrolabes, which are exhibited in our cases.

But not only these astrolabes, but these astronomical quadrants, these sundials, this series of instruments, in a word, will be able to give an idea of the manner in which, during the 16th and 17th centuries, mathematical and astronomical science was cultivated in Italy; and how numerous and renowned were the artificers of such apparatus.

Here is the great lens, made by Benedetto Bregans, of Dresden, and presented by him, about the year 1690, to the Grand Duke Cosimo III. About the beginning of this century it was made use of by your celebrated philosopher, Sir Humphry Davy, in the researches which he made in Florence, together with Faraday, who was at that time his assistant, on the chemical constitution of the diamond.

I must, however, tell you that Averani and Targioni had already succeeded, by means of the aforesaid lens, in vaporizing the diamond. We now come to the barometrograph of Felice Fontana, a well-known scientific man, born in Rovereto, and who was Director of the Royal Physical and Natural History Museum of Florence, founded by Peter Leopoldo. Not to impose upon the indulgence which you have so kindly granted me, I will omit describing to you the other experiments made by Fontana, who, in order to push forward meteorological science, constructed, upon the principle which you see here before you, a barometrograph, a thermometrograph, and other self-registering instruments. You will be able by the help of photography to understand how these instruments register themselves. The wooden stopper floating on mercury rises and falls with it, and transmits its motion to this pendulum, which in its upper portion has an arc covered with paper, on which at every hour, this needle, set in motion by clock-work, makes a mark. This same needle advances a little after every impression, and thus these marks form a curve upon the paper.

I come now to speak of Nobili, and shall, for the sake of brevity, mention only those objects which you see before you. Here is his first astatic galvanometer, made in the year 1825. And this is another instrument for hydro-electrical currents. Here is Nobili's magnetoscope for proving the presence of the very slightest magnetic influences; it is composed, as you see, of a system of astatic needles suspended to a thin thread of silk inside this crystal bell, surmounted by a tube which has on its upper part a graduated circle, with an index such as is used in scales "di torsione."

Here are three different models of thermo-electrical piles. In the construction of such instruments he attained great excellence. But I want particularly to draw your attention to this one, which is composed of thirty-six elements, bismuth antimony, and is provided with a conical reflector; it was of the greatest possible use to him in the experiments which he carried on, partly in collaboration with the celebrated Melloni, on the radiation of heat. All have heard of Nobili's coloured rings, obtained by means of the chemical action of the electric current; this "rosettone" was obtained by him in this wise. We have in Florence the complete chromatic scale constructed by him, and who has read his memoirs on the subject knows with what profundity he has treated this attractive argument.

And, although I cannot put the instrument itself before your eyes, allow me to remind you of his globe, round which there circulates an electrical current for imitating the magnetical phenomena of the earth. It was made by him in 1822, whilst it was only in 1824 that Dr. Birkbeck presented a similar one to the Royal Institution (of London); and, as he got his first idea from Barlow, it was henceforth known as Barlow's Globe.

With regard to the magneto-electrical machines of Nobili and Antinori, and to their publications on the phenomena of induction, I feel it my duty to add a few words. It is certain that your celebrated

Faraday, in whatever direction he investigated the mysteries of nature, he compelled her to disclose her secrets. He was the first likewise in this most important branch of science. But nevertheless, Nobili and Antinori, starting from an imperfect report of Faraday's works, which was communicated to the Academy of Sciences of Paris by M. Hachette, by dint of repeating Faraday's experiments, they not only succeeded in obtaining results equal to Faraday's, but, in some cases, produced effects that had never before been observed.

I well know that the English philosopher was somewhat displeased at Nobili's intrusion into the wide field of science which he had just thrown open; indeed, the illustrious Tyndall mentions this fact as characteristic of Faraday. I can perfectly understand his annoyance. But at the same time I hope that you will agree with me that the field of science is not merely open to all, but that it invites all to enter; that Faraday's renown is too firmly established for there to be any reason why some distinction should not be allowed to others; and that it was impossible for an investigator like Nobili not to throw himself eagerly into this new field of such marvellous and profound investigations. I beg you, therefore, to admit the fact that the first magneto-electrical machine was made by Nobili and Antinori. Here it is. If I quickly remove the keeper, armed, as you see, with this "rocchetto" of thread, of which one extremity is in contact with the magnet itself, the other with this spring *(molla)*, afterwards also in contact with the magnet; this second contact is immediately interrupted, and thus the inducted electrical current occasions, as you see, a spark. This instrument gave forth its first spark on the 30th of January, 1832.

Under No. 1298*c* of the Catalogue you may observe another example of a magneto-electrical machine made by Nobili and Antinori. It is exhibited by the illustrious Professor Dove of Berlin.

Latterly this same form of a magnetico-electrical machine has been returned to, for use in the lighting of mines (*accensione delle mine*) Then in order to get a more intense current, and a greater number of interruptions in a given time, Nobili and Antinori constructed their so-called "united magnets machine" (*macchina a calamite conjugate*); made up, in fact, of permanent magnets, among which, by means of a handle, or lever (*manubrio*), and of an eccentric, the keeper with its "rocchetto," swings, passing successively from the contact of the first magnet, to that of the second, and then back to the first again, and so on.

Then, by making use of the great natural magnet of the Regio Istituto Superiore di Firenze, a very large parallelopiped of about 50×65×78 centimetres in dimension, and fastening the keeper with isolated thread, they were easily able by detaching it, interrupting at the same time the inducted current, to obtain the spark and every other effect peculiar to electricity.

And with regard to Nobili, allow me to remind you that, besides his important researches in electro-physiology; on rotary magnetism (*di rotazione*); on electric-dynamical phenomena, &c., he founded his electrical condensator upon the principle of the extra current, upon which, shortly afterwards, so much light was thrown by your illustrious Faraday. It consisted of a "rocchetto" of thread interposed in the circuit of the electrical current, which acted in such a manner that the spark of opening (*d'apertura*) was greatly increased in intensity.

And now turning to Amici, I am sure that all will recognise in him a remarkable scientific man, of very remarkable mechanical genius, and one who has bestowed his name on many valuable and useful instruments. He was especially noted for his microscopes and telescopes, two of which are among the most honoured objects in the observatory of Arcetri. One of the objectives has a diameter of twenty-nine centimetres. The two microscopes which you now see before you, although not among the best of those made by Amici, are, however, the oldest. In one of them you may observe how the objective is made by means of a concave glass; among the accessories you will find various systems of camera lucida, some of them were imagined by him; and you can likewise see ingenious methods for concentrating light, especially on opaque bodies.

I cannot conclude this hasty review without reminding you of Matteucci, of whom science was, alas! too quickly robbed. The greater number of us have constantly present in our minds the weighty judgment pronounced upon his works by the most illustrious scientific men of Europe; and many venerable institutions inscribed his name on their lists of members. The fact that your Royal Society for the Advancement of Science granted him the Copley Medal frees me from the necessity of vindicating his claims to certain important discoveries, the originality of which have been contested by one who, from the perusal of Matteucci's writings, was imbued with the impulse to prosecute those researches upon which was afterwards founded his own fame. This is the galvanometer which he used in his principal experiments, and by means of which he discovered the muscular current, in the year 1844. It was rewarded by your Society with the Copley Medal.

And under No. 1742*b* of the Catalogue you will find another apparatus sent by France, which Matteucci and Arago used in their researches on the distribution of currents, inducted by magnetism in a revolving disc of copper.

I have the good fortune to be able to place before you the spectroscope made by Professor Donati, an astronomer of Arcetri; and it was by means of this instrument that in 1870, on the occasion of the total eclipse of the sun, he observed the luminous lines of hydrogen. In the last years of his too short career he constructed a new spectroscope for solar analysis, composed of twenty-five prisms, some of which were placed in the very tubes which have the fissure and the lenses of the telescope.

Finally, allow me to place before you the galvanic chronographic interrupter of Professor Felici, about which you have already heard something in the conference held by my excellent colleague, Professor Blaserna.

Before ending, allow me to express my regret at Italy being so inadequately represented at this great exhibition. A country in which, little more than ten years ago, youth was prohibited from tasting of the fountains of science or reading the works of the greatest of its citizens—of Dante, of Galileo—such a country is too new to comprehend suddenly the high importance of scientific rivalry, of these Olympiads of science. And yet it might have sent various objects of Galvani, Volta's piles, Brugnatelli's gilding, the electro-magnetic movers of Dal Negro, Melloni's apparatus, Belli's electrostatic induction machine, and many other ancient and modern instruments, to compete for the prize of merit with those of more fortunate nations.

Florence alone, the birthplace of physical science, and which has at all times cultivated it, has contributed largely with its treasures.

Let this be a proof of the existence of the spirit of science which has at all times inspired it, and of the goodwill of the directing Council of its Institution, which following in the wake of Galileo and of the Accademia del Cimento, places its trust in science and progress alone.

The PRESIDENT: In rising, as I now do, to offer our best thanks to our foreign friends who have just given us this account of these remarkable instruments, I wish to add, that it seems not without significance or happiness of accident that the last contribution to this Physical Section of the Conference should have been made by a foreigner. We are indebted very materially to our foreign friends for the success which, at all events up to this point, has attended this exhibition. We are indebted in the first place to the Governments who have given an impetus throughout all Europe to this undertaking. We are indebted to the foreign Academies who have taken up the movement on the part of the Governments; we are indebted, further, to the curators and the authorities connected with the museums who have so largely contributed to our collection; and we are indebted still further to the individual men of science who have been permitted to transport the collections of these instruments, and have put the crowning stroke upon their efforts by giving us the favour and instruction of their presence here. While, however, I am speaking of foreigners (and I cannot say too much on that score), we must not forget the efforts which have been made, and which originated in this country. We are indebted in the first place to the heads of the department under whose auspices this exhibition has been brought together. We are indebted also to the various men of science throughout the country who co-operated; but we must not forget those who have so largely contributed to removing the preliminary difficulties. I mean those few energetic and self-devoted individuals who have done perhaps more than any one else to insure the success of the undertaking. Many names might be mentioned, but I must not omit to name particularly Major Donelly, Major Festing, and Mr. Lockyer. To-day is, I am sorry to say, the last of the conferences in the Physical department, but I have no doubt that the success which has attended our meetings in this department will attend all the others to the close of the Conference. We have in these meetings, and in the collection which is covered in this building, a fresh illustration of the very old saying, that science is of no time and of no country. Science, it is true, is concerned with time and space, but within the range of those quantities it admits of and it will tolerate no distinctions. Science is the same whether it be pursued in these pleasant climes of Europe, whether it be followed out in the torrid zones of Central Africa, as Cameron and Livingstone and a host of others have done; whether it be studied in the deserts of Australia or in the forests of New Guinea; or whether it be followed out again as it has been done by foreign countries over and over again, and is now being by this country in the Arctic Expedition, of which we hope soon to hear tidings. Science is the same through all time—from the earliest dawn of intelligence, when the patriarchs went out to meditate in the fields at eventide, or when they studied in their fashion the moon and stars and the host of heaven. It is the same through all these days of which we have records in this museum, and it will be the same, I doubt not, until all feelings of wonderment and the like have been not superseded, but swallowed up by perfect knowledge. I beg now to congratulate the Section on the success which I have reason to hope has attended it.

Mr. DE LA RUE: Our Chairman has very briefly gone over the ground and described the commencement of this Exhibition, and also the valuable co-operation of the scientific gentlemen who have joined with the English philosophers in making these meetings interesting: but not only was the Loan Exhibition a bold undertaking, but so much of novelty attached to these Conferences that we were doubtful, even on the first day, whether they would be successful. That they have been eminently so is decided by the presence of so many ladies and gentlemen at our meetings; but the success does not come of itself. It required organization: it required a great deal of thought on the part of those who took part in the management, and to no one are our thanks more due than to Mr. Spottiswoode, who opened the *séance*. Throughout the whole of our meetings he has been daily employed in grouping subjects together, and in asking gentlemen to come forward to render them not only interesting but instructive. I ask you, therefore, to give your best thanks to the President of the Physical Section, Mr. Spottiswoode.

The PRESIDENT: I am much obliged to you for the vote of thanks which you have given to me for the very small part which I have taken. It has been a labour entirely of love, for the great readiness with which the gentlemen who have made communications have fallen in with such arrangements as I have thought it best to suggest has been very pleasing, and has made the whole thing a very easy business.

SECTION—MECHANICS (including Pure and Applied Mathematics and Mechanical Drawing).

President: Mr. C. W. SIEMENS, D.C.L., F.R.S.

Vice-Presidents:

Mr. F. J. BRAMWELL, F.R.S.

Mr. W. FROUDE, M.A., F.R.S.

M. le Général MORIN, Directeur du Conservatoire des Arts et Métiers.

Dr. WERNER SIEMENS.

M. TRESCA, Sous-Directeur du Conservatoire des Arts et Métiers.

Sir JOSEPH WHITWORTH, Bart., D.C.L., F.R.S.

May 17th, 1876.

DR. C. W. SIEMENS' OPENING ADDRESS.

In opening the proceedings of the Conferences regarding mechanical science, it behoves me to draw attention to the lines of demarcation which separate us from other branches of natural science represented in this Exhibition.

In the department of applied science we have collected here apparatus of vast historical interest, including the original steam cylinder constructed by Papin in 1690, the earliest steam-engines by Savery and by James Watt, the famous locomotive engine the "Rocket," by which George Stephenson achieved his early triumphs, as well as Bell's original marine engine, and a variety of models illustrative of the progress of hydraulic engineering and of machinery for the production of textile fabrics. In close proximity to these we find a collection of models illustrative of the remarkable advance in naval architecture which distinguishes the present day.

It would be impossible to deny the intrinsic interest attaching to such a collection or its intimate connection with the progress of pure science; for how could science have progressed at the rate evidenced in every branch of this Exhibition but for the great power given to man through the mechanical inventions just referred to. Yet, were mechanical science at these Conferences to be limited to the objects exhibited in the South Gallery (and separated unfortunately from apparatus representing physical science by lengthy corridors filled with objects of natural history), we should hardly find material worthy to occupy the time set apart for us. But, thanks to the progress of opinion in recent days, the barrier between pure and applied science may be considered as having no longer any existence in fact. We see around us practitioners, to whom seats of honour in the great academies and associations for the advancement of pure science are not withheld, and men who, having commenced with the cultivation of pure science, think it no longer a degradation to follow up its application to useful ends.

The geographical separation between applied science and physical science just referred to, must therefore be regarded only as accidental, and the subjects to be discussed in our section comprise a large proportion of the objects to be found within the rooms assigned more particularly to physics and chemistry. Thus, all measuring instruments, geometric and kinematic apparatus, have been specially included within our range, and other objects such as telegraphic instruments, belong naturally to our domain.

With these accessions, mechanical science represents a vast field for discussion at these Conferences, a field so vast indeed that it would have been impossible to discuss separately the merits of even the more remarkable of the exhibits belonging to it. It was necessary to combine exhibits of similar nature into subdivisions, and the Committee have asked gentlemen eminently acquainted with these branches to address you upon them in a comprehensive manner.

Thus they have secured the co-operation of Mr. Barnaby, the Director of Construction of the Navy, to address you on the subject of Naval Architecture, and of Mr. Froude to enlarge upon the subject of fluid resistance, upon which he has such an undoubted right to speak authoritatively. Mr. Thomas Stevenson, the Engineer of the Northern Lighthouses, will describe the modern arrangements of Dioptric lights, which mark a great progress in the art of lighting up our coasts. Mr. Bramwell has undertaken the important task of addressing you on the subject of Prime Movers, and Professor Kennedy upon the kinematic apparatus forwarded by Professor Reuleaux, of Berlin. M. Tresca will bring before us his interesting subject, the flow of solids. Mr. William Hackney will address you upon the application of heat to furnaces, for which he is well qualified both by his theoretical and practical knowledge. Mr. R. S. Culley, Chief Engineer of the Postal Telegraphs, will refer you to a most complete and interesting historical collection of instruments, revealing the rapid and surprising growth of the electric telegraph.

Measurement.—Regarding the question of measurement, this constitutes perhaps the largest and most varied subject in connection with the present Loan Exhibition. In mechanical science, accurate measurement is of such obvious importance, that no argument is needed to recommend the subject to your careful consideration. But it is not perhaps so generally admitted, that accurate measurement occupies a very important position with regard to science itself, and that many of the most brilliant discoveries may be traced back to the mechanical art of measuring. In support of this view I may here quote some pregnant remarks made by Sir William Thomson in his inaugural address, delivered in 1871 to the members of the British Association, in which he says, "Accurate and minute measurement seems to the non-scientific imagination, a less lofty and dignified work than looking for something new. But nearly all the grandest discoveries of science have been but the rewards of accurate measurement and patient long-continued labour in the minute sifting of numerical results. The popular idea of Newton's grand discovery is that the theory of gravitation flashed upon his mind, and so the discovery was made. It was by a long train of mathemetical calculation, founded on results accumulated through prodigious toil of practical astronomers that Newton first demonstrated the forces urging the planets towards the sun, determined the magnitude of those forces, and discovered that a force following the same law of variation with distance, urges the moon towards the earth. Then first, we may suppose, came to him the idea of the *universality of gravitation;* but when he attempted to compare the magnitude of the force on the moon with the magnitude of the force of gravitation of a heavy body of equal mass at the earth's surface, he did not find the agreement which the law he was discovering required. Not for years after would he publish his discovery as made. It is recounted that, being present at a meeting of the Royal Society, he heard a paper read, describing geodesic measurement by Picard, which led to a serious correction of the previously accepted estimate of the earth's radius. This was what Newton required; he went home with the result, and commenced his calculations, but felt so much agitated, that he handed over the arithmetical work to a friend; then (and not when sitting in a garden he saw an apple fall) did he ascertain that gravitation keeps the moon in her orbit.

"Faraday's discovery of specific inductive capacity, which inaugurated the new philosophy, tending to discard action at a distance, was the result of minute and accurate measurement of electric forces.

"Joule's discovery of thermo-dynamic law, through the regions of electro-chemistry, electro-magnetism, and elasticity of gases, was based on a delicacy of thermometry which seemed impossible to some of the most distinguished chemists of the day.

"Andrews' discovery of the continuity between the gaseous and liquid states was worked out by many years of laborious and minute measurement of phenomena scarcely sensible to the naked eye."

Here, then, we have a very full recognition of the importance of accurate measurement, by one who has a perfect right to speak authoritatively on such a subject. It may indeed be maintained that no accurate knowledge of any thing or any law in nature is possible, unless we possess a faculty of referring our results to some unit of measure, and that it might truly be said—*to know is to measure.*

To resort to a homely illustration of this proposition, let us suppose a traveller in the unknown wilds of the interior of Africa, observing before him a number of elevations of the ground, not differing materially from one another in apparent magnitude. Without measuring apparatus the traveller could form no conclusion regarding the geographical importance of those visible objects, which might be mere hillocks at a moderate distance, or the domes of an elevated mountain range. In stepping his base line, however, and mounting his distance-measurer he soon ascertains his distances, and observations with the sextant and compass give the angles of elevation and position of the objects. He now knows that a mighty mountain chain stands before him, which must determine the direction of the watercourses and important climatic results. In short, through measurement he has achieved perhaps an important addition to our geographical knowledge. As regards modern astronomy, this may almost be defined as the art of measuring very distant objects, and this art has progressed proportionately with the perfection attained in the telescopes and recording instruments employed in its pursuit.

By the ancients the art of measuring length and volume was tolerably well understood, hence their relatively extraordinary advance in architecture and the plastic arts. We hear also of powerful mechanical contrivances which Archimedes employed for lifting and hurling heavy masses; and the books of Euclid constitute a lasting proof of their powers of grappling with the laws regulating the proportion of plane and linear measurement. But with all the mental and mechanical power displayed in those works, it would seem strange that no attempt should have been made on the part of the ancients to utilise those subtle forces in nature, *heat* and *electricity*, by which modern civilisation has been distinguished, were it not for their want of means of *measuring* these forces.

Hero, of Alexandria, tells us that the power of steam was known to the Egyptians, and was employed by their priesthood to work such pretended miracles as that of the spontaneous opening of the doors of the temple whenever the burnt offering was accepted by the gods, or, as we moderns would put it, whenever the heat generated by combustion was sufficient to produce steam in the hollow body of the altar, and thus force water into buckets whose increasing weight, in descending, caused the gates in question to open.

Unfortunately for them, the Academia de Cimento of Florence had not yet presented the world with the thermometer, nor had Toricelli shown how to measure elastic pressures, or there would, at any rate, have been a probability of those clear-headed ancients applying the power of steam for preparing and transporting the materials which they used in the erection of their stupendous monuments, and for raising and directing the water used in their elaborate works of irrigation.

The Art of Measuring may be divided into the folowing principal groups :—

1st. That of linear measurement, the measurement of area within a plane, and of plane angles; comprising geometry, trigonometry, surveying, and the construction of linear measures, distance meters, sextants and planimeters, of which a great variety will be found within this building.

The subject of linear measurement will, I am happy to state, be brought before you by one whose name will ever be remembered as the introducer into applied mechanics of the absolute plane, and of accurate measure, I mean Sir Joseph Whitworth. It is to be regretted, I consider, that Sir Joseph Whitworth adopted as the unit of measure, the decimalized inch, instead of employing the centimetre, and I hope that he will see reason to adapt his admirable system of gauges, also to metrical measure, which, notwithstanding any objections that could be raised against it on theoretical grounds—that, namely, of not representing accurately the ten millionth part of the distance from one of the earth's poles to its equator—is, nevertheless, the only measure that has been thoroughly decimalized, and which establishes a simple relationship between measures of length, of area and of capacity. It possesses, moreover, the great practical advantage of having been adopted by nearly all the civilized nations of Europe, and by scientific workers throughout the world. Sir Joseph Whitworth's gauges, based upon the decimalized inch, are calculated to maintain their position for many years, owing to the intrinsic mechanical perfection which they represent, but the boon conferred by their author would be still greater than it is, if, by adopting the metre he would remove the last and only serious impediment in the way of the unification of linear measurement throughout the world. A discussion will probably arise regarding the relative merits of measurement *a but*, of which Sir Joseph Whitworth is the representative, and of measurement *a trait*, which is the older method, but is still maintained by the Standard Commissioners, both in this country and in France.

The second group includes the measure of volume or the cubical contents of solids, liquids, and gases, comprising stereometric methods of measurement, the standard measures for liquids, and the apparatus for measuring liquid and gaseous bodies flowing through pipes, such as gas meters, water meters, spirit meters, of which, likewise, a great variety, of ancient and modern date, will meet your eye, and upon which Mr. Merrifield will address you.

Another method of measuring matter is by its attraction towards the earth, or, thirdly, the measurement of weight, represented by a great variety of balances of ancient and modern construction. These may be divided into *beam weighing machines*, which appears to be at the same time the most ancient and the most accurate, into spring balances and torsion balances. The accuracy obtained in weighing is truly surprising, when we see that a mass of one ten-millionth part of a gramme suffices to turn the scale of a well-constructed chemical balance. Perfect weighing, however, could only be accomplished in a vacuum, and, in accurate weighing, allowance has to be made for the weight of air displaced by the object under consideration. The general result is that the mass of light substances is really greater than their nominal weight implies, and this difference between true and nominal weight must vary sensibly with varying atmospheric density.

Weighing in a denser medium than atmospheric air, namely, in water, leads us, fourthly, to the measurement of specific gravity, which was originated by Archimedes when he determined the composition of King Hiero's crown by weighing it in water and in air.

Among measures of weight may be noted a balance, which weighs to the five-millionth part of the body weighed, sent by Beckers Sons, of Rotterdam; another from Brussels, weighing to within a fourteenth-millionth part of the weight, in weighing small quantities; a balance formerly used by Dr. Priestly; and Professor Hennessy's standards, derived from the earth's polar axis, as common to all terrestrial meridians.

Next comes, fifthly, *the Measurement of Time*, which, although of ancient conception, has been reduced to mathematical precision only in modern times. This has taken place through the discovery, by Galileo, of the pendulum, and its application, by Huygens, to time-pieces in the 17th century. The most interesting exhibits in this branch of measurement are, from an historical point of view, the Italian, German, and English clocks of the 17th century; the time-keeper which was twice carried out by Captain Cook, first in 1776 and which, after passing through a number of hands, was brought back to this country in 1843; and an ancient striking clock, supposed to have been made in 1348—it has the verge escapement, which is said to have been in use before the pendulum. The methods employed in modern clocks and watches for compensating for variation of the thermometer and barometer, are illustrated by numerous exhibits, notably the astronomical clock, with Sir George Airy's compensation, which will form the subject of a special demonstration by Messrs. Dent and Co.

The measurement of small increments of time has been rendered possible only in our own days by the introduction of the conical pendulum, and other apparatus of uniform rotation, which alone convey to our minds the true conception of the continuity of time. Among the exhibits belonging to this class must be mentioned Sir Charles Wheatstone's rotating mirror, moved by a constant falling weight, by which he made his early determination of the velocity of electricity through metallic conductors; the rotative cylindrical mirror, marked by successive electrical discharges, which was employed by Dr. Werner Siemens in 1846, to measure the velocity of projectiles, and has been lately applied by him for the measurement of the velocity of the electric current itself, and the chronometric governor, introduced by him in conjunction with myself, for regulating chronographs, as also the velocity of steam engines under their varying loads; Foucault's governor, and a considerable variety involving similar principles of action.

Another entity which presents itself for measurement is, sixthly, that of *Velocity*, or distance traversed in a unit of time, which may either be uniform or one influenced by a continuance of the cause of motion, resulting in acceleration, subject to laws and measurements applicable both in relation to celestial and terrestrial bodies. I may here mention the instruments latterly devised for measuring the acceleration of a cannon ball before and after leaving the mouth of the gun, of which an early example has been placed within these galleries. Other measurers of velocity are to be found here, ships' logs, current meters, and anemometers.

In combining the ideas of weight or pressure with space, we arrive at, seventhly, the conception of work, the unit of which is the foot-pound or kilogrammetre, and which, when combined with time, leads us to the further conception of the performance of duty, the horse-power, as defined by Watt. The machines for the measurement of work, here exhibited, are not numerous, but are interesting. Among these may be mentioned Professor Colladon's dynamometrical apparatus, constructed in 1844; Richard's patent steam engine indicator, an improvement on Watt's, and Mr. G. A. Hirn's flexion and torsion pandynamometers.

Eighth.—The measurement of *Electrical Units*—of electrical capacity of potential—and resistance, forms a subject of vast research, and of practical importance, such as few men are capable of doing justice to. It may be questioned, indeed, whether Electrical Measurement belongs to the province of mechanical science, involving, as it does problems in physical science of the highest order; but it may be contended on the other hand that at least one branch of Applied Science, that of Telegraphy, could not be carried on without its aid. I am happy to say that this branch of the general subject will be brought before you by my esteemed friend Sir William Thomson, than whom there is no one more eminently qualified to deal with it. I may, therefore, pass on to the next great branch of our general subject, the ninth: *Thermal Measurement.*—The principal instrument here employed is the thermometer, based in its construction, either upon the difference of expansion between two solids, or on the expansion of fluids such as mercury or alcohol—(the common thermometer) or upon gaseous expansion (the air thermometer); or again, it may be based upon certain changes of electrical resistance, which solids and liquids experience when subjected to various intensities of heat. With reference to these, the air thermometer represents most completely the molecular action of matter which is the equivalent of the expansibility. I shall not speak of the different scales that have been adopted by Réaumur, Celsuis and Fahrenheit, which are based upon no natural laws or zero points in nature, and which are therefore equally objectionable upon theoretical grounds. Would it not be possible to substitute for these a natural thermometric scale? One commencing from the absolute zero, of the possible existence of which we have many irrefutable proofs, although we may never be able to reach it by actual experiment. A scale commencing in numeration from this hypothetical point would possess the advantage of being in unison throughout with the physical effects due to the nominal degree, and would aid us in appreciating correctly the relative dynamical value of any two degrees of heat which could be named. Such a scale would also fall in with the readings of an electrical resistance thermometer or pyrometer, of which a specimen has been added to this collection by myself.

When temperature or intensity of heat is coupled with mass, we obtain the conception of quantity of heat, and if this again is referred to a standard material, usually water, the unit weight of each being taken, we obtain what is known as specific heat. The standard to which measurements of quantity of heat are usually referred is the heat required to raise a pound of water one degree fahrenheit, or the cubic centimetre of water one degree centigrade.

The most interesting exhibits in this branch of measurement, are, from an historical point of view, the original spirit thermometer of the Florentine Academia del Cimento, and the photographs of old thermometers; the original Lavoisier Calorimeter for measuring the heat disengaged in combustion, Wedgwood's and Daniell's Pyrometers.

As illustrating modern improvement, may be instanced a long brass-cased thermometer showing the variation in the readings, when the bulb and when the whole thermometer is immersed; a thermometer with flat bulb to improve sensitiveness; a thermo-electric alarum, for giving notice when a given temperature is reached; an instrument for measuring the temperature of fusion by means of electric contact invented by Professor Himly; Dr. Andrews' apparatus for measuring the quantity of heat disengaged in combustion; Dr Guthrie's diacalorimeter for measuring the conductivity of liquids for heat, and a thermometric tube by Professor Wartmann for determining the calorific capacities of different liquids by the process of cooling.

Finally, Joule has taught us how to measure the unit of heat dynamically, and the interesting apparatus employed by him from time to time in the various stages of the determination of this most important constant in applied mechanics, are to be found, rightly placed, not among thermometers, and other instruments placed in the physical sections, but among the instruments required in the determination of three great natural standards—of length, time, and mass, and their combinations.

Another branch of the general subject is the *Measurement of Light*, which may be divided into two principal sections, that including the measurement of the wave-length of light, of different colours, and the angle of polarization, which belongs purely and entirely to physical science; and the measurement of the intensity of light by photometry, which, while involving also physical problems of the highest order, has an important bearing also upon applied science. The principal methods that have been hitherto employed in photometry are by the comparison of shadows, that of Rumford and Bouguer; by employing a screen of paper with a grease-spot, the lights to be compared being so adjusted that the spot does not differ in appearance from the rest of the paper, Bunsen's method; Elster's, by determining in combustion the amount of carbon contained in a given volume of a gas; and the one lately introduced by Professor Adams and Dr. Werner Siemens, by measuring the variation in the electrical resistance of selenium, under varying intensities of light.

Before concluding, I wish to call your attention to two measuring instruments which do not fall within the range of any of the divisions before indicated. The first is an apparatus designed chiefly by my brother, Dr. Werner Siemens, by which a stream composed of alcohol and water, mixed in any proportion, is measured in such a manner that one train of counter wheels records the volume of the mixed liquid; whilst a second counter gives a true record of the amount of absolute alcohol contained in it. The principle upon which this measuring apparatus acts may be shortly described thus:—The volume of liquid is passed through a revolving drum, divided into three compartments by radial divisions, and not dissimilar in appearance to an ordinary wet gas-meter; the revolutions of this drum produce the record of the total volume of passing liquid. The liquid on its way to the measuring drum passes through a receiver containing a float of thin metal filled with proof spirit, which float is partially supported by means of a carefully-adjusted spring, and its position determines that of a lever, the angular position of which causes the alcohol counter to rotate more or less for every revolution of the measuring drum. Thus, if water only passes through the apparatus, the lever in question stands at its lowest position, when the rotative motion of the drum will not be communicated to the alcohol counter, but in proportion as the lever ascends a greater proportion of the motion of the drum will be communicated to the alcohol counter, and this motion is rendered strictly proportionate to the alcohol contained in the liquid, allowance being made in the instrument for the change of volume due to chemical affinity between the two liquids. Several thousand instruments of this description are employed by the Russian Government in controlling the production of spirits in that empire, whereby a large staff of officials is saved, and a perfectly just and technically unobjectionable method is established for levying the excise dues.

Another instrument, not belonging to any of the classes enumerated, is one for measuring the depth of the sea without asounding line, which has recently been designed by me, and described in a paper communicated to the Royal Society. Advantage is taken in the construction of this instrument, of certain variations in the total attraction of the earth, which must be attributable to a depth of water intervening between the instrument and the solid constituents of the earth. It can be proved mathematically that the total gravitation of the earth diminishes proportionately with the depth of water, and that if an instrument could be devised to indicate such minute changes in the total attraction upon a scale, the equal divisions on that scale would represent equal units of depth.

Gravitation is represented in this instrument by a column of mercury resting upon a corrugated diaphragm of thin steel plate, which in its turn is supported by the elastic force of carefully tempered springs representing a force independent of gravitation. Any change in the force of gravitation must affect the position of this diaphragm and the upper level of the mercury, which causes an air-bubble to travel in a convolute horizontal tube of glass placed upon a graduated scale, the divisions of which are made to signify fathoms of depth. Special arrangements were necessary in order to make this instrument *parathermal*, or independent of change of temperature, as also independent of atmospheric density, which need not be here described. Suffice it to say that the instrument, which has been placed on board the S. S. "Faraday" during several of her trips across the Atlantic, has given evidence of a remarkable accordance in its indications with measurements taken by means of Sir William Thomson's excellent pianoforte wire-sounding machine ; and we confidently expect that it will prove a useful instrument for warning mariners of the approach of danger, and for determining their position on seas, the soundings of which are known.

Another variety of this instrument is the horizontal attraction meter, by which it will be possible to obtain continuous records of the diurnal changes in the attraction of the sun and moon as influencing the tides. This instrument belongs, however, rather to the domain of physics than to that of mechanical science.

These general remarks upon the subject of measurement may suffice to call your attention to its importance, several branches of which, those of *Linear*, *Cubical*, and *Electrical Measurement*, will now be dealt with.

The discussions which will follow these addresses will be carried on under circumstances such as have never before co-operated, namely, the presence of leading men of science of all civilised nations, who will take part in them, and the easy reference which can be had to the most comprehensive collection of models of scientific apparatus—both of modern and ancient—which has ever been brought together.

Mr. BRAMWELL, C.E.: I think it is our duty to move a vote of thanks to Dr. Siemens, our President, for his interesting and valuable address to our conference to-day. He has foreshadowed a large amount of work, which I hope will be faithfully fulfilled, and if so I am sure these meetings will be most useful to mechanical science.

Sir JOSEPH WHITWORTH: I shall be most happy to second that.

The motion was carried unanimously.

The PRESIDENT: I thank you very much for your expression of approval with regard to the address that I have just delivered. I found the subject was a very vast one, and that it was impossible to do full justice to it, but I hope that this address may be followed now by communications of a more specific kind, which will be both interesting and useful. I will now call on Sir Joseph Whitworth to read his communication

ON LINEAR MEASUREMENT.

Sir JOSEPH WHITWORTH: The two great elements in mechanics are the power of measurement and the true plane. The measuring machines which I have constructed are based upon the production of the true plane.

Measures of length are obtained either by line or end-measurement.

The English standard yard is represented by two lines drawn across two gold studs sunk in a bronze bar about 38 inches long, the temperature being at 62° fah.

There is an insurmountable difficulty in converting line-measure to end-measure, and therefore it is most desirable for all standards of linear-measure to be end-measure.

Line-measure depends on sight aided by magnifying glasses; but the accuracy of end-measure is due to the sense of touch, and the delicacy of that sense is indicated by means of a mechanical multiplier.

In the case of the workshop measuring machine, the divisions on the micrometer wheel represent ten-thousandths of an inch. The screw has twenty threads to an inch, and the wheel is divided into 500, which multiplied by twenty gives for each division the ten-thousandth of an inch.

We find in practice that the movement of the fourth part of a division, being the forty-thousandth of an inch, is distinctly felt and gauged. In the case of the millionth machine, we introduce a feeling-piece between one end of the bar to be measured and one end of the machine, and the movement of the micrometer wheel through one division, which is the millionth of an inch, is sufficient to cause the feeling-piece to be suspended or to fall by its gravity.

The screw in the machine has twenty threads, which number multiplied by 200—the number of teeth in the screw wheel—gives for one turn of the micrometer wheel the four-thousandth of an inch, which multiplied by 250—the number of divisions on the micrometer wheel—gives for each division one-millionth of an inch. The sides of this feeling-piece are true planes parallel to each other, and the ends both of the bars and the machine are true planes parallel to each other, and at right angles to the axis of the bar; thus four true planes act in concert. In practice we find that the temperature of the body exercises an important influence when dealing with such minute differences, and, practically, it is impossible to handle the pieces of metal without raising the temperature beyond 62°. I am of opinion that the proper temperature should be approaching that of the human body, and I propose that 85° fahr. should be adopted, and that the standards and measuring appliances should be made and kept in a room at a uniform temperature of 85° fahr.

In many workshops we hear the workmen speak in such vague terms as a bare sixteenth or full thirty-second, but minute and accurate measurement requires to be expressed in decimals of an inch.

In 1857, when President of the Institution of Mechanical Engineers, I read a paper on standard decimal measures of length, and I am happy to say that since that period the decimal system has been introduced to a certain extent in many engineers' works, but it is still far from being universal.

In the manufacture of our standard gauges, the workmen measure to the one twenty-thousandth of an inch, and these measures are as familiar and appreciable as those of larger dimensions.

As an illustration of the importance of very small differences of size, I have here cylindrical standards with a difference of the ten-thousandth of an inch. It is therefore obvious that a difference of one ten-thousandth of an inch is an appreciable and important quantity.

It will be at once conceded that the only scale of measurement which can be used for such small differences must be a decimal one.

For many years the decimal system has been in use at our works, taking the inch as the unit, and the workmen think and speak in tenths, hundredths, and thousandths of an inch.

It is of great importance to the manufacturer to have the means of referring to an accurate fixed measure, as it will enable him, at any time, to reproduce a facsimile of what he has once made, and so preserve a system of sizes of the fitting parts unaltered.

The great value of the workshop measuring machine is making difference gauges.

Every external diameter having to work in an internal diameter should have a certain difference of size; and close observation and experience can alone determine what this difference of size ought to be.

Take, for instance, a railway axle; if the bearing in which it has to work be too small the heating of the axle by rapid rotation will be the consequence; if, on the other hand, the bearing be too large, it will be sooner worn out.

It is therefore most important when rapid revolutions and great strains have to be undergone, that the proper difference of size, when once ascertained by experience, should be strictly adhered to.

In the manufacture of axles there should be two gauges used, the axle being made to the standard gauge, and the bearing bored out to fit a different gauge, which has to be as much larger as experience has found to be necessary, according to the conditions under which the axle has to work. Hence every manufacturer should be in a position to manufacture his own difference gauges.

Fifty years ago the thousands of spindles in a cotton factory had each to be separately fitted into the bolster in which it had to work. At the present time all these spindles are made to gauge, and are interchangeable.

It cannot be impressed too forcibly, both upon the student in mechanics and upon the workman, that accuracy of measurement is essential for good and efficient workmanship and that it tends to economy in all branches of manufacture, so as to have the parts interchangeable.

I will now endeavour to explain two or three things which I have on the table. I have said that the measuring machines are based on the production of a true plane. I have here a true plane of ten inches square. I will not go into the method by which these true planes were produced, which would take too long, but I may mention that in getting one true plane we have to get three, because if we had only two, one might be a little concave and the other a little convex, and still they might be made to fit, but if the third will fit the two, it must be a true plane. You will see that when I put this plane down on the other, they do not touch each other, but float on the air between them for a short time; and if I let one fall upon the other, you perceive the metals do not come in contact—there is a muffled sound. The plane is equally true when it is suspended as when it is supported. It is suspended by three points, and we also now lift them from those three points, so that when applied to a piece of work it remains true.

I will now refer to the workshop measuring machine. The divisions on the micrometer wheel represent ten-thousandths of an inch—that is if you move this wheel one division, the end of the machine is moved forwards one ten-thousandth of an inch. We find in practice that the movement of a fourth part of a division, being one forty-thousandth of an inch, is distinctly felt and gauged. I have here a small gauge one-fourth of an inch in diameter, and on setting this wheel to zero, I then regulate the machine with the small wheel at the other end, so that I can just feel this gauge go through. Now, if I move this wheel half a division, the gauge will not pass through. We get thus half a division, which would be one twenty-thousandth of an inch. No one would believe that so small a difference would produce such an effect as that. When everything is in nice order I can feel perfectly well the fifty-thousandth of an inch. In the case of the millionth machine we introduce a feeling-piece between one end of the bar to be measured, and one end of the machine, and the movement of the micrometer wheel through one division, which is the one-millionth, is sufficient to cause the feeling-piece to be suspended or to fall by its own gravity. This is a standard inch to be measured. We bring the inch bar up to one end, and introduce the feeling-piece. When the machine is adjusted, we find that moving the end one-millionth of an inch is sufficient to cause that feeling-piece to be suspended or to fall. As an illustration of the importance of very small differences of size, I have here cylindrical standards with the difference of ten-thousandths of an inch. This is a standard inch measure, and this is the corresponding one—internal gauge. You see that it is what we call a nice fit. These gauges are made of steel and case-hardened, so that they will last a long time, and nothing will cut the surface except a diamond or the grinding process by which they are made. This other gauge is one ten-thousandth part of an inch less, and you see the difference which it makes. The internal gauge slides down of its own gravity, and it is therefore obvious that the difference of a ten-thousandth part of an inch is an appreciable and important quantity. I have said that the great value of the workshop measuring machine is making difference gauges. These are a set of gauges which I made for Lord Hardinge in 1855 for the improvement of the Enfield rifle. The size is ·577. Each of these gauges are one five-thousandth of an inch less than the other, and you see how much more loosely one fits than the other. The value of these difference gauges is this: If you tell a man to bore a barrel to that size he is sure to make it wrong, he will get it too large; but if you give him this set of difference gauges he gets the bore first of all to the one-thousandth of an inch less than the standard, and then feels his way up to the standard. In making rifles we never made any allowance. We gave the men these difference gauges, and every barrel was made the same, so that all shot alike. There was no high gauge and low gauge as it is termed at Enfield. By the use of these difference gauges we were able to work up to a certain point, and they were all made alike. Then I have said that every external diameter having to work in an internal diameter should have a certain difference of size, and close observation and experience can alone determine what this difference of size ought to be. I may mention that in making these cylindrical gauges we do not go below one-tenth of an inch, and when we want a gauge less than that we make this class of gauge—flat gauges ; they are made with a tolerably good surface on each side, of steel and tempered, and they last pretty well. This is one-hundredth of an inch, and we can easily make this from the measuring machine. The largest of these gauges is one-tenth and the smallest is one-hundredth, and there are regular differences of sizes between. They serve the purpose of wire gauges. It is very important when you get so small a size that it should represent accurately what it professes to be. This is a little apparatus by which it is demonstrable that we can make the ends of this bar at right angles. It was once objected that we could not make the end of the bar at right angles to the faces. It is done by placing the standard in a solid block, with a V shaped channel to receive it, and the end of the standard and of the block is made a true plane. The standard is then turned round in its resting place, and again tried with the true plane. If it is not true it is again faced up until it becomes so, and when it corresponds to the true plane in each of the four positions it is evident that the end is at right angles to the four sides. This is a rod which is another application showing the sense of touch. It was made for ascertaining after a gun had been proved or fired a certain number of times whether there was any appreciable wear or whether one part expanded more than another. At one end there is an inclined plane, which moves out three little feeling-pieces. The bar is introduced into the bore of the gun and is moved endways, and there is no difficulty at all in feeling the ten-thousandth part of an inch. There is a little roller fixed on the muzzle of the gun on which it works, and the attendant can feel quite easily a difference of the ten-thousandth part of an inch.

M. Tresca recognised the fact, that, in the actual state of things, the exactness of one-millionth of an inch is the utmost limit which it is possible to appreciate : and that Sir J. Whitworth's process is the only one which allows of this appreciation of small quantities.

In the proceedings which are being carried on by the members of the International Commission on the Mètre, it will be as much as they can do if they succeed in appreciating, by the most delicate optical means, a quantity four times as great, viz., the ten-thousandth part of a millimètre (millimètre=·03,937 inch).

Sir J. Whitworth's process is then the most exact which is as yet known, and whenever short distances are concerned, it can be adopted with perfect security.

The "Mètre" Commission has nevertheless adopted the "mètre à traits" as being more easily comparable. In it the true length is in reality merely defined by the medium of the thickness of a line, which is not to be less tha five-thousandths of a millimetre in breadth. The fact is, that for long distances, the direct measurement between two trials (touches) is much less precise, unless recourse is had to a complemental standard of the same perfection as those of Sir J. Whitworth ; but then the true measurement is further complicated by two contacts, and it is absolutely necessary, in measures of precision, to take particular precautions with regard to all that concerns the influence of the temperature and of flexion.

In order to satisfy these two conditions in the best way possible, a proposal of mine has recently been adopted. It consists of a form of set section ruler having the shape of an X, in which the neutral fibre is apparent. As the international mètres must be "mesures à traits," recourse will exclusively be had to optical processes, in order to compare them; proceedings which alone can guard against contacts and the wear of the faces, which might result rom frequent use. We could not, however, in any case, hope to reach the degree of exactness attained by Sir J. Whitworth's English inch standard, which he has made with so much care and trouble, and which must be considered, until the definitive adoption of the metrical system, as the type of the most perfect precision, which mechanical means permit of attaining.

Mr. CHISHOLM : The question whether a measure of length was best defined by the whole length of a bar, or by the distance between two points or lines marked upon it, was carefully considered by the

Standards Commission for restoring the British Standards destroyed in 1834.

The ancient standards of length, the yard of Henry VII. and the yard of Queen Elizabeth, were both end-standards.

The Parliamentary standard yard, legalised in 1824, and destroyed in 1834, was defined by two dots marked on a brass bar.

The ancient Prussian standard of length was a line-standard, but in 1838 there was substituted for it an end-standard, constructed by Bessel. He gave as his reasons—

1. That if a flexible bar be supported on two points, the extreme length of the bar from the centre of one end to the centre of the other end is not sensibly altered by its flexure, whilst the distance between two points or lines upon the upper surface may be sensibly altered.

2. That the principle of end-measure is more convenient than that of line-measure for the production of copies, in other words, for comparisons.

As to the first of these points, the Standards Commission remarked in their report, that Bessel himself admitted this objection would be removed if the lines were engraved on surfaces depressed to the middle of the thickness of the bar. And they stated, moreover, that the tendency to an alteration of the apparent length, either at the surface or mid-depth of the bar, might be obviated by proper adjustment of the points of support, and still more surely by supporting the bars at numerous points on lever frames, with equal supporting forces at all the points, or by floating the bar in quicksilver.

As to the second point, the Commission considered that line-measures had been invariably adopted for measuring British geodetic bases; and their use led the Commission to form a high estimate of the convenience of the line principle, and to consider that a standard which was intended to apply to them should be constructed on the same principle. They considered, also, that the construction of defining lines on a standard bar was more simple and easy than that of defining ends; and that the tendency of an end-standard to an alteration of length was by wear in only one direction, whilst a line-standard was practically invariable.

The end-standards are connected with spherical ends, the radius of the spherical curve being half the length of the bar.

That although a *single* comparison by end-measure was, perhaps, more accurate than a single comparison by line-measure, yet there was no doubt that by repeating the comparisons, unexceptionable accuracy might be given to observations of line measures.

Upon full consideration, the Commission unanimously preferred to adopt a line-measure for the new standard of length, the defining line to be marked at the mid-depth of the bar.

Specimens of both the line-standard and end-standard yards, constructed under the superintendance of the Standards Commission, and of the lever-supports, are now exhibited.

The same question was also fully considered by the International Metric Commission at Paris, and a similar conclusion was arrived at for the construction of the international standard metres.

There can be no question of the advantages of the form of end-measures, adopted by Sir Joseph Whitworth, with the defining ends constructed as true planes normal to the measuring axis; and more particularly to his application of the principle to guages, to which, indeed, the principle of line-measure is clearly inapplicable. But this mode of application of the principle of end-measure is not so well adapted to a standard unit of length, such as our standard yard, from which secondary standards of multiples and parts of this unit are to be derived. If a measure of length, whether an end-standard or a line-standard, is to be subdivided into parts, these parts can only be marked upon it by defining lines.

There are now in existence no less than ten standard Egyptian measures of the home of the Pharaohs. These are standard cubits, end measures, divided by lines into palms and digits. A model of the best specimen of these ancient standard cubits is now exhibited.

It appears to me, however, that one great practical objection to an end-measure, as used by Sir Joseph Whitworth, with a contact or feeling apparatus, consists in alterations of its length by variations of temperature, and in the difficulty, if not impossibility, of determining the extent of this influence. In comparisons made with a micrometer microscope, on the other hand, whether of line-measures or of end-measures, the actual temperature of the measuring axis of the bar may be maintained nearly constant, and may be ascertained, and allowance made for its effect. There is really no practical difficulty in the accurate comparison of line-measures with end-measures, and could easily be shown if time allowed.

It is also of the utmost importance to determine with precision the rate of dilatution by heat of a standard bar, and such determination can be practically made more easily and accurately from observations of the defining lines of a standard bar.

It may be desirable here to examine the extent of alteration of a measure of length arising from variations of temperature and to show it in the following tabular form, in regard to the materials of which standards of length are ordinarily made :—

Material.	Co-efficient of line. Expansion For 1° Fahrenheit.	Variation of length of 36 in. For 1° Fahrenheit.
		in.
Brass	0. 00000956	0. 000344
Bronze	0. 00000947	0. 000341
Wrought Iron ...	0. 00000611	0. 000220
Cast Iron	0. 00000550	0. 000198
Cast Steel	0. 00000575	0. 000207
Platinum	0. 00000476	0. 000171

It may here be seen that an iron standard yard alters its length sensibly—that is to say, one ten-thousandth of an inch, with a variation of temperature of half a degree Fahrenheit. At the same time, an iron guage one inch in diameter will not alter the length of its diameter to the extent of one ten-thousandth of an inch from dilatution with less than an increase of 36° Fahr.

The conclusion to be drawn from a consideration of all these circumstances is, that whilst for practical uses, where mechanical accuracy is required, end-standard measures may be satisfactorily used, line-standards are preferable for primary standards where the highest scientific accuracy is needed.

The CHAIRMAN : I am afraid we cannot go into a full discussion of this very important and intricate subject, or of the relative merits of measurement *a but*, and measurement *a trait*. If the commissions which have been sitting in this country and in France upon this subject for months have not arrived at a definite solution, it is to be feared, that we, pressed as we are for time, shall not be able to do so, although we have had the advantage of hearing three gentlemen on this subject, who rank first as authorities. I think M. Tresca puts it very fairly, that both methods have their advantages. The measurement *a but*, as we have seen it explained by Sir Joseph Whitworth, seems to me most adapted to the determination of very slight differences of linear measure, provided the piece of substance measured is short, as is the case in measuring thickness, whereas for the measurement of long articles, the other method seems to possess advantages of its own. Sir Joseph Whitworth will say a few words in reply to what has been advanced, and then I am afraid we shall have to close this discussion, as we have a great deal of business on hand.

Sir JOSEPH WHITWORTH: I will just say that the line-measure previous to 1851, had engaged the attention of a number of gentlemen—the Reverend Mr. Sheepshanks being one—who were appointed the Commission, and were engaged for several years upon it. They had rooms in Somerset House, and were going to get an Act of Parliament for these standards which they recommended. The original standard as was observed by Mr. Chisholm had been destroyed when the Houses of Parliament had been burned down. In the meantime I was making my experiments on end-measure, and I exhibited at the Exhibition of 1851 a standard yard measure, and showed there, that the differences in length by small increments of heat was very great. In the machine that I exhibited, which was the standard yard, I showed repeatedly that when the feeling-piece at the end was adjusted, you could not touch the bar with your finger nail, without the feeling-piece being suspended. You can make, of course, any number of standards, and supposing we had a room at the temperature of the human body, we could then handle these things; but the commission determined that the degree of heat should be 62° Fahrenheit, and of course you cannot touch anything with your hands without making it longer. The way in which I supported the standard yard was objected to before this commission, and it was said, it was better to support it on rollers, but I may say that the gentlemen who were engaged in determining what the standard yard should be, though they were scientific gentlemen of the greatest eminence, did not know how to use their hands. They recommended that the standard inch, which I considered was very important, should be our end-measure, and also the standard foot. I agree with my friend Dr. Siemens, that if we could only have one measure it would be a great thing for all countries, and I would go into the matter at once if the standard yard and our standard inch would only divide into the metre, but unfortunately it will not; and as all our screws and all our machinery are made to the standard inch, of course those standards must be used until everything is worn out. Perhaps it may be advisable in the end to adopt the metre, but it is quite impossible in our time. I have shown you the importance here of very small differences, but the most important of all, perhaps, is with reference to screws. A perfect screw is one of the most important things in mechanism, and it would be impossible to alter the distance of the thread of our screws, otherwise I fully admit it would be a great advantage to all countries if we could have only one measure, and I am sure I would do anything I could to promote it.

The CHAIRMAN: I am sure you will all agree with me in passing a vote of thanks to Sir Joseph Whitworth, for his interesting communication.

The vote of thanks was passed unanimously.

The PRESIDENT: The next paper on our list is by Mr. Merrifield,

ON SOLID MEASUREMENTS.

Mr. C. W. MERRIFIELD, F.R.S.: I have been desired to give you an account of solid measurement, and I must ask you to excuse any shortcomings, by telling you that it was not until Monday that it was proposed that I should undertake this task, and it was only yesterday that I was aware I should have anything to say about fluid measure. It has also been suggested to me that, besides solid measurement, I should touch on a subject you have not yet had before you, which would otherwise have been omitted from the programme, namely, surface measurement. That necessarily precedes solid measurement, because, unless a measure of solid contents be mere replacement or displacement, we must first be able to measure the surfaces, and from them only can we measure solids; at least, that is the usual course pursued by geometers, and the only course to which geometrical measurement applies. I must first say a word or two with reference to the accuracy of this measurement. The only tangible idea we have of infinitesimals, and the only clear idea of a boundary is, to my mind, a solid boundary, and from that comes one of the great material advantages which you have heard described by Sir Joseph Whitworth in accurate end-measurement. The next to that in facility of perception probably is the boundary of a plane by straight lines, or of a line by cross lines, for there is no such thing observable by us as any reasonable approach to a point. The accuracy of lines is not so easily felt as a surface. A surface, if highly polished, probably presents a greater approach to an accurate boundary than anything we can have in the way of lines. On lines there are more or less, roughly—especially in regard to material lines, of which I am now speaking—two sides to the line. They are not generally well defined, but we can average the line by finding its middle, and in that consists the accuracy of line-measurement. When we come, however, to lines carelessly drawn perhaps, or ill-defined, to which we cannot give great precision, we find we are dealing with a very indefinite thing indeed. For instance, if you take a hard pencil, and draw as fine a line as you can with it on a piece of paper, and put it under a microscope, you will find a series of dots scattered about an irregular breadth, more or less as if you had been dropping seed out of a hopper in going across a field. If you do the same with a mathematical pen you will then find under the microscope that you have got something equivalent to a cart track on a road, with a rut on each side. Still, if it is done with great care and precision, you are able to get a very definite middle to that line. Now, with surface measurement, as ordinarily presented, we have a very different thing. The surfaces we have to measure, are generally defined in a very different way, either by an edge, which we have not ourselves had the means of making accurate, or by a line which has been drawn for us with more or less accuracy, generally with very small accuracy, because there is no very good means of drawing with any degree of accuracy any line but a straight line or a circle. Therefore, in this measurement of areas we have not as yet felt the necessity of any instruments at all approaching in precision either the solid measures used and so greatly perfected by the Warden of the Standards, or the still more accurate linear-measures, whether those of the French or English Commission of Standards, or Sir Joseph Whitworth's end-measures.

To return to the subject of ordinary plane measurement. The first idea of measurement of plane surface is to reduce it to one or more square units, that is, having obtained its linear dimensions, and taken account of the irregularities of its area, to find out the number of square units contained in it. That is generally done geometrically by cutting it up into strips of some definite shape, either by parallel lines, or else by lines radiating from a point, and then these strips are each separately measured with great ease and with tolerable accuracy; and the characteristic of the measurement is such that the error on each strip, although noticeable when altogether, is still very small and diminishes very much more rapidly than the number of strips into which the area is divided. Every person accustomed to the quadrature of curves is well aware of that. That is to say, if you use any tolerably accurate mode of measuring, and cut up an area, either physically or mathematically, into ten strips, you have a certain definite error of, say, perhaps one-tenth of a square inch. But if you cut it up into 100 pieces, instead of reducing the error by ten, you generally, if your arrangements are well made, reduce it by 100. Now when we come to solid quantities, the ordinary method of dividing them is to cut them up into slices. You add the slices together to make the solid, in the same way as you add the solids together to make the surface. But that is hardly the case in general application. We do not cut these cylinders before us into strips, but we are obliged to have resort to means of replacement, this means being, for various reasons, far less accurate than any of the modes of linear-measure we have just described. But yet in the hands of a gentleman like Mr. Chisholm we can get a degree of accuracy which few persons unaccustomed to consider the thing would dream of—probably far more accuracy than we ordinarily meet with in surface measurement.

Now, there are also certain other beautiful little instruments called *Planimeters*, by which a task, which at first sight would hardly be considered conceivable, has been successfully accomplished, namely—that by simply running a pen point round any irregular closed curve, of which we wish to measure the area, a little wheel records the area that the pencil goes round. That seems conceivable enough if we had to deal either with a circle or an ellipse, but it seems almost inconceivable when we have to deal with curves with any amount of irregularity. The most useful of these, as at present arranged, is probably Amsler's Planimeter. The principle of this, although really very beautiful, is a little intricate, and has not always been well understood. It depends upon this principle—that if you take a bar of definite length, and give it a small motion, then you may measure the surface swept over by that bar by simply multiplying the length of the bar into the travel of the middle point resolved at right angles to the bar. I can explain this perhaps more clearly by the aid of a diagram. The way that principle is made use of in Amsler's planimeter is this. He puts a small roller on the bar in the direction of its own length, and if the bar moves the wheel simply slides, but if transversely it rolls. If the bar moves in any intermediate direction it both rolls and slides, but the wheel records only that component of the motion which is at right angles to the bar. That of course only applies to small motions of the bar, but as it applies to every small motion it must be true for the sum of them as well as for every part. Supposing you have an irregular curve to measure, you make one end of the bar follow the curve and let the other end reciprocate along a certain curve. Its longitudinal motion will have no record taken of it, but the motion at right angles to the bar will be accurately counted by the little wheel. As it moves in one direction, the wheel will run one way, and as it goes round the other, it will go the other way; and the difference of those two will be recorded when the bar gets into its original position: and that, read to a proper scale, and multiplied into the length of the bar, will give the difference between the line swept out and the true area all round; consequently the difference between the readings of the wheel in its first position and when it gets back to its first position will give the area. It is quite immaterial what this curve may be so long as one end of the bar moves backwards and forwards along the same curve, no matter what; the reading will be exactly the same. Those who have used it will be aware that the wheel is not put at its middle point as I have put it, but it may be proved geometrically that so long as you measure a closed curve with a bar, one end of which is always outside the curve, it is immaterial upon what part of the bar you put the little wheel. It is, of course, material, if you are only going to measure a part of the curve, but not when you measure the whole curve. That, then, contains the whole principle of Amsler's Planimeter, that the surface swept out by the bar is measured by the travel of the middle point, resolved at every instant at right angles to the bar, and the success of the instrument depends on the little wheel recording that transverse component only. The other points are mere details of mechanism. One form of the instrument is shown here. Practically, these are not generally made of a size to be used for any other than small diagrams, but they really measure with great accuracy. In fact, when I was in the School of Naval Architecture we used to try it upon squares, and we invariably found that it measured a square with at least as great accuracy as we were able to draw it, and I need hardly tell you that in surface measurements it is not necessary to obtain an accuracy that goes beyond the drawing.

There is another kind of Planimeter founded on a principle that is well known to mechanicians in the form of a continuous indicator. Suppose there is a disc turning on its centre and we have another little disc at some point of it in rolling contact with it, the travel of this disc will depend not only on the rapidity with which the primary disc moves, but also largely on its distance from the centre, and, moreover, supposing that the travel of the primary disc is uniform, the travel of the follower will be exactly proportionate to its distance from the centre. That circumstance is taken advantage of in a little machine I have here. The disc you see the edge of is the following disc, and the one which is horizontal when the instrument is in its proper position is the driving disc. The bar that you see with a little wire upon it is always kept parallel to a base line, and a tracing point follows exactly the curve. Now, when this following disc is on the line of centres it records nothing, but as it moves away from the line of centres there is a specific ordinate to which the travel of the following wheel is proportional, consequently the operation of the instrument is to add this exactly as we do in ordinary algebraic integration. In that way we get the surface recorded with as great accuracy as the drawings are framed. In point of fact, I have very little doubt its performance is quite equal to that of any drawings that can be submitted to it. A modification of that is to be found in Professor James Thomson's instrument down stairs. I do not propose to describe that, because it it really very much more than an instrument for measuring an area, and may be applied to a great many things, but in its simplest form it is a case of measurement of area, on exactly the same principle as this, with the exception that the sliding motion is entirely got rid of, being replaced by the motion of a sphere which rolls both on a sphere and on a cylinder. The motion that it records depends upon the distance of the point of contact with the sphere from the centre of the disc on which it rolls. In principle it is exactly the same as this, and the principle of both is the same as in ordinary integration, namely, the ordinary rule that the area equals the integral of $y\,dx$.

With regard to the measurement of solids, the general principle of solid measurement in ordinary use is simply that of displacement or replacement, namely, that a standard measure is made, and all other measures are compared with that by actually pouring some kind of material, either corn, or some dry material of that kind, or water, from one to the other. The most obvious measurement of solids by displacement, is by filling a vessel up to a certain edge with a liquid, and dipping the solid into it either entirely, or up to some measured point, and ascertaining the quantity that overflows. We have, I am afraid, no exemplification of either of those modes of displacement here in their simplest form. With regard to some instruments I have seen, I may mention that it is necessary to observe great caution in the form of the lip by which the overflow takes place if you wish to secure any accurate measurement, for a difference in the form will make it measure differently, and there is also great difference in the viscosity or molecular character of different liquids, and scarcely any two liquids measure alike. It would not be fair, however, to say that is the only mode of measuring solids. Of course, practically, the first thing you have to do in order to compare solid-measure with linear, is to construct very carefully a solid, either hollow or not, upon geometrical principles, so as to connect it with linear-measure, and the most ordinary mode of doing that is by making cylinders of which you can measure the height with very great accuracy, and in which you can secure a very considerable degree of accuracy in form, and also are able with accuracy to measure the diameter. Taking the three things together, I do not think it can be said, that these measures at all approach in accuracy the linear-measures. There are various reasons why they should not. In the first place, they depend not only on three different sets of linear-measure, each one of which requires to be got with accuracy, but also depend on accuracy of form, and upon the possibility of measuring round forms, which is scarcely so great as with the linear-measure. They also depend on having square ends—that is, on the bottom or top being set perfectly square to the edges. Still a very considerable degree of accuracy is to be obtained by these methods.

With the purely geometrical modes of measurement by means of ordinates I need not detain you, but I should be neglecting a very large part of the subject, if I were to fail to call your attention to some very important economical measures, namely, timber measuring and cask gauging. In timber measuring, practically, one's first idea of measuring a spar is to consider it as a cylinder. Of course any spar, as the tree grows, is considerably thinner at one end than at the other, but the correction for that is not very difficult to apply, and is ordinarily applied by geometry, with which I need not trouble you here, particularly as we have before us no instruments for doing it. Sliding rules are used for the purpose, but they are rather calculating than measuring instruments. With regard to cask gauging, however, it may be done with as much accuracy as the subject admits of. It has been made the subject of very elaborate calculation by Hutton and others, and great perfection has been obtained in the instruments with which the measurements are taken, considering the mode of applying them. The instruments I am now showing are those manufactured by Messrs. Dring & Fage for the use of the English Customs and Excise officers. This is the calliper with which the length of the cask is measured. Of course the staves generally stick out an inch or two beyond the head of the cask, and consequently the length is measured by two rules sliding one upon another, which fit over the head of the cask, and enable you to get a fairly accurate measurement of the length. The calliper for obtaining the diameter of the cask is also here. It is constructed on the same principle as the first. In getting the diameter of the cask we have first to get the bung diameter, the diameter at the widest point, and also at the head, and these are quite different. I only mention for the sake of those who are not acquainted with round measure, that it is necessary that the callipers should be applied in several places: you have either to turn the cask or turn the calliper round the cask, because you very seldom find that the cask is a perfect circle, and for the measurement you have to take the average. Then when you have got the diameter of the cask at the bung and at the head, for the internal diameter you have to allow for the thickness of the planks and the curve of the cask. Supposing that you know the shape of the cask, you would then have the elements for a tolerably accurate measure of it. But, in the first place, casks are not always in the same shape; and no person has yet been able, so far as I can discover, to make up his mind as to what the proper shape for a cask ought to be. Those difficulties, however, are very small, and are not very important. For fiscal purposes it has been found quite sufficient to take the head diameter and the bung diameter. Where much more accuracy is required—that is to say, where any large casks have to be specially measured, it is necessary to take a diameter intermediate between the bung and the head; the actual measurement is then done on paper and also by the slide rule. Here is one of the slide rules ordinarily used. It is a good long one, and that is an advantage, because it is read much more easily. The scales laid down upon it are the ordinary A B C and D scales of a slide rule. These serve generally for ordinary calculation, which merely requires the scale to a half radius on one side of the rule and a whole radius on the other. Besides this there is another method for measuring casks in a rough way, and that is the diagonal measure. Here is a diagonal staff. It is simply thrust through the bung, and put to the edge of cask, and then, on the supposition that all casks are of the same size; the volume of the cask, of course, varies with the cube of the diagonal. There is also a scale on the slide rule for that purpose. The ullage of a cask is also of some importance. It means the measuring a cask which is partially full. At the same time, as the measurement is chiefly done for fiscal purposes, it is not a matter of much consequence, in measuring a cask perhaps one-third full, whether there is a pint this way or the other. Therefore the measurement is very rude. What is really done is to dip a stick into it, and to see how much of it is wetted, and then by comparing that with the known measurements of a standard cask you find the quantity in the cask. There is no attempt at an accurate geometrical measurement of ullage.

What remains for me to say is on the measurement of gases and liquids, and on that I shall be very short. I shall first speak of gas meters; for although the engineers present are pretty certainly aware of the construction both of dry and wet gas meters, some of my hearers would perhaps like to have them described in a general way. The ordinary wet gas meter is a mere cylindrical box about half full of water, in which there are four vanes fastened to a horizontal spindle. The gas flows in at the bottom, and as it flows in it turns the vanes round, then delivers the gas into the upper part, from whence it goes into the house through the supply pipe. Therefore, for each quarter revolution a definite measure of gas passes in, and consequently passes out. There is a counter fixed on the spindle which counts the number of revolutions, and as that is done to a scale, it is read off in cubic feet or hundreds of cubic feet. I have not gone into any details, and I dare say the rough sketch I have made does not represent it with anything like accuracy, but that is the principle. It is not very satisfactory for many reasons. It regulates with tolerable truth, supposing the meter is perfectly and honestly arranged; but it has the disadvantage of wetting the gas, which makes it burn badly for one thing, and it also tends to deposit water in the joints of the pipes and to give trouble in that way. The dry meter is very much like the cylinder of a steam engine, only differently arranged. If you can imagine a steam engine with the piston fixed and the cylinder moveable, it would be very like it, only instead of being made of rigid materials there is a bag on each side of a fixed diaphragm. The gas first comes into the bag on one side, and when it is full up to a certain point it presses the end of it against a stop which shuts it off, and at the same time opens an out-flow cock into the house. But at the same time that it shuts off the one bag it opens an inflow cock into the other bag so that the other side begins to fill. Thus at every half stroke one bag gets filled, and the other emptied, and at the opposite stroke the first gets emptied and the second filled. Each stroke is recorded on a counter which thus registers the quantity of gas passing through the meter. This is in several respects a much better arrangement than the wet meter. I think it is a little more expensive to make, and gas companies generally prefer wet meters. There is a practical reason generally attributed to them for that, although perhaps unfairly. That I cannot speak about with certainty, but it is generally supposed that if anything goes wrong with the meters the dry meter makes an error in favour of the consumer, while the wet meter makes an error in favour of the gas company, and consequently the gas companys go in for wet meters, and will not let you have dry meters if they can help it; but if you take my advice, go in for dry meters whenever it is possible.

Now, with respect to water meters. There are various kinds of water meters. First, there is an absolute water meter, in which the water is received into a receptacle, much in the same way as in the dry gas meter, and delivered in the same way, a reciprocating motion, and an arrangement giving the means of measuring it; just as if you passed water instead of steam through a steam-engine, the number of strokes of the steam-engine would give the quantity of steam passed into the engine, and out again. If you want absolute measurement, that is doubtless the best you could arrange for, but it gives you an intermittent current, and destroys the head of water; consequently, in all modern water meters the arrangement depends on either the turbine or the screw. To put it in the simplest form, if you had a screw, like a screw propeller, fixed in the middle of a pipe, and revolving, the number of revolutions of the screw would depend on the velocity of the water going past it, and that would give you a rough measure of the amount, but only a very rough measure, for there is a certain amount of friction caused by turning the screw, and besides that a part of the water escapes through the interstices of the screw without acting upon it, and in that way the screw does not record the quantity of water with any great degree of accuracy. Not only that, but it does not record it with the same degree of accuracy at different speeds, and therefore you are not able to apply an easy or certain correction for it. An improvement of that has been introduced by a more complicated form of screw called the turbine or Barker's Mill, where the water is got under more complete control by directing it against vanes properly arranged and proportioned, and in that way a very accurate water meter can be obtained for all ordinary speeds. There is no doubt that if you were to pass the same stream of water through at the rate of half-a-mile an hour, and then, at some other time, at the rate at which water is driven through the hose of a fire-engine, there would be considerable errors, but with Dr. Siemens' turbine arrangement, under ordinary circumstances, with considerable variation of velocity, there are practically very small limits of error. It really is a turbine or Barker's Mill arrangement, with proper directive vanes to give compensation

All water meters more or less destroy the head of water; that is, make a considerable difference between the velocities at which the water enters the pipe and leaves it, and, consequently, where the stream is used to develope power, any water meter is a very serious drawback, but where no power is wanted from it, that is immaterial.

This machine before us, which is called Siemens' alcoholimeter, is one used to determine both the quantity of liquid that passes out of a still, and also the quantity of alcohol that it contains, and it is used largely in Russia. There are two distinct parts, first, a water meter of tolerably simple construction, not unlike what I said before about gas and water meters; a copper disc revolves in a box upon a horizontal shaft, with three vanes at equal angles, and the water comes in at the centre, and drives always in one direction, delivering at the circumference like a water wheel or vertical turbine. So far, it merely measures the quantity of liquid that passes, and the lower scale upon the instrument measures the volume of liquid that flows out of the still. That may be, of course, either wholly alcohol, or wholly water, or any mixture of the two. To indicate the quantity of alcohol there is another counter, and the way it is set to work is this:—There is a weight at the end of a lever, calculated either to represent a certain weight of pure alcohol, or of proof spirit, in such a manner, that if there were no spirit at all a second counter would always stand at zero; while if it were entirely pure spirit, it would count the same as the lower one, which shows the quantity of liquid passing. The spirit which passes is generally considerably below proof. When pure water passes, this upper counter is not affected, but when, on the contrary, pure alcohol is passing, a bob comes down and pushes a second lever, with a curved edge over, and this second lever regulates the difference between the two counters. You may say that its edge ought to be a straight line, and so it could be, but for the fact that one pint of water and one pint of alcohol do not make a quart, but rather less, and consequently there is a slight curvature in it. What happens in the two receptacles is this. The liquid is weighed in one and measured in another. One scale represents the measure, and the other represents the joint effect of the weight and the measure; consequently, one scale records the total quantity of liquid that passes, and the other represents either the absolute quantity of pure alcohol passing through it, or of proof spirit, according to the way in which the instrument is contrived and set. I am assured it works with great accuracy, and being put under a glass case and locked up, the Russian excisemen have not half the trouble with the distillery that our excisemen have. The distiller cannot tamper with it in any way without being found out. The pipes coming from the still are completely in sight; he cannot tap them without the hole being discovered, and he cannot tamper with this machine without showing signs of it. I think I have now said all I need say with regard to these measurements, and I hope I have made myself clear to you.

The PRESIDENT: I have only now to call upon you to pass a vote of thanks to Mr. Merrifield for his very able and lucid explanation of the different modes of measuring solids. He undertook this task only upon two days' notice, and is therefore entitled to our special thanks.

The vote of thanks having been passed unanimously,

The PRESIDENT: I will now call on Sir Wm. Thomson to give us some explanation upon that most difficult and important subject of electrical measurements. I am afraid Sir William will not be able to go as fully into the subject as it deserves to be gone into, inasmuch as our time is exceedingly limited, but I am sure that whatever falls from Sir Wm. Thomson will convey information to all, and that we shall profit by it.

Professor Sir W. THOMSON, L.L.D., F.R.S.: The beginning of electrical measurements, are, I believe, the measurements of Robinson in Edinburgh, and of Coulomb in Paris of electrostatic forces. The great results which followed from those measurements illustrated how important is accurate measurement in promoting thorough scientific knowledge in any branch of physical science. The earlier electricians merely describe phenomena attractions and repulsions and flashes and sparks, and the nearest approach to measurement which they gave us, was the length of the spark under certain circumstances, the other circumstances on which the length of the spark might depend being left unmeasured. By Robinson's and Coulomb's experiments was established the law of electrostatic force, according to which two small bodies, each electrified with a constant quantity of electricity, exercise a mutual force of attraction or repulsion, according as the electricity is similar or dissimilar, and which varies inversely as the square of the distance, when the distance between two bodies is varied.

In physical science generally, measurement involves one or other of two methods, a method of adjustment to a zero, or a what is called a *null* method, and again, a method of measuring some continuously varying quantity. This second branch of measurement was illustrated in Coulomb and Robinson's experiments, where the law according to which the electric force varies, where the distance between the mutually influencing bodies varies continuously, was determined. The other mode of experimenting in connection with measurement, is illustrated by another exceedingly important subject, bearing upon electrical theory, and that is the evanescence of electrical force in the interior of a conductor. Both kinds of measurements were practiced by Cavendish in a very remarkable manner, and I look forward with great expectation to the results we are soon to have of Cavendish's work. One most interesting result which will follow from the Cavendish laboratory in Cambridge, from its director Professor Clerk-Maxwell, and from the relationship thus established between the physical laboratory of the University of Cambridge, and its director on the one hand, and the munificent founder of the institution, the Duke of Devonshire, on the other hand. The Cavendish manuscripts still remain in that family, and are, I believe, at present in the possession of the Duke of Devonshire, and have been by him put into the hands of Professor Clerk-Maxwell for the purpose of having either the whole, or extracts from, published, which may be found to be of scientific interest at the present day. The whole of them, no doubt, had great scientific interest at one time. A large part of these manuscripts, I believe, will be found to be excessively interesting even now, and from something I heard a few days ago from Professor Maxwell, when he was here on the opening day of this exhibition, I learnt that much more than was even imagined is to be found in these manuscripts, and particularly the whole branch of electrical measurements worked out, from the measurement of electrostatic capacity. The very idea of measuring electrostatic capacity in a definite scientific way is, as it now turns out, due to Cavendish. A great many years ago, in 1846 or 1847, when the Cavendish manuscripts were in the hands of Sir Wm. Snow Harris, at Plymouth, I myself found one paper, out of a box full of unsorted manuscripts which startled me exceedingly. It contains the description of an experiment and its result, measuring the electrostatic capacity of an insulated circular disc. That is one of the cases in which the theory founded by Cavendish and Coulomb as developed in the hands of the mathematicians who followed, allowed the result to be calculated *a priori*, and I found the result agreed within, if I remember rightly, one-half per cent. of Cavendish's measurements. When I mention these cases of the measurement of electrical force by Coulomb and Robinson, which has led to the true law of force and of the measurement of electrostatic capacity, a subject which is the least known generally, and held to be the most difficult, I have said enough to show that we must not in this century claim all the credit of being the founders of electrical measurement.

The other main method of experimenting in connection with measurement to which I have referred is illustrated also by Cavendish's writings, that is the seeking for a zero. It is very curious, that while Coulomb and Robinson by direct measurement of a continuously varying quantity discovered the law of the inverse square of the distance, Cavendish, quite independently, pointed out by very subtle mathematical reasoning that the law must either be the inverse square of the distance, or must vary in a determinate manner from the law of the inverse square of the distance if in a certain case, which he defined, either a perfect zero of electric force is observed, or if instead of a perfect zero any particular amount of electric force is observed. It is quite clear from Cavendish's writings that he believed that perfect zero would be found when the experiment should be made, but with a caution characteristic of the man and also proper to his position as an accurate philosopher and mathematician he never would state the law absolutely. He had that scrupulous conscientiousness which prevented him from guessing at the conclusion which no doubt he arrived at. His mind was probably a great deal quicker than are many other minds in which the conclusion is jumped at and given as if it were proved, but he conscientiously avoided stating it as a conclusion, and held it over until exact measurement should prove whether or not it was to be concluded.

The subject of measurement in this case of a null method pointed out by Cavendish was this. If in the interior of a hollow electrified conductor the electric static force upon a small insulated and electrified body is exactly zero, then the law of variation of the electric force must be according to the inverse square of the distance. On the other hand, if a certain attraction of a small positively electrified body towards the sides of the supposed hollow electrified conductor is observed, then the force varies according to a law of greater variation of the distance than the inverse square of the distance ; and *vice versa* if a small body electrified in the opposite way to the electrification of the conductor seems to be repelled from the sides, then the law of diminution of force with the distance will be something less than would be calculated according to the inverse square of the distance. The case supposed is an insulated electrified body—an infinitely small body—charged with electricity opposite to that of the electrified body. If this small body, then, put into the interior as a test, exhibits attraction towards the sides, the law of variation of the force will shew a greater diminution according to the distance than according to the inverse square of the distance, and *vice versa.* It was left for Faraday to make with accuracy the concluding experiment which crowned Cavendish's theory. Faraday found by the most thoroughly searching investigation that the electrical force in the circumstances supposed was zero, and supplied the minor proposition of Cavendish's syllogism. Therefore the law of force varies with the inverse square of the distance. This result was obtained with far less searching accuracy by Coulomb and Robinson, because their method did not admit of the same searching accuracy. On this law is founded the whole system of electrostatic measurement in absolute measure. Mathematical theory lays down the proper static unit—that quantity, which if a quantity equal to it is possessed by two bodies, those two bodies react upon one another with unit force at unit distance. On this is founded the system of absolute measurement in electrostatics.

Cavendish's other experiments, and series of experiments—because I believe Professor Clerk-Maxwell is to edit a whole series of experiments measuring electrostatic quantities—led to the general system of electrostatic measurement in absolute measure.

But now there is another great branch of electrical measurement, and that is the measurement of electro-magnetic phenomena. Our elementary knowledge of electrostatics was complete, with the exception of this minor proposition of Cavendish's syllogism, and the great physical discovery of Faraday of the peculiar inductive quality known as the electrostatic inductive capacity of dielectrics. With these two exceptions the whole theory of electrostatics was completed in the last century. It was left for us to work out the mathematical conclusions from the theory of Cavendish, Coloumb, and Robinson; and it was not until after the end of the last century that the existence of electro-magnetic force became known. Orsted made the great discovery in 1820 of the mutual connection between a magnet and a wire in which an electrical current is flowing, and the remarkable developments which were very speedily given to that discovery by Ampère, led to the foundation of the other great branch of electrical science, and pointed to the subject of electro-magnetic measurement, upon which I must say a word or two now.

I think the principles of the mathematical theory of the interaction of wires containing currents mutually between one another, and again their mutual action upon magnets, was fully laid down by Ampère in consequence of Örsted's discovery. The working out of the accurate measurement of currents, and generally of the system of measurement founded on these principles, was founded altogether in Germany. The great work of Gauss and Weber on terrestrial magnetism belongs strictly to this subject. I believe Gauss first laid down the system of absolute measurement for magnetic force. The definitions and mathematical theory, of Poisson and Coulomb as to magnetic polarity, and the magnetic force founded on it, was worked out practically by Gauss, and made the foundation of the whole system of magnetic measurement followed in our magnetic observatories. This was an immense step in science, and one of great importance, giving, not merely definite measurement, but measurement on a certain absolute scale, which, even if all the instruments by which the measurements were made were destroyed, would still give us a perfectly definite result. It was, for the first time in physical science, worked out in consequence of Gauss' foundation of the system for terrestrial magnetism. That, then, is really the beginning of absolute measurement in magnetic science, and for electro-magnetics and electrostatic science. Gauss and Weber carried on the work for terrestrial magnetism together, and Weber carried on by himself, I believe, during Gauss' lifetime and also after his death—the system of absolute measurement in electrostatics. One most interesting result, brought out by Weber, is that the electric resistance of a wire, in respect of electro currents forced to flow from it, is to be measured in terms of certain absolute units, which lead us to a statement of velocity in units of length per unit of time, as the proper statement for the electro-magnetic measure of the resistance of a wire. It would take too long to occupy your attention on matters of detail if I were to explain minutely how it is that resistance is to be measured by velocity. It seems curious, but you will form a very general idea of it in this way. Suppose you have two vertical copper bars and a little transverse horizontal bar, placed so as to press upon those two bars. Let the plane of those two bars be perpendicular to the magnetic meridian; place then a little transverse bar, like one step of a ladder, across the two vertical bars. Let this bar be moved rapidly upwards; being moved across the line of the horizontal component of the earth's magnetic force, it will, according to one of Faraday's discoveries, experience an inductive effect, according to which one end of it will become positively electrified, and the other negatively. Now, let the two bars upon which this presses be connected together: then the tendency I have spoken of will give rise to a current. That current may be made, as in Orsted's discovery, to cause the deflection of a galvanometer needle. Now, you will see how resistance may be measured by velocity. Let the velocity of the motion of this little bar, moved upwards in the manner I have described, be such as to produce in the galvanometer a deflection of exactly 45°. Then the velocity, which gives that deflection, measures the resistance in the circuit, provided always the galvanometer be arranged to fulfil a certain definite condition as to dimensions. The essential point of this statement is that the result is independent of the magnitude of the horizontal force of the earth's magnetism. The galvanometer needle is attracted by the horizontal magnetic force of the earth. Let us suppose that to be doubled; the attracting force in the needle is doubled, but the inductive effect is doubled also, and, therefore, the same velocity which causes the needle of the galvanometer to be deflected 45° with one amount of magnetic force of the earth, will cause the needle to be deflected by the same number of degrees, with a different amount of magnetic force of the earth. Thus, independently of any absolute measurement of the terrestrial magnetic force, we get a certain velocity which gives a certain result. Thus, it is that velocity is the proper measure of the resistance of a metallic circuit to the flow of a current through it. Going now to electrostatics, — the resistance of an insulator to the transmission of electricity along it, may be measured in a curious manner in connection with the velocity. It may be measured by the reciprocal of a velocity, or in other words, the conducting power of a wire may be measured, with reference to the electrostatic phenomena, by a velocity. Thus, imagine a globe in the centre of this room, at a great distance from the walls. Imagine that globe to be two metres in diameter and one metre in radius; let it be electrified, and hung on a fine silk thread, perfectly dry, so as to insulate perfectly. There we have a perfectly insulated globe in the middle of this room. Now if you apply an excessively fine wire, say a wire one ten-thousandth of an inch in diameter, to the globe, and bring that to a plate of metal connected with the walls of the room, or you may suppose the walls of the room to be metallic, so that we may have no confusion owing to the imperfect conductors. When you apply this very fine wire connecting the insulated globe with the walls of the room, the globe instantly loses its electricity. By instantly, I mean in so short a time as would be impossible to measure by any method we could apply—I mean a time as small as one-millionth of a second—the globe would lose its electricity, if we had connected to it ten or twenty yards of the finest wire we could imagine. But suppose the wire a million times finer (if we can suppose that) than we can apply, the same thing would happen. Or take a cotton thread, and get such a globe as I have been imagining, surrounded with metallic walls, that moist cotton thread will gradually diselectrify it; in a quarter of a minute the globe will have lost perhaps half its electricity, in another quarter half of the remainder, and so on. If the resistance of the conductor I have supposed is constant, the loss will follow the compound interest law—so much per cent. per second of the charge will be lost. Now imagine a conductor of perfectly constant resistance to be put between the ideal globe and the walls of the room, and imagine the globe to be connected with one of these electrometers—of which I shall say a word in conclusion—by an excessively fine wire going into the instrument, and suppose the electrometer to indicate a certain degree—a potential, as we now call it—that subject of electric measurement really discovered by Cavendish in his measurement of electric capacity. Now suppose, then, we are measuring the electric value—the potential of the charge in the globe by an electrometer, then we shall see the electrometer indications decreasing and the potential gradually going down according to the logarithmic, or compound interest law, in the circumstances I have supposed. But instead of this being carried out, let us suppose the following conditions, which we can imagine, although it would be impossible for any mechanician to execute it. Let the globe by some imaginary means be gradually diminished in its diameter. Suppose, in the first place, the insulation to be exceedingly perfect, and suppose the resistance of the conducting wire to be enormously great, so that in the course of a minute or two there is but little loss of potential. Now let this globe, which is supposed to be shrinkable or extendable at pleasure, be shrunk from the metre radius to 90 centimetres radius, what will the effect be? The effect will be that the potential will increase in the ratio of 90 to 100. Shrink the globe to half its dimensions the potential will be double, and so on. That follows from the result of the mathematical theory that the electrostatic capacity of a globe is numerically equal to its radius. Now, while the globe is charged let its radius be diminished. Let the globe shrink at such a speed that the potential shall remain constant. There, then, you can imagine a globe losing a constant quantity of electricity per unit of time, because it is kept now at a constant potential. A globe kept by this wonderful shrinking mechanism at a constant potential will lose a constant quantity of electricity per unit of time, losing in equal times equal quantities; and the globe going on shrinking and shrinking so as to keep a constant potential, the velocity with which the surface approaches the centre measures the conducting power of the wire in absolute electrostatic measurement. So, then, we have the very curious result that according to the electrostatic law of the phenomena we can measure in terms of electrostatic principles the conducting power of a wire by the velocity. Although I have put an altogether ideal case to you, it would be very wrong for me to allow you to suppose that this is an ideal kind of measurement; in point of fact, we measure regularly in electrostatic measurement the capacity of the Leyden jars in that way, and in future when anyone goes to buy a Leyden jar of an optician, let him tell you to give him one of one or two metres or whatever it may be, and require him to find out how to produce it. I give that as a hint to anyone interested in electrostatic apparatus, or in the furnishing of laboratories. There is no likelihood that he will understand what you mean, but perhaps if you teach him a little he will soon come to understand it, and I hope in ten years hence, in every optician's shop where Leyden jars are sold, there will be a label put on each jar saying the capacity is so many centimetres. It could be done to-morrow. We have all the means of doing it, only we have not the knowledge.

The relation between electrostatic measurement and electro-magnetic measurement is very interesting, and here from the supposed uninteresting realms of minute and accurate measurement we are led to the depths of science, and to look at the great things of Nature. These old measurements of Weber led to an approximate determination of the particular velocity at which the electro-magnetic resistance is numerically equal to the electrostatic conducting power of a wire. The particular degree of resistance of a wire which shall be such that the velocity which measures the resistance in electro-magnetic measure shall be the same as the velocity which measures the conducting power in electrostatic measure, was worked out by Weber, and he found that velocity to be just about 300 kilometres per second. I unhappily have British statute miles in my mind, through the misfortune of being born thirty years too soon, and I remember the velocity of light in British statute miles. That used to be considered about 192,000 miles per second, but more recent observations have brought it down to about 187,000. Now I think the equivalent of that in metres is about 300 kilometres per second, and that was the number found by Weber. Professor Maxwell gave a theory leading towards a dynamical theory of magnetism, part of which suggested to him that the velocity for which the one measure is equal to the other in the manner I have explained should be the velocity of light. This brilliant suggestion has attracted great attention, and has rendered it an object of intense interest, not merely for the sake of accurate electro-magnetic and electrostatic measuring, the measuring with great accuracy the relation between electrostatic and electro-magnetic units, but also in connection with physical theory. It seems that the more accurate an experiment up to the present time, the more nearly does the result approach to being equal to the velocity of light, but still we must hold opinion in reserve before we can say that. The result has to be much closer than it is shown by the experiments already made. But you can all see by the mere mention of such a subject how intensely interesting the pursuing of these investigations further must be. I believe Maxwell at present is making a measurement of this kind on a different plan from any that have been yet made. I have now spoken too long, or I should have described something of the methods already followed in this department, but they are already fully published, and can easily be referred to.

Now with respect to accurate measurement—theory was left far behind by practice, and I need not to be reminded by the presence of our President how very much more accurate were the measurements of resistance in the practical telegraphy of Dr. Werner Siemens and his brother, our President, than in any laboratory of theoretical science. When in the laboratory of theoretical science it had not been discovered that the conductivity of different specimens of copper differed at all, in practical telegraphy workshops they were found to differ by from thirty to forty per cent. When differences amounting to so much were overlooked —when their very existence was not known to scientific electricians, the great founders of accurate measurement in telegraphy were establishing the standards of resistance accurate to one-tenth per cent. Dr. Werner Siemens and our President were among the first to give accurate standards of resistance, and the very first to give an accurate system of units founded upon those standards. This Siemens unit is still well known, and many of the most important measurements in connection with submarine cables are stated in terms of that unit. By a coincidence, which in one respect is a happy one, although there is something to be said on the other side, the unit adopted by Messrs. Siemens founded on the measurement of a certain column of mercury produces and reproduces in the accurate resistance coils—the Siemens unit approaches somewhat nearly to the unit which in Weber's system would be 10^{10} or a thousand million centimetres per second. This is so far convenient that measurements in Siemens units are very easily reduced to the absolute measure. The committee of the British

Association, of which our President was one, and I also had the honor to be a member, proposed a method of measurement which was carried out chiefly by Professors Clerk-Maxwell, Balfour Stewart and Jenkins, who laid down what is called the British Association unit to which the name, according to the advice of Mr. Latimer Clark, of "Ohm," was given in commemoration of one of the great founders of electro-magnetic science, Ohm being the man who gave us the first law of currents in connection with electro-motive force, it was considered appropriate that his name should be given to the electric unit, but I may mention as a matter of great importance and interest in physical science, a revision of the measurement of the British Association unit is being undertaken. There is now an endeavour to measure with the greatest possible accuracy what is the value of the "Ohm" in terms of the absolute scale of centimetres per second. It will certainly come within a small percentage of being exactly a thousand metres per second. One per cent. away from that amount, it may, but that it is two or three per cent. or four per cent. or one-third per cent. is of course possible as anyone may judge by looking at the difficulties that will have to be met with in making the experiments. I will just say in connection with the electro measurement that it touches on another point of measurement, that of heat. Joule in a quite independent set of experiments which I can only name, showed another way of arriving at similar results, and Joule's electro-magnetic experiments taken in connection with other experiments of his on the dynamical equivalent of heat, show some disagreement from the British Association measurement of their unit of resistance. There is something to be reconciled here. Joule on the one side holds that the British Association unit, the Ohm, is a little too much or rather too little, I forget which, but on the other side, in Germany, Kohlrausch holds the Ohm to be a little on the other side of the exact thousand million centimetres per second. I believe if you eliminate doubt by the method of averages, Kohlrausch and Joule's experiments would show the British Association to be very nearly right, but I do not approve of that method of removing doubts, and we shall not be satisfied until both Joule and Kohlrausch are satisfied.

I will now mention a number of experiments with electrometers which, I am afraid, are of little interest to anyone in the world, but myself. Here is the first attempt at a quadrant electrometer, but it is exhibited below. It is well known now to many electricians, and a descriptive pamphlet also accompanies it. I really do not know, considering that the British Association report on electrometers has been republished in connection with the whole series of their reports, that I need go into detail with respect to any of these instruments. This is the very first portable electrometer, and I will tell you how it came into existence. I had one that I was very proud of, I am ashamed to say in these days. I was proud of its smallness, and how easily it could be carried up to the top of Goatfell and back; and there was one before then, the highest character of which was, that it was heavier than a rifle; but, that was in the days of what Lord Palmerston called the "rifle fever," and I was touched a little with that at the time, being a rifle volunteer; and I found that my electrometer weighed a pound less than my weapon. It only weighed thirteen lbs., and the rifle weighed fourteen lbs. I had that at Aberdeen, but it is not now to be found, although it has been searched for, or it would have been exhibited. Part of it, the stand that was on the top of it, is below. The next that followed, was this one. I got down the weight to about one-half, and I was perfectly satisfied then, and this one has gone up the Goatfell a great many times; but it is fully described in my book, and in the paper I have referred to. I was showing it with great pride on one occasion to Professor Tait, and I said to him: "You should get one like that." He said, "I will wait until you can get one that you can put into your pocket. Get one the size of an orange, and then I will have it." That literally was the origin of this electrometer. I felt rather challenged by what he said, and in the course of my next run up to Glasgow, Mr. White, who is so indefatigable in making new things, and who has so admirable an inventive capacity, helped me in my endeavour, and we had something like this one. In the course of a month, this very electrometer was got into action. This is the first attracted disc electrometer. It differs from the portable electrometers now known merely in some minor details; the moveable disc turns round with a micrometer screw instead of moving up and down in a slide. In all other respects, it is the same, except the awkward arrangement for placing the pumice, which with my great care, did not lead to any accident, but with almost any other person led to the instrument being destroyed by the sulphuric acid placed on it getting shaken down into the instrument below. The more convenient arrangement of the pumice is now made, but that is the only alteration. The mechanical arrangement of the disc is only changed in the portable electrometer as it now exists, and two specimens have been sent out to the Arctic expedition. Just one word of practical advice, with respect to the electrometers. I have been continually asked how to keep them in order, and have frequently heard complaints that these will not hold; that they do not retain the charge. In each of these electrometers there is a porous jar, the heterostatic system being adopted in each of them. It is necessary the insulating should be perfect, and then it all depends afterwards on the cleanness of the surface of the glass. If you will allow me to use the phrase of Lord Palmerston, with regard to water, when he said that, "Dirt is matter in its wrong place," and to consider that water, or any moisture on the face of the glass—which ought to be perfectly dry—is in its wrong place, and is, therefore, dirt, you will understand what I mean. If there is no dirt on the glass it is certain to insulate well. But then how to get the glass perfectly clean? In the first place, wash it well with soap and water. If you like you may try nitric acid, and then pure water, or you may wash it with alcohol, and then with pure water. I have gone through almost incantations to get perfect cleanness of the surface of the glass, but I doubt much whether I have got any result which I could not have got with soap and water, and then running pure water over the surface of the glass. After it is done, wash it well, somehow or other. You may use acids, or alcohol, if you like; but I think you will generally find that soap and water, and enough clean water, to end with, will answer as well as anything. Then shake it well, and get it well dry, but do not use a duster, however clean, to dry it. Shake the moisture off, and take a little piece of blotting paper, and suck up very carefully any little portion of water which may remain by cohesion, but do not rub it with anything that can leave shreds or fibres; that is dirt. The finest cambric will leave on the glass what will answer Lord Palmerston's definition. When you have got the glass clean of everything except water, then dry it, and you will sure to find it answer. The way to dry it, and to keep it dry, is to have the sulphuric acid in the proper receptacle. Each of these instruments has a receptacle for sulphuric acid, which must be freed from volatile vapours by a proper process; boiling with sulphate of ammonia suffices. The sulphuric acid need not be chemically pure, but it must be purified from volatile vapours, and it must be very strong. I believe, oftener than from any other cause, these instruments fail to hold well because the sulphuric acid is not strong enough, and frequently, when an electrometer has failed, by putting in stronger acid, the defect has been perfectly remedied.

The PRESIDENT, in rising to move a vote of thanks to Sir William Thomson for his very profound observations on a subject which is, perhaps, one of the most difficult in physical science, said: he would only allow myself to make one criticism, namely, that Sir William Thomson had dealt more with the labours of others than with his own. We all know that he has worked more than any other living man to bring theory and practice into one focus regarding this subject of applied electricity. There are many apparatus with which his name is connected, and will always be connected, such as electrometers for the measurement of currents, and apparatus for working long submarine lines. These he has not dwelt upon, except in a very cursory manner. Regarding the electrometer, I heard Sir William Thomson give a lecture at the Royal Institution, some years ago, when he brought some Scottish electricity to London in this very instrument. These electrometers have had not only a theoretical importance, but are of practical uses to the electrician, especially one variety of them, which is used largely by those who apply electricity to practical work. Without occupying your time longer, after listening to so lucid an explanation, I beg to call upon you to pass a hearty vote of thanks to Sir William Thomson.

The vote of thanks having been passed,

The PRESIDENT said: I will now call on M. Tresca to give us a discourse upon his very remarkable experiments on the flow of solids. This is a subject which is peculiarly the work of M. Tresca, of whom we have all of us more or less heard, but we have not all of us had the author himself to give us a lucid explanation of them, as I am sure he will do now.

On the Fluidity and Flow of Solid Bodies.

M. Tresca: Gentlemen, when ten years ago, after many careful experiments, I made use of, and commented on the scientific expression, Flow of Solid Bodies, my first communications were not received without some shadow of incredulity. I, therefore, feel it my duty to mention with gratitude the names of Mr. Tyndall, Mr. Fairbairn, and Mr. Scott Russell among the scientific men, who at the very outset, interested themselves in this subject. I should wish to thank them in your language, but I am afraid that I am not sufficiently familiar with it, and I therefore rely on your indulgence for allowing me to address you in French.

The question of the flow of solid bodies has been a great success; it is, thanks to it that I now hold a position to which I should never have dared to aspire, and which allows me to represent French science at this Exhibition. We have been most desirous, I can assure you, to afford you heartily all the help that lay within our power.

The principal fact connected with the flow of solid bodies was very simple. If a resisting mass, enclosed in a wrapper, be submitted to an exterior action of sufficient power, it will exert in every direction a greater or less pressure, and if a hole be made in the wrapper, the matter will escape through this aperture, forming a jet, which is made up of different portions of the mass. When the latter is homogeneous, and the shape of the mass is regular, and the hole is in a certain symmetrical position with regard to it, the mode of final distribution may be deduced from the mode of initial distribution, and the first experiments made will determine the kinematical conditions of such a flow.

Thus, in the formation of a cylindrical block by the superposition of a certain number of slabs of lead, it was found that each one of the slabs penetrated by turn into the jet, and formed there a concentric tube when the aperture itself was concentric.

Why were we so astonished to find, on cutting, according to a meridian plane the so formed jet, that it was composed of as many continuous tubes as there were slabs, until the complete exhaustion of the matter which supplied food for its formation? Could, indeed, the perfectly geometrical position of the natural phenomena be explained in any other way? It is always as you see, when the regularity could not be more perfect, it appears to us like the most absolute evidence. The indefinitely reducible molecular formation, could not, assuredly, be better justified than by means of these parallel walls, which preserve to a microscopic thickness natural tissues, and an equality of appearance and movements which are very remarkable.

These experiments in concentric flow have, with regard to the relative displacements, been submitted to calculations, and we are now able, in such a deformation, to lay down the trajectory passage of each one of the molecules of the mass, and establish with certainty the final places which they will occupy, compared to those in which they were at first; and also, by the same means, the transformed of any line or of any surface to their first position.

These experiments became far more convincing when the shape and position of the apertures are varied; tubes still replace the slabs, but the relative thickness of these tubes, the juxtapositions and the convolutions which result are just as instructive with regard to the clearness of their shapes and the distribution of the pressure in the whole extent of the mass which is escaping. But this question will occupy our attention more especially in the study of "punching."

If a punch be driven into a plate of metal, it will propel before it the material of which the plate is composed, which, at a given moment begins to form a protuberance on the opposite site, and finally detaches itself in the form of a cylinder of the same diameter as the punch, and to which the very characteristic name of "debouchure" has been given. Under certain thicknesses the "debouchure" preserves a height equal to the thickness of the plate, but this is no longer the case when the thickness of the plate is very high.

A "debouchure" one centimeter high has been made by a block five centimeters thick. If the block be formed of several plates laid one over the other, all these plates will be represented in the "debouchure." The lower plates, pretty nearly, by their original thickness; the superior plates by a kind of convex lens, of which the curved side has for basis the flat side which has remained in contact with the punch, and which has kept, according to the common axis, almost the whole of its original thickness; the intermediate plates by a number of cups pressed one into another, which reach the same flat side, and the lower parts of which have variable thicknesses, which become extremely thin at a certain point of the axis of the "debouchure."

This point is the principal centre of the flow which has been produced from the axis to the circumference by the pressures caused and transmitted by the punch to the very interior of the mass. When this flow has been able to take place, the virtual resistance of the partition to be punched through was necessarily greater than the latteral resistance; but these two resistances will, on the contrary, be of the same force when the "debouchure" begins to detach itself, and the effort necessary to continue the punching will go on, after this, quickly decreasing to the end of the operation.

The mode of action of these resistances, which we can determine from the facts themselves, has furnished us with the true knowledge of the action of the transmitted pressures, and we have been able to formulate the laws of the transmission of pressures in a solid mass in course of deformation, in a zone more or less extended which we have named the zone of activity. And, finally, by proving the force necessary to be exerted at the moment when the "debouchure" separates itself, we have been able to determine the true co-efficients of transversal cuttings. They are as follows, for each square centimeter:

	Kilog.
Lead	182
Tin	209
Alloy of Lead and Tin	230
Zinc	900
Copper	1893
Iron	3757

It is worthy of remark that these numbers come very near those of the resistance to rupture by extension for each of the metals experimented on.

The results of the calculations, with regard to the transmission of mechanical work, from one layer to another next to it, allow us to lay down the mathematical law for the distribution of pressures, not only for punching, but also for the various methods of deformation previously studied. The formula arrived at by this method have been subsequently justified by M. de Saint Venant and by M. Lévy; quite lately M. Boussinescq, of the Faculty of Lille, announced to us that he has reached the same conclusions by purely mechanical theories, and from this moment, therefore, they can be considered as belonging to science in the most complete manner.

Another kind of research has occupied us for a long time ; it is that connected with the planing of metals. The shaving which the tool carries away, and which up to that time had not even attracted the attention of practitioners, offers many remarkable peculiarities. Not only does it often twist itself spirally, which is the result of the unequal shortenings of the different files of molecules of which it is composed, but speaking generally, it can be calculated by how much it will be shortened in a given case. It is usually reduced about a fifth of its original length under ordinary circumstances ; and we can also consequently tell by how much its transversal dimension will increase during the same time, its density not having varied to any sensible extent. This shaving, which is the "transformée of the prism" carried off by the graver, increases in thickness in an inverse ratio to the variations of its length. With regard to those complex convolutions, in which the definite forms no longer depend on those of the moulds, of the drawing-frames, or of the punches of which we have spoken above, but merely on the exterior and interior forces which are brought into play, we have not yet been able to characterize them with certainty, except in a small number of cases. Here, however, is a model, which shows how matter, driven back by a tool of a peculiar and symmetrical shape, passes from a square section to a triangular section of much greater dimensions, under the influence of pressures transmitted in passing through geometrical mediums, traced according to a theoretical diagram and belonging to analytically defined surfaces. Many a discovery will be made in molecular mechanics by following up these phenomena, which have been so entirely overlooked up to the present time, and which are so closely allied to the properties of matter.

Smelting is also one of the means that can be employed in this attractive study of molecular mechanic. A bar of iron is simply a mass of agglutinated filaments placed in juxtaposition, which proceed individually from a determined nucleus. In proportion as the number of the filaments is increased, and the mass formed by their juxtaposition is stretched out, they become more and more tenuous, and it is easy, by special means of oxydation, to discover, in a manufactured piece of iron, all the accidents of the various processes which have been used for its formation. You can, indeed, in smelting, make these filaments expand or concentrate them ; you can unite or separate them ; but you cannot make them disappear ; and we have here another mode of investigation, which must be made use of for the study of the interior convolutions which correspond to certain exterior changes of shape.

When once all the relative displacements, as well as the force necessary to produce them are known, there will be no great difficulty in determining the best way of carrying them out, and of calculating the mechanical work which it will demand.

In one ot our smelting experiments, made on a large scale upon brightened platinum, an accessory phenomenon presented itself, which attracted the whole of our attention, and which is so closely connected with the deformation of solid bodies, that you will allow me, I am sure, to say a few words about it, although the experiments which have originated from it are not yet finished. It is, moreover, a great satisfaction to me to be able to communicate to a gathering of English scientific men the first results of these experiments, before any publication whatever of them has been made.

On the 8th of June, 1874, we only pointed out to the Academy of Sciences the principal fact : when the bar of platinum, at the moment of smelting, had already cooled down to a temperature below that of red heat, it has happened several times, that the blow of the stamp hammer which occasioned, at the same moment, in this bar, both a local depression and a lengthening, heated it anew along two inclined lines, forming upon the sides of the piece the two diagonals of the depressed portion, and this re-heating was so great, that the metal along these two lines was brought back to the red hot temperature, plainly enough for its form to be distinguished. These lines of greater heat even remained luminous for several instants, and presented the appearance of the two jambs of the letter X. In some cases we have been able to count simultaneously as many as six of these X, produced one after the other, as the bar under operation was being moved, in order to draw it out by degrees to a certain part of its length.

The explanation of these luminous traces could not for an instant be doubtful ; the lines of greater slipping were, at the same time, the zones ot greater developed heat, and we had before ûs a perfectly definte thermo-dynamical fact. If this fact had never been observed before, it was clearly owing to the circumstances necessary for its manifestation never having been all united together in so favourable a manner. Brightened platinum requires for its deformation a great amount of work. Its surface is unalterable and almost translucent when the metal is heated to a red hot temperature ; it is but a moderately good conductor of heat ; its calorific capability is somewhat feeble ; all conditions rendering the phenomena visible in the smelting of platinum whilst it has passed unnoticed in the case of all the other metals.

But, although anticipating this explanation, we, nevertheless, felt it our duty to prove it by more direct experiments, of which I shall now speak, and which constitute the chief novelty—may I say the chief interest—of this communication.

Given a metallic bar at the ordinary temperature—if two of its lateral surfaces be coated with wax or tallow, and is then subjected to the action of a single stroke of a ram, the wax will melt opposite the depression which is produced ; and it is proved that, in certain cases, this melted wax takes the shape of the two arms of the letter X, which we have noticed on the platinum ; in many cases, the jambs are curved, having their convex sides in front. This happens when the heat has spread to a greater extent, and the wax has melted in the whole of the space between them.

The prism which has this line for its base, and the width of the bar for its height, represents a certain bulk and a certain weight, and if it be admitted that the whole of it has been raised to the temperature of the melted wax, this rise in the temperature must represent a certain quantity of heat, or, by its mechanical equivalent, a certain quantity of interior work, which is directly proved by the experiment.

By comparing this converted work to the work furnished by the fall of the ram, we find a co-efficient of mechanical return, which is not less than 70 per cent. We do not consider this number definitive ; it depends upon the conductibility of the metal, the solidity of the apparatus used in carrying out the experiment, the cleanliness of the surroundings of the melted surface. But what I wished to point out, Gentlemen, is, that we have come back to M. Joule's first methods, and that our labours on the flow of solid bodies are already bringing us back to the verification of some thermo-dynamical statements.

We had, at the very beginning of these researches, expressed our opinion, that the phenomena observed in glaciers are, in more than one respect, due to the mechanical deformation of the ice which is subjected to the action of the immense burdens which its plasticity allows it to transmit in all directions.

Quite recently we were called in to help M. Daubrée in his work on the schistosity of rocks. The experiments made on this subject do not let the slightest doubt remain as to the ease with which the very least relative motion, caused in any mass whatever, were it even absolutely homogeneous, would bring about perfectly manifest sliding tendencies. The best way to make this most perfectly obvious is to scatter in the mass little spangles of mica, which range themselves in a line as exactly as may be observed in the micaschists following the very plan of each one of the relative motions. In this specimen, for example, which is the result of the flow of a piece of clay earth, through the aperture of a rectangular drawing-frame, not a spangle can be seen in the transversal fractures, whilst they all show themselves perfectly ranged, one after another, on the clefts which are easily produced longitudinally by taking advantage of the schistosity which earth preserves as distinctly after its desiccation as in the schists of the different formations.

On the other hand, the crushing of blocks of earth or of lead between compression moulds which allow the matter to escape in one direction only, has given us strikingly true representations of the straightening of certain strata, their convolutions with schistose qualities that are distributed throughout the body, and finally real upheavings, of which the bends and the rents are equally satisfactory imitations.

We are therefore quite authorised in saying that the study of the flow of solid bodies, has already thrown some light upon geological phenomena, or more generally, upon the phenomena of inorganic nature. The question is, whether this will likewise be the case, in the future, with regard to organic phenomena. The matter, however, cannot yet be investigated scientifically: we only know that vegetable growth is carried on by the formation of cells which always arrange themselves in groups according to a tubular order, that certain animal developments, such as the horny tissues, assume, during their growth, forms which are invariably the same, and which seem to be derived from their first disposition. Just as if, in either state, the final arrangement were carried by means of the circulation of nutritive fluids into casings formed by the tissues, and already completed and prepared to receive them.

However distant this similarity may be, which is, to a certain extent, the predominant characteristic of all organic formations—I was going to say, the mechanical characteristic of the phenomenon of the development of all objects in the vegetable and animal kingdom—it is nevertheless by no means unreasonable to suppose that the day will come when these phenomena will be explained, at least in their essentially mechanical characteristic, and independently of all other purely physiological and the more frequently predominant actions, by considerations similar to those which you have been kind enough to listen to to-day.

[*It has been thought desirable that this important communication should also be given in the original French*].

SUR LA FLUIDITÉ ET L'ÉCOULEMENT DES CORPS SOLIDES.

Lorsqu'il y a dix ans, à la suite d'expériences soigneusement répétées, je pensai devoir associer, dans le langage de la science, ces deux mots : écoulement des corps solides (flow of solid bodies), mes premières communications ne furent pas accueillies sans quelque nuance d'incrédulité, et c'est pour moi un devoir de reconnaissance de citer ici M. Tyndall, M. Fairbairn et M. Scott Russell parmi les savants qui les ont tout d'abord considérées avec intérêt. Je voudrais les remercier dans votre langue, mais elle ne m'est pas assez familière et je compte sur votre indulgence pour me permettre de vous entretenir en Français.

Cependant, la question de l'écoulement des corps solides a fait fortune, elle m'a valu des suffrages que je n'aurais pas osé ambitionner, et qui me permettent aujourd'hui de saluer votre exposition au nom de la science Française, qui a tenu à vous prêter à son sujet une large et sympathique collaboration.

Le fait principal de l'écoulement des corps solides était bien simple : lorsqu'une masse résistante, enfermée dans une enveloppe, est soumise à une action extérieure d'une intensité suffisante, elle transmet, dans tous les sens, des pressions plus ou moins grandes, et si la paroi est percée d'un orifice, la matière s'échappera par cet orifice, en formant un jet aux dépens des différentes parties de la masse. Lorsque celle-ci est homogène, lorsque la forme de la masse est régulière, lorsque l'orifice est placé par rapport à elle dans certaines conditions de symétrie, le mode de répartition finale peut être déduit du mode de répartition initiale, et les premières expériences faites décident les conditions cinématiques d'un tel écoulement.

C'est ainsi qu'en constituant un bloc cylindrique par la superposition d'un certain nombre de plaques de plomb, on a pu reconnaître que chacune de ces plaques venait à son tour pénétrer dans le jet et y former un tube concentrique, lorsque l'orifice est lui même concentrique.

Pourquoi avons-nous éprouvé tant d'étonnement lorsqu'en coupant, suivant un plan méridien, le jet ainsi formé, nous l'avons trouvé composé d'autant de tubes continus qu'il y avait de plaques, jusqu'à l'épuisement complet de la matière qui en avait alimenté la formation ? La précision toute géométrique des phénomènes naturels pouvait elle vraiment se traduire d'une autre façon ? Toujours est-il, comme vous le voyez, que la régularité ne saurait être plus complète, qu'elle nous apparaît avec le caractère de l'évidence la plus absolue. La constitution moléculaire indéfiniment réductible, ne saurait assurément être mieux justifiée que par ces parois parallèles, conservant jusqu'à l'épaisseur microscopique, qui sont si souvent l'apanage des tissus naturels, une égalité d'allure et de direction bien remarquable.

Ces expériences d'écoulement concentrique ont été aussitôt, au point de vue des déplacements relatifs, soumises au calcul, et nous pouvons, aujourd'hui, dans une telle déformation, assigner sa trajectoire de chacune des molécules de la masse et fixer avec assurance sa position finale par rapport à celle qu'elle occupait dans la masse primitive ; par les mêmes moyens aussi, les transformées de toute ligne ou de toute surface définie dans sa position première.

Ces expériences deviennent bien autrement probantes lorsqu'on varie la forme et la position des orifices ; ce sont toujours des tubes qui répondent aux plaques primitives, mais les épaisseurs relatives des parois de ces tubes, les juxtapositions et les contournements qui en résultent sont aussi instructifs, au point de vue de la netteté de leurs formes et de la répartition des pressions dans toute l'étendue de la masse qui s'écoule. Cette question, d'ailleurs, va nous occuper d'une manière plus spéciale dans l'étude du poinçonnage.

Lorsqu'un poinçon pénètre dans une plaque métallique, il chasse devant lui la matière de cette plaque, qui commence, à un moment déterminé, par former une protubérance sur la face opposée, et qui finit par s'en détacher, sous forme d'un cylindre de même diamètre que le poinçon, et que l'on désigne sous le nom bien caractéristique de débouchure. Au dessous de certaines épaisseurs, la débouchure conserve une hauteur égale à la dimension de la plaque en épaisseur, mais il n'en est plus ainsi lorsque la plaque est beaucoup plus épaisse.

Une débouchure de 1 centimètre de hauteur a été fournie par un bloc d'une épaisseur de 5 centimètres. Si le bloc est formé de plaques superposées, toutes les plaques sont cependant représentées dans cette débouchure : les plaques inférieures, à peu de chose près, par leurs épaisseurs primitives ; la plaque supérieure par une sorte de lentille plan-convexe, dont la face courbe a pour base la face plane, restée en contact avec le poinçon, et qui a conservé, suivant l'axe commun, la presque totalité de son épaisseur primitive ; les plaques intermédiaires par autant de gobelets emboutis les uns dans les autres, aboutissant à la même face plane, et dont les fonds ont des épaisseurs variables, d'une minceur extrême sur un certain point de l'axe de la débouchure.

Ce point est le centre principal de l'écoulement qui s'est produit de l'axe à la circonférence, sous l'influence des pressions déterminées et transmises par le poinçon à l'intérieur même de la masse. Lorsque cet écoulement a pu se produire, la résistance verticale de la cloison à poinçonner était nécessairement plus grande que la résistance latérale ; les deux résistances seront, au contraire, du même ordre lorsque la débouchure commencera à se détacher, et l'effort à faire pour continuer le poinçonnage ira ensuite en diminuant rapidement jusqu'à la fin de l'opération.

Ce jeu des résistances, que nous pouvons conclure des faits eux-mêmes, nous a donné la vraie connaissance du jeu des pressions transmises, et nous avons pu formuler dès lors les lois de la transmission des pressions dans une masse solide en voie de déformation, dans toute une zone plus ou moins étendue, que nous avons appelée la zone d'activité. Enfin, en constatant l'effort à exercer au moment où la débouchure se sépare, nous avons pu déterminer les véritables coefficients de cisaillement, qui sont les suivants, par centimètre carré :

	Kilog.
Plomb	182
Etain	209
Alliage de plomb et d'étain	239
Zinc	900
Cuivre	1893
Fer	3757

Il est à remarquer, que ces chiffres se rapprochent beaucoup de ceux de la résistance à la rupture par extension, pour chacun des métaux expérimentés.

Les résultats du calcul de la transmission du travail mécanique d'une couche à la suivante ont permis d'établir la loi mathématique de la répartition des pressions, non-seulement pour le poinçonnage, mais encore pour les différents modes de déformation précédemment étudiés. Les formules indiquées par cette méthode ont été justifiées ultérieurement par M. de Saint-Venant et par M. Lévy. Tout récemment, M. Boussinescq, de la Faculté de Lille, nous annonçait qu'il les avait retrouvées par des considérations exclusivement mécaniques, et elles peuvent dès lors être considérées comme acquises à la science de la façon la plus complète.

Un autre genre de recherches nous a longuement occupé; ce sont celles qui se rapportent au rabotage des métaux. Le copeau que l'outil enlève et qui, jusqu'alors, n'avait pas même attiré l'attention des praticiens, offre des particularités bien remarquables. Non seulement il se vrille souvent en hélice, ce qui est le résultat des raccourcissements inégaux des différentes files de molécules qui le composent, mais, à un point de vue plus général, on peut dire de combien il se raccourcira dans une circonstance donnée, environ des quatre cinquièmes de la longueur primitive dans le cas les plus ordinaires de la pratique, et de combien, par conséquent, sa dimension transversale augmentera en même temps, sa densité n'ayant pas varié sensiblement. Ce copeau, qui est la transformée du prisme enlevé par le burin, augmente en épaisseur en raison inverse de sa variation en longueur. L'examen de ces déformations complexes, dans lesquelles les formes définitives ne dépendent plus de celles des moules, des filières ou des poinçons dont nous avons parlé précédemment, mais seulement des forces extérieures et intérieures qui sont mises en jeu. Nous n'avons pu encore les caractériser sûrement que dans un petit nombre de circonstances ; mais voici cependant un modèle qui montre comment la matière, refoulée par un outil d'une forme particulière et symétrique, passe d'une section carrée à une section triangulaire très agrandie, sous l'influence des pressions transmises, en passant par des intermédiaires géométriques tracés d'après une épure et appartenant à des surfaces analytiquement définies. Plus d'une découverte sera faite en mécanique moléculaire par la poursuite de ces phénomènes, absolument négligés jusqu'ici et qui touchent de si près aux propriétés mêmes de la matière.

Le forgeage est aussi un des moyens qui peuvent être employés dans cette étude si attrayante de la mécanique moléculaire. Une barre de fer n'est rien autre chose qu'un paquet de filaments juxtaposés et agglutinés, provenant individuellement d'un noyau déterminé. A mesure qu'on augmente le nombre des mises, et qu'on étire davantage la masse formée par leur juxtaposition, ces filaments deviennent plus ténus, et il est facile, par des modes spéciaux d'oxydation, de reconnaître dans une pièce fabriquée tous les accidents des différents procédés qui ont été employés pour sa production. Vous pouvez bien, par le forgeage, épanouir ces filaments ou les concentrer, les réunir ou les séparer, mais vous ne pouvez les faire disparaître, et c'est encore là un des moyens d'investigation qui doivent servir à l'étude des déformations intérieures, correspondant à tel ou tel changement de forme extérieure. Une fois que chacun des déplacements relatifs est connu, ainsi que l'effort nécessaire pour le produire, on est bien près de pouvoir déterminer le mode d'exécution le meilleur et d'évaluer le travail mécanique qu'il exige.

Dans l'une de nos expériences de forgeage, faite sur une grande échelle et sur du platine iridié, il s'est présenté un phénomène accessoire qui a nécessairement appelé toute notre attention et qui se rattache si intimement à la déformation des corps solides, que vous me permettrez d'en dire quelques mots, quoique les expériences qui en dérivent ne soient pas encore terminées ; c'est pour moi, d'ailleurs, une grande satisfaction d'en faire connaître, avant toute publication, les premiers résultats devant une assemblée de savants Anglais.

Le 8 Juin, 1874, nous avons seulement indiqué à l'Académie des sciences le fait principal : lorsque la barre de platine, au moment du forgeage, s'était déjà refroidie jusqu'au-dessous de la température rouge, il est arrivé plusieurs fois que le coup de marteau pilon qui determinait simultanément, dans cette barre, une dépression locale et un allongement, se réchauffait suivant deux lignes inclinées formant sur les côtés de la pièce les deux diagonales de la partie déprimée ; et ce réchauffement était tel que le métal était, suivant ces deux lignes, ramené assez franchement à la température rouge pour qu'on pût en distinguer très-nettement la forme. Ces lignes de plus grande chaleur restaient même lumineuses pendant quelques instants et présentaient l'aspect de deux jambages de la lettre X. Dans certaines circonstances, nous avons pu compter simultanément jusqu'à six de ces X produits successivement, les uns à la suite des autres, à mesure que l'on déplaçait la pièce en travail pour l'étirer de proche en proche sur une partie de sa longueur.

L'explication de ces traces lumineuses ne pouvaient faire aucun doute : les lignes des plus grand glissement étaient aussi les zones deplus grande chaleurdéveloppé, et nous étions en présence d'un fait de themodynamique parfaitement défini. Si ce fait n'avait pas été observé encore, cela tenait évidemment à ce que les circonstances nécessaires à sa manifestation ne s'étaient pas trouvées, toutes ensemble, réunies d'une façon aussi favorable. Le platine iridié exige pour sa déformation une grande somme de travail ; sa surface est inaltérable et presque translucide lorsque le métal est porté à la température rouge ; il est médiocrement conducteur de la chaleur ; sa capacité calorifique est assez faible ; toutes conditions pour que le phénomène fût rendu sensible dans le forgeage de ce métal, alors qu'il était resté inaperçu avec tous les autres.

En anticipant cette explication, nous avions cependant pour devoir de la justifier bientôt par des expériences plus directes, dont nous avons maintenant à vous entretenir, et qui constituent la principale nouveauté, oserai-je dire le principal intérêt de cette communication.

Une barre métallique étant donnée, à la température ordinaire, si après l'avoir enduite de cire ou de suif sur deux faces latérales, on la soumet à l'action d'un seul coup de mouton, la cire fond en regard de la dépression produite, et l'on constate que cette cire fondue affecte, dans certains cas, la forme de deux branches de l'X que nous avions observées sur le platine; dans d'autres cas, les jambages sont courbes et présentent en regard leurs convexités; c'est qu'alors la chaleur s'est disséminée davantage et que la cire s'est fondue dans tout l'intervalle qui les sépare.

Le prisme qui a ce tracé pour base et pour hauteur la largeur même de la pièce représente uncertain volume et uncertain poids, et si l'on admet qu'il a été tout entier porté à la température de la cire fondue, cette élévation de température représente une certaine quantité de chaleur, ou, en raison de l'équivalent mécanique, une certainé quantité de travail intérieur qui se trouve directement constatée par l'expérience. En comparant ce travail transformé au travail fourni par la chute du mouton, on trouve un coefficient de rendement mécanique qui n'est pas inférieur à 70 pour 100. Nous ne considérons pas ce chiffre pour définitif; il dépend de la conductibilité du métal, de la stabilité de l'installation, de la netteté des contours de la surface fondue; mais ce que je voulais vous dire, messieurs, c'est que nous voilà revenu aux premières méthodes de M. Joule, et que nos travaux sur l'écoulement des corps solides nous ramènent déjà à quelques constatations thermodynamiques.

Nous avions, dès l'origine de ces recherches, exprimé la pensée que les phénomènes observés dans les glaciers sont, sur plus d'un point, dus à des déformations mécaniques de la glace soumise à l'action des charges immenses que sa plasticité lui permet de transmettre dans tous les sens. Il nous a été donné, tout dernièrement, d'aider M. Daubrée dans son récent travail, sur la schistosité des roches.

Les expériences faites à ce sujet ne laissent plus subsister le moindre doute sur la facilité avec laquelle le plus petit mouvement relatif, distribue dans une masse quelconque, fût-elle absolument homogène, détermine des sens de glissement tout à fait manifestes. Le meilleur moyen de les mettre en plus parfaite évidence consiste à disséminer dans cette masse de petites paillettes de mica qui s'alignent, aussi exactement qu'on l'observe dans les micaschistes, suivant le plan même de chacun des mouvements relatifs. Dans cet échantillon, par exemple, qui résulte de l'écoulement d'un pain de terre argileuse par l'orifice d'une filiere rectangulaire, on n'aperçoit aucune paillette dans les cassures transversales, tandis qu'elles se montrent toutes, parfaitement rangées les unes à la suite des autres, sur les fentes que l'on détermine facilement dans le sens longitudinal, en profitant de la schistosité que conserve la terre après sa dessication, aussi nettement que dans les schistes des différentes formations.

A un autre point de vue, l'écrasement de blocs de terre ou de plomb produit entre des plans de compression, en permettant à la matière de s'échapper dans un seul sens, nous a donné des représentations, frappantes de vérité, du redressement de certaines couches, de leurs contournements avec des caractères de schistosité, distribués comme ils le sont dans la nature, et, enfin, de véritables soulèvements dont les courbures et les déchirures sont d'une imitation tout aussi satisfaisante.

Nous sommes ainsi très-fondé à dire que l'étude de l'écoulement des corps solides a déjà jeté quelque jour sur les phénomènes géologiques ou plus généralement sur les phénomènes de la nature inorganique. En sera-t-il de même dans l'avenir en ce qui concerne les phénomènes organiques ? C'est là une question qui ne saurait être encore envisagée dans le domaine scientifique : nous savons seulement que les végétaux croissent par l'accession des cellules, qui se groupent toujours suivant une disposition tubulaire, que certaines excroissances animales, telles que les tissus cornés, affectent aussi, pendant leur développement, des formes qui sont invariablement les mêmes, et qui semblent dérivées de leur première disposition, comme si, dans l'une et dans l'autre condition, la disposition finale se trouvait déduite, par voie de circulation des fluides nourriciers, dans les enveloppes formées par les tissus déjà confectionnés et préparés pour les recevoir.

Quelque éloignée que soit cette similitude, qui est en quelque sorte le caractère dominant de toutes les formations organiques, j'allais dire le caractère mécanique du phénomène de développement des végétaux et des animaux, il n'est pas irrationnel de supposer qu'un jour viendra où ces phénomènes s'expliqueront, au moins dans leur partie essentiellement mécanique, et indépendamment de toutes les actions plus purement physiologiques et le plus souvent dominantes, par des considérations analogues à celles dont vous avez bien voulu nous permettre de vous entretenir aujourd'hui.

The PRESIDENT: You have already by your acclamations expressed your gratification at hearing the remarkable explanation given to us by

M. Tresca. He has worked out a subject which was very little known —hardly conceived of, in fact, in any form whatever—before, he took it up, and commencing by giving us the results of his experiments, he has drawn one deduction after another, and his subject has extended into one which promises to be of very great importance indeed in mechanical science. I cannot follow him into physiology, because I am not sufficient physiologist to say whether our nails are forced out of us in the manner M. Tresca has described; but it is my want of apprehension, no doubt, rather than his want of perfect conception which is at fault. M. Tresca, by his investigation, throws down as it were the boundary line between solids and liquids, a solid, according to his view, being only a liquid with greater viscosity, and this is an enlargement of our general conception of matter which cannot fail to be of practical importance in mechanical science. With these few observations, I beg to call upon you to pass a hearty vote of thanks to M. Tresca.

Mr. J. SCOTT-RUSSELL, F.R.S. : I beg to second the vote of thanks, and, in saying so, allow me to say, from my own observation, that M. Tresca has given to this meeting to-day, in comparatively few words, the whole result of some nearly twenty years of continuous thought and continuous experiment devoted to this subject, in the most admirable and methodical way. I have watched his progress for that number of years with the deepest interest, and I cannot tell you how profound a gratification it was to me to find that all this was to be brought before you to-day. And allow me to say this : that in addition to washing away the curious and narrow prejudice in which our minds are bound as to the radical difference between a solid and a fluid, and we must get rid of this prejudice in order to go further in the matter of science—he has paved the way for enormous improvements in the important arts of metallic manufactures. He has given us the key to make out of every kind of metal, on the first occasion in which it is lignified, every kind of body to which we may wish it afterwards to be converted. Whereas we, in our clumsy way, up to the present time, take a body of metal, melt it, and then cool it; and then to make a little change in it, heat it again, and cool it again; and then to make another little change in it, heat it again, and again cool it. But he now tells us that you have only to communicate to a little bit of melted metal, the first day it comes to the state of metal, what you want it to come to, and you can put it into that shape of all sorts of sizes, at once, with one heating instead of several. I think the mere getting this idea, founded as it is on careful investigation, may be of the greatest value to us. I therefore wish to say how warmly we should appreciate the benefit M. Tresca has conferred upon us to-day by his very humorous, and I may say enthusiastic and inspiring lecture.

SECTION—MECHANICS (including Pure and Applied Mathematics and Mechanical Drawing).

Monday, May 22nd, 1876.

Dr. SIEMENS, President, in the chair.

The PRESIDENT: Ladies and Gentlemen, we now re-open the Conference on mechanical subjects, and the first paper on our list is that by Professor Kennedy, on Reuleaux's Collection of Kinematic Models. I may take this opportunity of stating that it is the intention of the department to organise a system of explanations at stated intervals of the exhibits in the building. A list will be published stating the days in the weeks at which exhibitors or their representatives, or gentlemen qualified to do so, will give full explanations of certain exhibits, and by this means it will be possible to extend the knowledge to be obtained from this collection to a fuller extent than could be realised by discussing these questions at these Conferences only. We are naturally limited here to time, and can only sketch out the general outline of the very large and interesting collection with which we have to deal. I will now call on Professor Kennedy.

Professor Alex. B. W. KENNEDY, of University College, London, then delivered the following address

ON THE COLLECTION OF KINEMATIC MODELS, By Professor Reuleaux, of Berlin:

Most of the models, of which a small number are upon the table before you, are a portion of an educational collection designed by Professor Reuleaux, the Director of the Royal Polytechnic Academy in Berlin, for use in the classes of that Institution, the rest have been sent to the Exhibition by Messrs. Hoff and Voigt of Berlin, and Messrs. Bock and Handrick, of Dresden. They have been designed chiefly to illustrate Reuleaux's Theory of Machines, a theory which differs in some very important respects from any treatment of mechanisms hitherto adopted. Looking at the models by themselves, some of them seem extremely intelligible, and some very much the reverse. I shall endeavour, in the very limited time at my disposal, to point out the leading ideas which run through the whole, and connect the very familiar mechanisms with those more complex ones which, although differing only in degree from the former, appear at first sight so entirely dissimilar.

In the old books upon machinery, such as that of Ramelli for instance (1588), each machine was taken up by itself, and treated as a whole from beginning to end. One kind of pump after another, for example, may be described without any recognition of their essential identity, or the use of any single word to express the concept *pump*. Each machine is described separately as an apparatus which raises water from such a place, in such a way, and delivers it at such a place. The complex idea which we cover by the word pump, had not yet found a place in the writer's mind.

Presently it was found, of course, that machines were not all different from beginning to end, but consisted of various combinations and repetitions of similar elements ; these elements in time became more distinctly recognised, and were called *mechanisms*. Each machine accordingly was not now described as a whole, but was analysed into the mechanisms of which it consisted, and these received separate treatment. Very much valuable matter has been written upon machinery from this point of view. In our own country, Professor Willis especially gave most valuable contributions to the science of machinery on this basis. Hitherto, however, we have stopped at this point. We have obtained each mechanism "somehow," but have not yet troubled ourselves as to *how* it was invented, or what the elements were. We have, that is, analysed the machine into mechanisms, but we have not yet analysed the mechanisms themselves. We are all familiar with the interesting and valuable work which has been done in the way of examining the motions of particular pieces or members of a mechanism after it has been presented to us, but it cannot be denied that in all cases the mechanism itself has been in the first place taken as a whole.

Professor Reuleaux has attempted to perform the final analysis to which I have alluded, and to discover of what elements mechanisms consist, and how these elements have been combined. He starts from what appears to be the fundamental principle of every machinal combination or arrangement; that each particular part of the combination must have at each instant only one definite motion relatively to every other part. If, in any machine, there be a piece which at any one instant can move in two directions, there is obviously some defect in the machine. Engineers are familiar with the many devices that have to be employed in connexion with certain mechanisms to carry them across "dead-points," &c. This condition, that it shall be impossible for any point in a machine to move at any instant in any direction other than that intended, is a universal one. The way in which it is satisfied is by giving to certain portions of those pieces which form the machine suitable geometric forms. These forms are arranged in *pairs*, in such a way as to be reciprocally *envelopes* one of the other. The one piece, then, so envelopes the other, on account of the forms given to both, that each can move only one way relatively to the other at any instant. The motion of such pieces is called *constrained* motion. Let me take a very simple case—a screw and nut. These two pieces are formed in such a way that the nut can move only in one way at any instant relatively to the screw, and the screw in one way relatively to the nut. If the pitch of the screw be made zero, we have simply a pair of solids of revolution, or "revolutes," having such profiles at the end as to prevent any axial motion; and again we have a pair of mutually enveloping forms, whose relative motion is absolutely constrained. If, on the other hand, the pitch be made infinite, we have a pair of prisms, in which the only possible relative motion is axial. In the first case, the constrained relative motion is twisting; in the second, it is turning (twisting without any translation); and in the last case, where the pitch is infinite, it is simply sliding (twisting without any turning). You recognise, in these three pairs of bodies, the geometrical forms which are used in ninety-nine cases out of a hundred to constrain the motions of machinery. Such bodies as these Reuleaux calls *pairs of kinematic elements*, and —when they have the peculiarity that one entirely encloses the other,— *lower* pairs or *closed* pairs.

It is not essential, however, that one should enclose the other. There are before you a number of examples of *higher* pairs, the bodies of which mutually constrain each other without complete *enclosure.*

I may briefly point out the nature of one of these (Fig 1.) The one element is an equilateral curve-triangle, which possesses the property that any pair of parallel tangents to it are at the same distance apart. The second element is a square, the side-length of which is equal to this distance. The two elements always touch each other in four points, the normals to these points intersect in one point (as O), and at any instant motion can take place about this point only. The two bodies form a pair of elements in exactly the same sense as before, but with one important difference. In the lower pairs the paths of all points in the moving element are *similar*. Every point in the moving element of the twisting pair, for instance, moves in a helix of the same pitch, all points in that of the turning pair in concentric circles, &c. Here, however, the paths of all the points are different. Some of

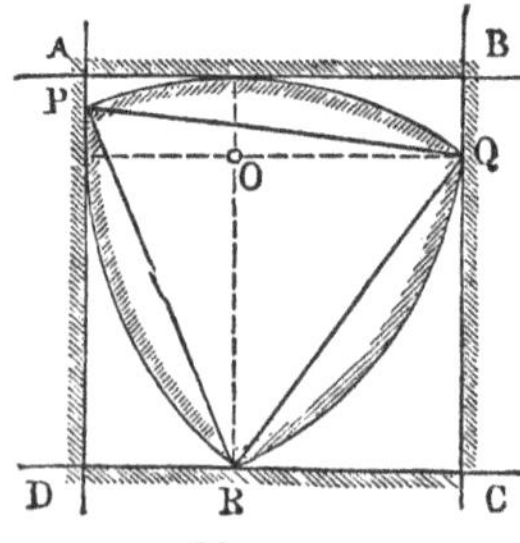

Figure 1.

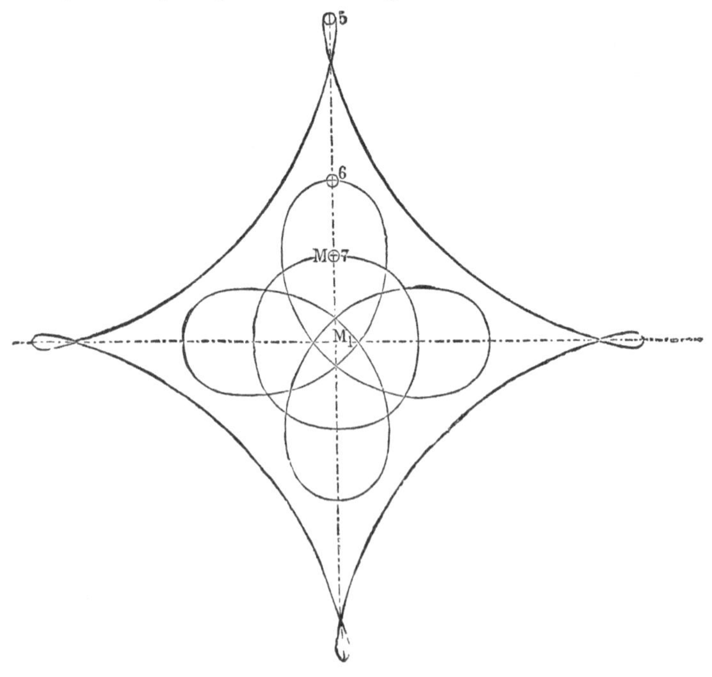

Figure 2.

these paths are very beautiful curves, as you will see if you can follow the motion of the pointer in the model (see Fig. 2). These paths are, in this case, combinations of epi-and-hypotrochoidal arcs. I cannot here go more fully into their nature, I merely show them to illustrate the fact that *enclosure* is not an essential characteristic of pairs of elements having constrained motion.

If two elements be joined together rigidly by a body of any form whatever, we have what is called a *kinematic link*. If I take, for instance, this nut, and fasten it to this open cylinder, I have such a link. This solid cylinder, connected to the screw, gives another link, and so on. By pairing together a number of links we get a combination

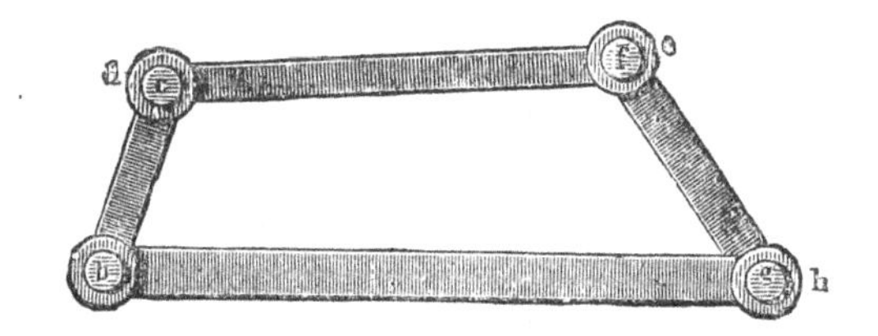

Figure 3.

which Reuleaux has called very happily a *kinematic chain*. In the particular chain which I have here (Fig. 3) there are four links, each being a bar rigidly connecting two elements, and these elements belong in each case to a turning (closed) pair of elements. Before me on the table are a number of other chains similarly constituted, and containing both turning and sliding pairs of elements.

I must now direct your attention to a matter which is of the greatest simplicity, but of equally great importance. If I take any pair of elements in my hands, and move it, you see at once that although the motions of each element, *relatively to the other*, are perfectly determinate, the *absolute* motions are perfectly indeterminate. The elements may move anywhere in space. In the kinematic chain there is just the same peculiarity. It is none the less important that it is almost absurdly obvious. The motions of every link relatively to every other, in other words the relative motions of the links, are absolutely determined ; the absolute motion of the whole chain, or, what is the same thing for our purpose, the motion of the links relatively to this table, is left entirely indefinite.

The conversion of the relative motion into absolute motion (in the restricted sense in which we have used this expression) is a very simple matter. In the case of the pair of elements, all that is required is that one element should be *fixed*, that is prevented from moving relatively to any portion of space which is, for our purposes, stationary—it may be to a room (as here), or to a railway carriage, or a ship, &c. The same method applied to the kinematic chain enables us to convert the relative motions of the links into absolute motions. We must, that is to say, fix or make stationary one *link* of the chain.

A combination of kinematic links, therefore, whose absolute motions are unconstrained, is a kinematic chain. The same combination, when one of its links is fixed, forms what is universally known as a *mechanism*. By fixing, for instance, the link *a h* (Fig. 3), we obtain a mechanism similar to the beam and crank of an ordinary beam engine, *b c* revolves, while *f g* swings to and fro. But the chain has four links, and it is obvious that I may fix any one of them. The combination, that is the kinematic chain, remains the same, but the nature of the mechanism may be entirely altered. Suppose, for instance, the link *b c* be fixed, we obtain a mechanism which you see at once differs entirely from the last, and which you will recognise as the common drag link coupling, *a h* and *d e* both revolving as cranks about the fixed centres *b* and *c*.

A mechanism is, therefore, a kinematic chain of which one link is fixed. Two links cannot be fixed simultaneously without making the whole chain immovable. Any one link, however, can be fixed, and thus from any chain we can obtain as many mechanisms as it has links. I shall endeavour to show you, by a few illustrations, what a wonderful insight this gives us into the nature of some familiar mechanisms. I will only mention in passing what may, perhaps, be new to some, that this chain (Fig. 3), with which we are so very familiar, is not moveable because the axes of all its pairs are parallel, but because they all intersect in one point. In this particular case

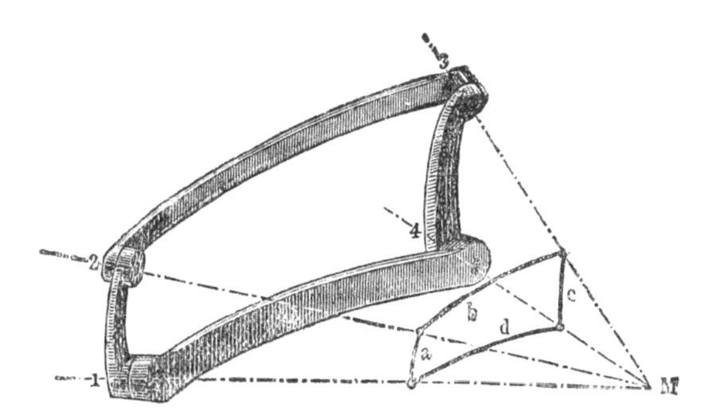

Figure 4.

that point is at an infinite distance, but the chain moves equally wel if the point of intersection be at a finite distance, as is illustrated by Fig. 4. Upon this mechanism many engines (disc engines, &c.), and other machines have been based. Hooke's universal joint is a familiar illustration of it. The complex constructive forms of these, however, make the recognition of their real nature almost impossible without the aid of some such system as Reuleaux's analysis.

In order to obtain the constrained motion of a closed pair, it is not necessary that both elements should be constructed as fully as in the cases we have hitherto looked at. Grooves might be cut down the sides of a pin, for instance, without affecting its motion in an eye. Professor Reuleaux has made some investigations, which I can only mention here, on the extent to which this process can be carried. Without further proof, I have no doubt you will recognise at once that by the use of the slot and sector of Fig. 5 instead of the pin and eye

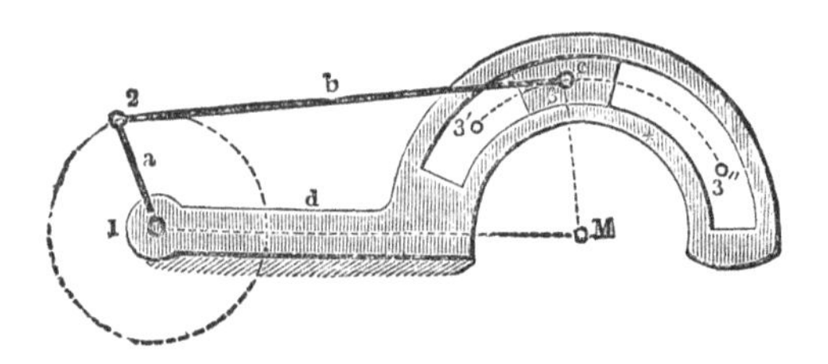

Figure 5.

g h (Fig 3), no change has been made in the chain. The complete cylinder pair has been replaced by a sector and a slot concentric with it, but of totally different diameter; the motions remain absolutely identical. We obtain thus a very convenient method of altering the length of the links of the chain, for the link *f g* has now taken the form of the sector *c* (Fig. 5). To make that link longer, herefore, all we have to do is to give the sector a larger radius. If we make the link infinitely long, the sector becomes a prism, working in a straight slot of which the axis passes through the centre of the pin 1 (Fig 5), and we obtain this very familiar chain (Fig 6), the driving

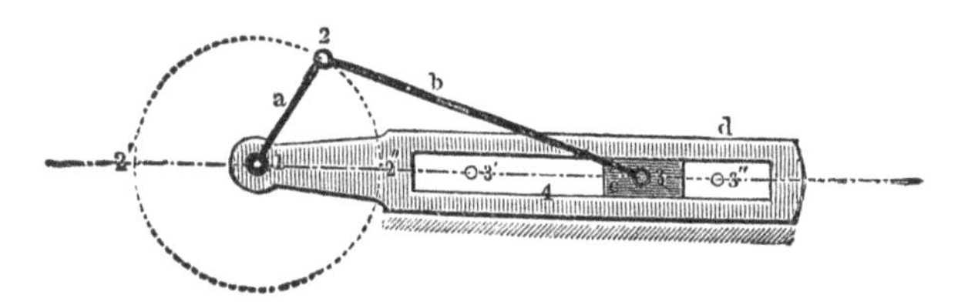

Figure 6.

mechanism of the common steam engine. It is derived from the former merely by increasing the lengths of two of its links, making them equal and infinitely long. I particularly wish to draw your attention to this chain, both because it is very familiar, and because by its inversion, that is by fixing one or another of its links, we get such very notable results. I have already fixed one link (Fig. 6), and you have seen the common steam engine driving train. If I fix another link, say

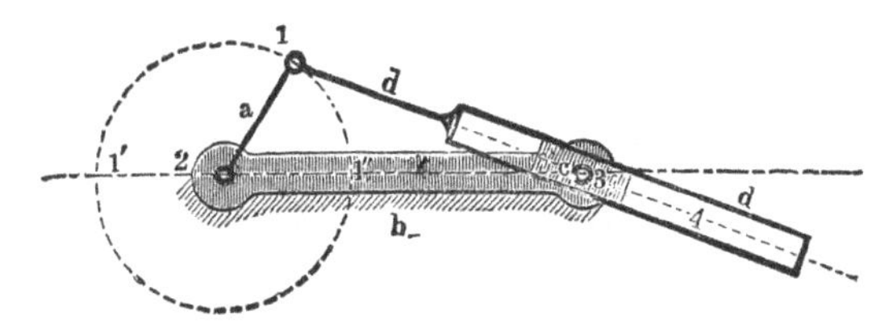

Figure 7.

the connecting rod, we have a mechanism which I think you will at once recognise as the driving train of the common oscillating engine (Fig 7). This appears even more distinctly if we reverse the sliding pair 4, making the link *d* carry the full prism, and the link *c* the open prism (Fig. 8). The link *c* becomes the steam cylinder, *d* (which was the

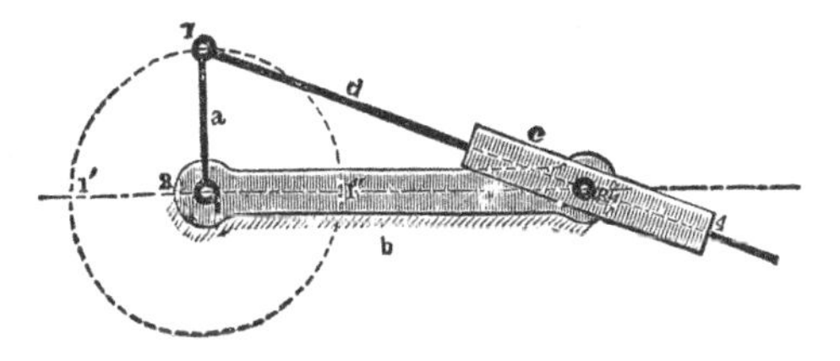

Figure 8.

frame in Fig. 6) the piston and rod, and *b* (the connecting rod in Fig. 6), the framing of the engine.

This intimate connexion between the driving mechanism of the direct-acting and oscillating engines was never, I think, recognised until Reuleaux pointed it out. They are the *same chain*, the only kinematic difference between them being in the link which is fixed. But we can go further; we can fix, for instance, the crank. We get thus a mechanism doubtless familiar to most of my hearers (Fig. 9),

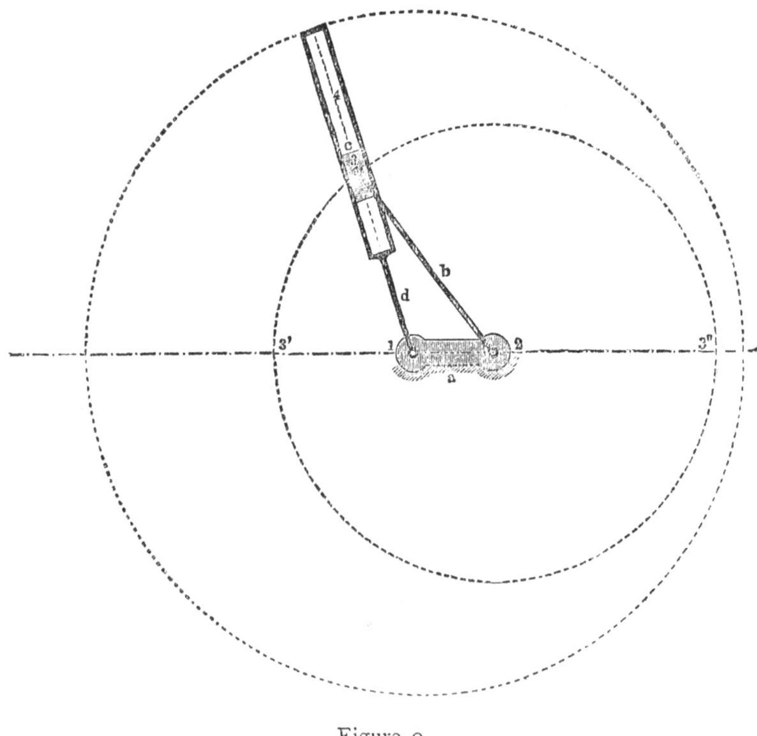

Figure 9.

and which has often been described and applied. Sir Joseph Whitworth, for instance, has used it as a quick return motion. If the link *b* revolve with a uniform velocity, it gives to d the varying velocity which has been so often utilised. The mechanism is obtained from the same chain as before, simply by choice of a different link for the stationary one. In general it is constructively disguised in such a way as to be only recognisable with difficulty, but when put in this schematic form it can easily be analysed into its kinematic elements.

From the same chain, we may obtain one more mechanism by fixing the block (Fig. 10). This mechanism has been seldom used,

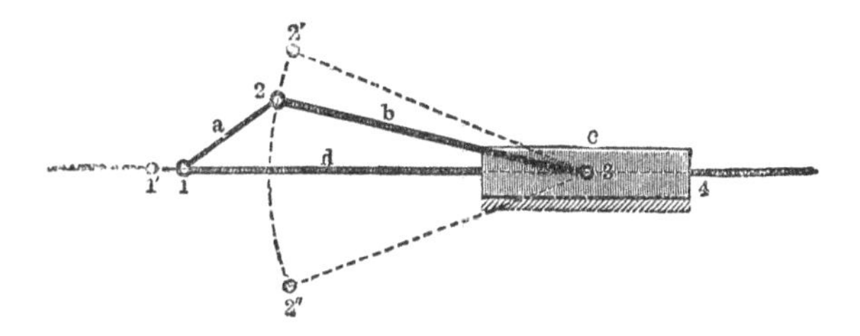

Figure 10.

but it is occasionally employed. The motion of the link a is now characteristic.

We must now proceed to look at a few other leading ideas illustrated by these models. First, we may notice as long as the *form* of the elements of the pairs remains unchanged, their relative size is a matter of indifference. We have already seen this incidentally in comparing Figs. 3 and 5. By utilising this principle we can obtain a great number of very different-looking forms of one and the same mechanism. Models of many of these are on the table before me: some of them are constantly used by engineers, others have been seldom or never applied. The ordinary eccentric is a familiar illustration of this "expansion of elements." It differs from the train shown in Fig. 6 only in the relative size of two of its elements. The turning-pair at 2 (Fig. 6) is made so large as to extend beyond the pair 1; all the motions in the train remain, however, as before.

In order to illustrate the way in which the system of kinematic analysis, which I have sketched to you, can be applied to actual machines, Professor Reuleaux has examined analytically an immense number of those unfortunate devices called rotary engines, on which so much ingenuity and excellent brainwork has been wasted. In his "Theoretische Kinematik," he gives illustrations of between sixty and seventy of these machines, analysing every one of them into one or other of the three chains which we have been examining (Figs. 3, 4, and 6). The inventor has generally called his machine "rotary" because of some notion that there were more rotating parts, or more direct rotation, than in the ordinary engine. While there are a few engines in which this is the case, in the vast majority it is entirely a mistake. Here is one (see Fig. 11) which has been invented over and over again. I believe it was invented for the first time in 1805, by a Mr. Trotter; it was re-invented in 1831, 1843, 1863, 1866, 1870, 1872, 1873,

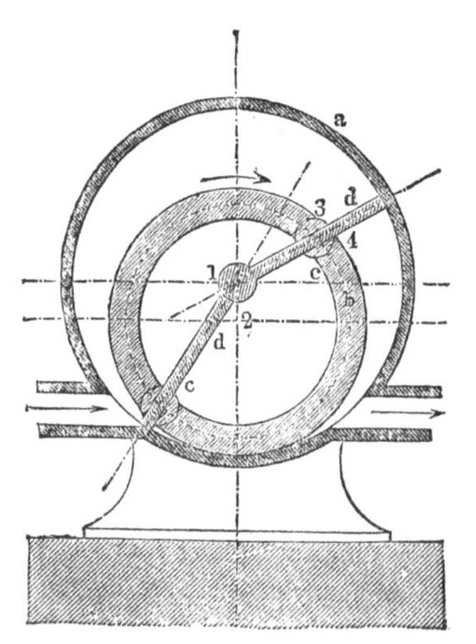

Figure 11.

and possibly at many other times. One gentleman, whom I see here, told me he had invented it himself twenty years ago! After all, it is absolutely identical with the chain of Fig. 6. The link fixed is the crank, so that, as a mechanism, it is represented by Fig. 9. I am sorry my time will not allow me to prove this, but by analysing the pairs of elements which it contains it shows itself at once. Many other rotary engines of which there are models on the table are of the same kind. Here is one which was exhibited at the Exhibition of 1851, and which attracted a good deal of notice at the time—Simpson and Shipton's machine. It is really nothing more than the mechanism shown in Fig. 10, although its constructive form so disguises its real character. I might go through a great many more in the same way, but time compels me to leave this part of my subject.

In Reuleaux's work, to which I have alluded, and of which I have just completed the English translation, he uses a method which, while it is by no means original with him, has never been formerly developed to the same extent and in the same way. I must try in a few words to indicate the general nature of this method. If we have any plane figure moving in a plane, its motion, at any instant, may always be considered as a motion about one particular point (it may be at a finite or an infinite distance), and this point is called the *instantaneous centre* for the motion of the figure. The body may continue to move about the same point, in which case the instantaneous centre becomes a *permanent* centre ; but in general the motion of the body in successive instants is about *different* points, each being for the time, the instantaneous centre. The locus of the points which thus become instantaneous centres for the motion of any figure is some curve, and is called by Reuleaux *Polbahn*, for which I propose the use of the word *centroid.* If we have any two figures A and B, having a definite relative motion, and make the relative motion of B to A absolute by fixing A, we can, by moving B, find as many points in the centroid as we wish, so to construct the curve. This centroid remains, of course, stationary, like the figure A, and is called the centroid of A. By fixing B and moving A relatively to it, we can in the same way obtain the centroid of B. We have then two curves, one connected with each figure, and these possess certain properties which are of great value in the study of mechanism. As the figures move these curves *roll* upon one another, and their point of contact is always the instantaneous centre of motion for the time being. It is, of course, impossible for me to *prove* these or any other properties of these curves here; many gentlemen here must be quite familiar with the proofs. I must content myself with simply showing you, by way of illustration, models of mechanisms in which the centroids are constructed in such a way that their rolling can be distinctly seen. By the aid of these centroids we can treat a great number of kinematic and dynamic problems in an extremely simple and beautiful manner, and can treat complicated and simple problems by one general method, instead of using different methods. Centroids have not, so far as I know, been used in English text books hitherto, but have not unfrequently found more or less extended use in German and French works, principally in static and kinematic problems. I have found them, however, even more useful in Kinetics. Among the more recent books in which I have noticed them, I may mention Dwelshauver's Dery's "Cinématique," Schell's "Theorie der Bewegung und der Kräfte," and Pröll's "Graphische Dynamik."

Considering the constrained motion of *bodies* instead of plane figures only, the instantaneous centre becomes an axis, and the centroidal curve a ruled surface, the locus of all the axes. These surfaces are called *axoids*. For bodies having complane motion, they are cylinders; for bodies having motion about a fixed point, cones; and for general motion in space, general ruled surfaces, in general non-developable, of which successive generators *twist* upon each other. Poinsot was, I believe, the first to mention these surfaces.

I shall only mention one more point in connection with the subject before us. Professor Reuleaux has devised, for the purpose of aiding his kinematic work, a system of notation founded upon his analysis, which can be used to represent these mechanisms in a perfectly simple manner. Of course, to write down a description of them is a long matter, but hitherto we have not known their real nature analytically, and therefore could do nothing else. Now that we do know it, we may treat them exactly as we do chemical compounds, where instead of writing down the whole names of everything, we use short symbols for known elements. This kinematic notation I can do no more than mention. It seems to me very original, and I have myself found it very useful in the analysis of really complicated machinery. Mechanisms of great apparent complexity come out, often, in forms of really wonderful simplicity, and one's work is made at the same time much more easy and much more satisfactory.

These models are very beautiful, but they are also necessarily somewhat expensive. I have here a couple of wooden ones, however, which will serve to show that they can be constructed and used in schools without any very extraordinary cost. These two models, and also the stand for carrying them, have been made for me by M. Paul Nolet, of University College.

I must now thank you for the kind attention with which you have listened to my sketch of the nature of this beautiful collection of models, and I should like also to take this opportunity of thanking Herr Kirchner, of the Berlin Akademie, for the trouble he has taken in preparing and arranging them for me.

The PRESIDENT : I am sure we have passed a very agreeable half hour in listening to Professor Kennedy's explanations of these most beautiful and instructive models. I think this new branch in mechanism, kinematics, will be extremely useful in helping mechanical engineers in devising improved means to an end, because by combining motions into systems, we can resort to any combination that seems likely to answer our purpose, without having to go to the fountain head of our own brains to originate it. I regret that we have so little time to go with Professor Kennedy into the details of the various contrivances put before us, but I hope that Professor Kennedy will give further explanations under the new arrangement which I have had the pleasure of announcing. I will now call upon you to pass a vote of thanks to Professor Kennedy for his interesting communication.

Dr. MANN : I wish to ask Professor Kennedy if the work he referred to just now was a translation.

Professor KENNEDY : Yes.

The PRESIDENT : I will now call on Mr. Barnaby, the Chief Constructor of the Navy, to give us his communication on Naval Architecture. I will take this opportunity of announcing that we shall have to make some little change in our programme. General Morin arrived yesterday from France, and has to leave again in a couple of days. As we should all like to hear him on the subject to which he has given such very great attention, that of ventilation, it has been thought desirable to take his paper at two o'clock to-day.

On Naval Architecture

In nominating for the distinguished position which I occupy this morning in an International Assembly, the Director of Construction of a War Navy, the Committee have exposed themselves to some possible objections. I can find their defence not only in my own personal fraternal and kindly feelings towards our guests, but also in the general aspect of war machinery to the eyes of an Englishman. He would show his newest ship of war or his newest gun to his foreign guest with the same sort of feeling that he would exhibit the trappings and weapons of the policeman or the special constable. They are the means by which he hopes to make himself the terror of evil doers, and he does not reckon his guest among them. Still the Admiralty has not exhibited a single model of a ship of war, nor any warlike apparatus. Every modern ship of war equipped for sea is in itself a series of laboratories full of scientific instruments and apparatus, but they are of such a character that they could not be well transferred bodily to these galleries, and it is impossible to represent them by models or drawings. But, thanks to the labours of the Committee, the Exhibition in this department of Marine Architecture is not without interest.

Notwithstanding that ships and apparatus for modern naval warfare are represented in such a very incomplete manner in this splendid Exhibition, a state of things which, perhaps, no one much regrets, I should not be justified in ignoring ships of war in an address on Naval Architecture.

There is a model in the Exhibition of some of the floating batteries originated by the late Emperor, Napoleon the Third, of France, built in 1854-56, by Messrs. Green, of Blackwall, and there is also a model of the first English seagoing ironclad, the "Warrior," designed under the superintendence of Mr. Isaac Watts.

The "Warrior" is a very remarkable ship, not because she is a powerful fighting engine, for that she cannot be said to be, but because in the material of construction and in the disposition of her armour, she is precisely what experience has shown to be the best; and our latest designs, the "Ajax" and "Agamemnon" differing from her, as they do widely, in proportions, in thickness of armour, in power of guns, in the mode of propulsion, and in internal arrangements, are still like her in that they are built of iron and use wood for only secondary purposes, and are like her also in having no side armour at the bow or stern, watertight decks under water, with cellular divisions over them, taking the place of armour. This latter feature has been subsequently elaborated at various times and with different degrees of completeness and success by Mr. Reed; Mr. Michael Scott; by the present Minister of Italian Marine, Signor Brin; and by myself and colleagues. Our newest design for seagoing battle-ships of 8,400 tons displacement, *i. e.*, the "Ajax" and "Agamemnon," differs broadly from this first ship in the following respects :—

1. The length of the "Warrior" is six and a-half times her breadth, and the "Ajax" and "Agamemnon" are only four and a half times.

2. She has four and a-half inches armour and 9-ton guns, while the latter ships, although of less dimensions, are to have eighteen inches of armour and 38-ton guns.

3. She obtained fourteen and a-third knots speed with engines indicating 0·6 of a horse-power per ton of displacement. In these new designs we shall be content with a knot less speed, obtained by means of a horse-power per ton of displacement.

4. This speed is obtained in the "Warrior" with great draught of water, a single screw, and fine lines; it will be obtained in the "Ajax" and "Agamemnon" with two screws, a light draught of water, and full lines.

5. In the "Warrior" there is only one means of propelling and one means of steering, and the steering gear is exposed to shot. In these latest ships there are two distinct isolated means of propelling, two means of steering, and the steering gear is completely protected.

6. In the "Warrior" there is only a single bottom, except for a small breadth of the ship; in the latest ships, and in nearly all the ironclads, there is a complete double bottom.

7. In the "Warrior" there is no fire from behind armour either ahead or astern; in the latest designs the protected fire ahead and astern is as powerful as that on the beam.

8. In the "Warrior" the consumption of fuel per horse-power indicated is four and a-quarter pounds; in our recent ships it is less than three pounds.

The ships which I have thus compared with the "Warrior" are to be constructed on the same principle as the "Inflexible" but will have a knot less speed than the "Inflexible," thinner armour (one and a-half feet instead of two feet) and 38-ton guns.

These changes have been made in several steps extending over several years. I do not know precisely to what extent Mr. Scott Russell influenced the general conception of the design of the "Warrior," but I believe he did so to a considerable extent. I know that the influence he has exerted while the development of iron shipbuilding has been going on in the Royal Navy has been far greater than has ever been recorded or acknowledged.

It ought to be recorded by me here, I think, that the Exhibition contains models of the three grandest ships of war in the navy of the German Empire. The "König Wilhelm," and the "Kaiser" and "Deutschland," now forming part of the squadron commanded by Admiral Batsch, were designed by an English architect, and built in English dockyards, and I may also mention that China, which has now an important naval establishment at Foochow, where ships of war and marine engines are constructed under the superintendence of a French officer, has recently had built for her in England two of the Rendel gunboats, with 25-ton guns, the models of which are furnished by the builders, Messrs. Laird. The Thames Company send a model of the "Vasco da Gama," the first ironclad for the famous navy of Portugal.

I should like now in a few brief sentences to discuss and dismiss the question of armour-plating, and pass on to matters of less national moment, but of more general interest.

There are people who say that it would be a gain to the world, and notably to England, if some one would find the way to get rid of armour-plating. Their minds are distressed with a condition of things of which they cannot pretend to foresee the probable end and issue, for the contest between the armour and the gun is of this character.

I do not myself pretend to foretell at what point on the road we are travelling we shall be obliged to turn off, but I do not doubt that wherever it may be, and in whatever direction the new road may lie, naval armaments will not become less costly or less subject to the progress of mechanical invention. The introduction of artillery, of steam, of armour-plating, and of torpedoes into naval warfare have all tended to increase the power of civilisation, of wealth, and of mechanical skill to assert themselves against the power of mere numbers and of brute force, and the nations represented in this assembly are the last in the world that should regret that naval battles may be fought successfully by the few against the many, provided that the few have on their side, in addition to the universal qualities of animal courage and endurance, the rarer possessions of wealth, of science, and of mechanical skill.

I would add that I do not advocate—indeed I have always opposed—the protection of men in detail by armour. I would protect them in the mass by protecting their ship, their signalling and steering instruments, and their fighting power as a whole; but I would not attempt to protect each gun and each man. My own ideal of a fighting ship is not the "Inflexible," the "Alexandra," the "Temeraire," or the "Ajax," and "Agamemnon," although I am primarily responsible for their designs, but it is the ships now building in Scotland, the "Nelson" and "Northampton," which represent in my view the best disposition of the offensive and defensive powers.

In these ships the central part is armoured up to a shot-proof deck four feet out of water. The ends are without side-armour, but have an under-water deck protecting everything that is vital. There is a high battery with numerous heavy guns; the battery is protected from end-on fire, and the bow and stern guns, which fire in line with the keel, and are the most powerful of any, are protected from broadside fire also. The intermediate guns have in front of them a thin side incapable of being splintered, and each gun's crew is cut off from the next by a splinter-screen or traverse. This broadside of guns can be loaded and laid in a close engagement under the shelter of the bow or stern armour, and may be fired by electricity without exposing the crew. The ships are propelled by two screws; the propelling machinery is divided into compartments separated from each other by water-tight bulkheads, and there is for a time of peace good sail-power. In a time of war only the lower masts would stand, and the ships could then carry a very large supply of fuel. The cost of each of these ships for hull and engines, exclusive of fittings and rigging, is about £350,000.

To turn now to the Loan Collection more particularly I would remark that in it there are examples of the beginnings of things which have assumed gigantic proportions.

There is a model and a drawing of the first steamboat in Europe advertised for the conveyance of passengers and goods, and there is also the original engine made and fitted on board that vessel.

That vessel was the "Comet," built in Scotland in 1811-12 for Mr. Henry Bell, of Helensburgh, the boat being designed and built by Mr. John Wood, at Port-Glasgow. The "Comet" was the first steam-vessel built in Europe that plied with success in any river or open sea. The little vessel was forty-two feet long and eleven feet wide. Her engine was of about four horse-power, with a single vertical cylinder. She made her first voyage in January, 1812, and plied regularly between Glasgow and Greenock at about five miles an hour.

There had been an earlier commercial success than this with a steam vessel in the United States of America, for a steamer called the "Clermont" was built in 1807, and plied successfully on the Hudson River. This boat, built for Fulton and Livingstone, was engined by the English firm of Boulton and Watt.

The reason for this choice of engineers by Fulton appears to have been that Fulton had seen a still earlier steamboat for towing in canals, also built in Scotland, in 1801, for Lord Dundas, and having an engine on Watt's double-acting principle, working by means of a connecting rod and crank, and single stern wheel.

This vessel, the "Charlotte Dundas," was successful so far as propulsion was concerned, but was not regularly employed because of the destructive effects of the propeller upon the banks of the canals. This brings me to another interesting model. The engine of the canal boat just spoken of was made by Mr. William Symington, and he had previously made a marine engine for Mr. Patrick Miller, of Dalswinton, Dumfriesshire. This last-named engine, made in Edinburgh in 1788, marks, it is said, the first really satisfactory attempt at steam navigation *in the world*, and the veritable engine is in this exhibition. It was employed to drive two central paddle wheels in a twin pleasure-boat (a sort of "Castalia") on Dalswinton Loch. The cylinders are only four inches in diameter, but a speed of five miles an hour was attained in a boat twenty-five feet long and seven feet broad. The models, &c., which I have referred to, cover the period from 1788 to 1812.

There is also a model of the first steam vessel built in a royal dock-yard. She too is called the "Comet." She appears to have been built about the year 1819, and was engined by Boulton and Watt. This ship had two engines of forty horse-power each, to be worked in pairs on the plan understood to have been introduced by this firm in 1814.

In 1833-4 iron appears to have first come into use for the construction of steamships, and there are models of the "Rainbow," the "Nemesis," and other vessels built soon after this date. In 1838 the "Sirius" and "Great Western" commenced the regular Atlantic passage under steam. The latter vessel, proposed by the late I. K. Brunel, and engined by Maudslay, Sons, and Field, made the passage at about eight or nine knots per hour. There is an excellent model of the engines of the "Great Western" in the collection. One year earlier, *i.e.* in 1837, Captain Ericsson, a scientific veteran who is still among us, towed the admiralty barge with their lordships on board from Somerset House to Blackwall and back at the rate of ten miles an hour in a small steam vessel, driven not by paddles, but by a screw. Messrs. Laird supply a model of the "Robert L. Stockton," built by them in 1839, from Capt. Ericsson's designs. This vessel had the rudder before the screw, arranged as in the modern fast launches built by Mr. Thorneycroft. This firm exhibits a model of a proposed screw ship-of-war of a still earlier date, 1836. The screw did not come rapidly into favour with the Admiralty, and it was not until 1842 that they first became possessed of a screw vessel. This vessel, first called the "Mermaid" and afterwards the "Dwarf," was designed and built by the late Mr. Ditchburn, and engined by Messrs. Rennie. Her model is in the exhibition.

In 1841-3 the "Rattler," the first ship-of-war propelled by a screw, was built for and by the Admiralty under the general superintendence of Mr. Brunel, who was also superintending at the same time the construction of the "Great Britain," built of iron. The engines of the "Rattler," of 200 nominal horse-power, were made by Messrs. Maudslay. They were constructed, like the paddle-wheel engines of that day, with vertical cylinders and overhead crank shaft, with wheel gearing to give the required speed to the screw. The "Rattler," built of wood, does not exist now, but the "Great Britain," built of iron is still at work.

The next screw engines made for the Royal Navy were those of the "Amphion" 300 nominal horse-power, made in 1844 by Miller and Ravenhill. In these the cylinders took the horizontal position, and they became the type of screw engines in general use. This ship had a screw-well and hoisting gear for the screw.

In 1845 the importance of the screw propeller for ships of war became fully recognised, and designs and tenders were invited from all the principal marine engineers in the kingdom. The government of that day then took the bold step of ordering at once nineteen sets of screw engines of the following firms, viz., of Messrs. Maudslay, Sons, and Field, four sets for the "Ajax," "Edinburgh," "Niger," and "Desperate;" of Messrs. Seaward and Capel four sets, for the "Blenheim," "Hogue," "Conflict," and "Termagant;" of Messrs. Jno. Penn and Sons two sets, for the "Arrogant" and "Encounter;" of Messrs. Boulton and Watt three sets, for the "Eurotas," "Horatia," and "Vulcan;" of Messrs. Rennie two sets, for the "Forth" and the "Seahorse;" of Messrs. Napier two sets, for the "Dauntless" and "Simoom;" of Messrs. Fairbairn one set for the "Megœra;" and of Messrs. Scott and Sinclair one set for the "Greenock."

Of these, the four last-named vessels and the "Desperate" and "Termagant" had wheel gearing. In all the rest the engines were direct acting. The steam pressure in the boilers was from five to ten pounds only above the atmosphere, and if the engines indicated twice the nominal power, it was considered to be a good performance.

The most successful engines were those of the "Arrogant" and "Encounter" of Messrs. Penn. They had a higher speed of piston than the others, and the air-pumps were worked direct from the pistons, and had the same length of stroke. These engines developed more power for a given amount of weight than other engines of their day, and were the forerunners of the many excellent engines on the double-trunk plan, made by this firm for the navy.

The engines with wheel-gearing for the screws were heavier, occupied more space, and were not so successful as the others, and no more of that description were ordered for the Royal Navy.

Up to 1860 neither surface-condensers nor super-heaters were used in the Royal Navy. The consumption of fuel was about four and a-half pounds per one horse-power per hour.

In 1860 a step was taken in the Royal Navy which receives illustration from a beautiful set of drawings contributed by the firm of J. Elder & Co.

Three ships, the "Arethusa," "Octavia," and "Constance," were fitted respectively by Messrs. Penn, Messrs. Maudslay, and Messrs. Elder, with engines of large cylinder capacity to admit of great expansion, with surface-condensers and super-heaters to the boilers.

Those of the "Arethusa" were double-trunk, with two cylinders; those of the "Octavia" were three cylinder engines; and those of the "Constance," illustrated by the drawings referred to, were compound engines, with six cylinders; the two former were worked with steam of twenty-five pounds pressure per square inch, and the latter with steam of thirty-two pounds pressure. All these engines gave good results as to economy of fuel, but those of the "Constance," were the best, giving one indicated horse-power with two and a-half pounds of fuel.

But the engines of the "Constance" were excessively complicated and heavy. They weighed, including water in boilers and fittings, about five and a-half hundred weight per maximum indicated horse-power, whereas ordinary engines varied between three and a-half and four and three-quarter hundred weight.

For the next ten years engines with low pressure steam, surface-condensers, and large cylinder capacity were employed almost exclusively in the ships of the Royal Navy. A few compound engines, with steam of thirty pounds pressure, were used in this period with good results as to economy, but they gave trouble in some of the working parts.

Compound engines, with high pressure steam (fifty-five pounds), were first used in the Royal Navy in 1867, on Messrs. Maudslay's plan, in the "Sirius." These have been very successful. In the Royal Navy the compound engine is now generally adopted. They are rather heavier than the engines which immediately preceded them, but they are about twenty-five per cent. more economical in fuel, and taking a total weight of machinery and fuel together, there is from fifteen to twenty per cent. gain in the distance run with a given weight. We are now introducing wrought iron largely in the framing in the place of cast iron, and hollow propeller shafts made of Whitworth steel. By these means the weight is being reduced, and it is to be hoped that a still further reduction may yet be made by the use of high class materials in the engines, and steel in the boilers. That there is room for improvement is indicated by the marvellous results obtained by Mr. Thorneycroft, of Chiswick, and others, by means of high piston speed, forced com bustion, and the judicious use of steel As much as 250 indicated horse-power is obtained with a total weight of machinery of eleven tons, including water in boilers and spare gear. The ordinary weight of a sea-going marine engine of large size, with economical consumption of fuel, would be four or five times as great.

Mr. Brotherhood, with his three-cylinder air-engine worked at high speed, and with a pressure of forty atmospheres, indicates, as he informs me, forty-three horse-power, with a weight of engine (not including air or air reservoir) of only forty-three pounds. There is in this Loan Collection an illustration of the three-cylinder plan, as patented by Mr. Willans, in the shape of a complete launch engine.

I have dwelt longer upon steam machinery than I am probably entitled to do, because Mr. Bramwell will go over the ground in his address as prime mover. But I feel so strongly the great debt of gratitude which naval architecture in England owes to her marine engineers, that although I am not myself a marine engineer, I could not help seeing that a discourse on English naval architecture without the marine engine would be ridiculous.

There are three or four men of modern times who have done much for naval architecture, as it concerns the hulls of ships, but England owes at least as much, in my judgment, to Mr. John Penn, the late Mr. Joshua Field, the late Mr. John Elder, and the late Mr. Thomas Lloyd, for her position in the world as to steam navigation.

In the matter of the forms of ships I regret to say that opinions differ outside the Admiralty as to whether ships intended for high speed should be about as broad as they are long, or should have a length any number of times, up to twelve times their breadth.

Mr. Scott Russell's experiments, made in 1832 and subsequently, have long influenced, to a considerable extent and favourably, the forms of steamships, but there does not appear to be any general agreement among the designers of fast ocean-going ships, as to the best proportions and form for securing the least resistance under the average conditions of an ocean voyage. Splendid results are obtained in the Atlantic service—ships averaging over fifteen knots an hour on the whole passage.

For the Royal Navy all our proportions and forms are subjected to the investigation which Mr. Froude is able to give by his beautiful experimental apparatus at Torquay. We do not believe that the resistances of the ships will accord exactly with those of their models, but we are satisfied that the behaviour of the models as to resistance approximates very closely to that of the ships in smooth water. But even when we have obtained the best form for resistance in smooth water, it is quite certain that that will not be the best form for service among waves; we are still open to the corrections of practice and experience at sea.

The Admiralty are building fourteen-knot ironclad ships with only four and a-quarter beams in the length; and shallow draft, nine--knot unarmoured vessels, with only three and a-quarter beams in the length. The Admiralty experiments at Torquay are also directed towards the discovery of the causes of the enormous loss of power in propelling machinery indicated by the fact that only about thirty-seven to forty per cent. of the maximum power of the steam delivered in the engines is useful in propelling the ship. It is proposed to continue in H.M.'s ships tried at the measured mile, the experiments commenced by Mr. Denny, of Dumbarton, with this object.

The present condition of the case appears, according to Mr. Froude's estimate, to be that, calling the effective horse-power (that is, the power due to the net resistance) 100, then at the highest speeds the horse-power required to overcome the induced negative pressure under the stern consequent on the thrust of the screw is 40 more; the friction of the screw in the water is 10 more; the friction in the machinery 67 more; and air-pump resistance perhaps 18 more; add to this 23 for slip of screw, and we get that, in addition to the power required to overcome the net resistance $= 100$, we need $40 + 10 + 67 + 18 + 23$, making in all 158, *i.e.*, at maximum speeds the indicated power of the engines needs to be more than two and a-half times that which is directly effective in propulsion.

I regret that the present practice of naval architecture has a side about which I would rather say nothing, but which is yet, in my judgment, so important that I do not feel justified in passing it by in silence. That dark side is the tendency of keen competition among owners, and consequent lowness of freights, to make them content with ships built without that regard to their safety in the matter of division into compartments, which I hold to be in the highest degree desirable.

There can be no doubt that ships built of iron are more liab'e to fatal damage by collision or grounding than those built of wood, unless they are properly divided into compartments. If so divided they are safer than those built of wood. I regret to say that with few exceptions iron passenger steamers are getting to be worse instead of better cared for in this respect.

There is no law enforcing the existence of bulkheads, or regulating the height to which they should be carried, or what doors there should be in them, either in the regulations of the Board of Trade or in those of the Surveyors to Lloyds, excepting that there must be one bulkhead at each end of the ship in a steamship and at one end in a sailing ship. A ship may be 500 feet long, and there may be over 400 feet of the central ship practically in one compartment.

In some ships the bulkheads are sufficiently numerous, but they do not extend above the deck which is nearly level with the water when the ship is laden, so that if one compartment fills, the ship sinks far enough to bring the tops of all the bulkheads under the water, and then the bulkheads are of no more value in keeping the ship afloat than if they did not exist.

In others the bulkheads are neither sufficiently numerous nor sufficiently high out of water.

In others, even where there are good bulkheads, there are doorways cut in them which destroy their integrity and there are no water-tight doors to close them.

It may be supposed that the power of the pumps is so great, that with even a large leak the water would not be allowed to rise up to the top of a bulkhead dividing the compartments. But pump-power is a poor resource. It is very difficult to provide, and very rare to find, pump-power in even the finest ships sufficient to cope with a hole one square foot (I might almost say half a square foot) in area ten feet under water.

It is quite true that by great care in management such ships can be worked for years without losses of life, but the loss of a single large ship with hundreds of women and children on board is such a terrible event, that the security afforded by compartments sufficient in number and in height to provide against immediate sinking, if any one is filled, ought to be sought for. That this arrangement is consistent with commercial success is proved by the practice of some of the best

English owners. I know too well how dependent the ship is, even then, upon her water-tight doors; but such care bestowed upon them, as is given by some owners and builders, would get over that difficulty.

I have been supplied at my request with sketches of some admirable arrangements designed by Messrs. Harland and Wolff, of Belfast, and fitted in some ships recently built by them.

Description of Watertight Doors, &c., as fitted in the s.s. "Britannic" and "Germanic."

No. 1.—This floating door works on pivots and is plaeed at the entrance of the screw shaft tunnel on the after side of the bulkhead, and in case of the water rising in the tunnel, closes itself by the buoyancy of the air-chamber attached to it. Chains are also fixed on the foreside and extending to the engine-room platform on the middle deck for closing by hand if any obstruction occurs.

No. 2.—Any irruption of water into the after boiler and engine space would float and close the lower door of this pair, and at the same time lower the upper one; these two also have chains led to the middle deck grating in the stokehold for closing by hand.

No. 3.—This floating door is fixed on the same bulkhead at No. 2, but on the opposite side, and closes in case of water getting into the forward boiler space, by a similar self-acting arrangement to No. 1.

No. 4.—This door is opened by a rack and worm from the middle deck, and can also be closed *instantaneously* from the stokehold by means of the handle A, which works on an eccentric, throws the worm out of gear with the rack, and lets the door drop, any shock being prevented by the pistons above cushioning cn the air in the cylinders, which does not commence until the two hangers T T come in contact with ends of pistcn rods S S.

I bring also an Admiralty contribution towards this good object, in the shape of a model of a door which cannot be left open because it never is open, but it nevertheless serves for passage. We are pro posing to fit it in fore and aft bulkheads where ventilation is not necessary, but where the integrity of the division is of the utmost importance.

I do not wish it to be supposed that I complain of the action of the Lloyd's surveyors in this matter. They can only enforce what the general sense of owners accepts as necessary. The excellent and careful details of construction exhibited by the London surveyors is an illustration of their skill and faithfulness, but they are powerless here. So also is the Board of Trade so long as foreign ships are in the same condition, and owners and builders are satisfied with the existing state of things. I should not be faithful to the profession which I represent if I did not ask owners and builders whether it is not high time that there was an alteration.

In the matter of the material employed in shipbuilding I believe that the iron in ordinary commercial use is better than is commonly supposed. Still it is not sufficiently good or uniform in quality to come up to the Admiralty tests, and tested iron is a special manufacture costing from £4 to £8 a ton more than that known as ship plates.

The fact that even this excess in cost is sometimes exceeded for bottom plates in ships of war brings up the cost of first-class boiler plate iron to near that at which excellent steel can be produced.

Steel is now being made suitable for shipbuilding and boiler making at a reasonable cost by both the Siemens' and the Bessemer processes. There are two ships of war now building for the Admiralty at Pembroke, of Siemens' steel in all parts of the hull and in the boilers, and nearly all the other ships building have certain portions made of a similar material produced by the Bolton Steel Company by the Bessemer process. The Admiralty is also using a more costly material produced by Sir Joseph Whitworth for cylinder liners and for propeller shafts, and the bodies of the Whitehead torpedoes are made of the same material, viz., fluid compressed steel, having about fifty tons per square inch of tensile strength and twenty per cent. of ductility, whereas the steel preferred for ships and boilers has only twenty-eight or twenty-nine tons of tensile strength, and from twenty per cent. to twenty-five per cent. ductility. This latter material corresponds in quality with that produced by the works of Messrs. Schneider and used by the French Government.

It is perhaps right that I should record that all vessels now building in the Royal Navy have iron or steel frames. Wood is employed for that purpose no longer. Timber is still largely used in some other national navies, but I think unwisely. We have vessels as much as 220 feet long, and with thirteen knots speed, having wood planking in two thicknesses on such iron frames, but with no iron skin. The inner thickness of wood is secured to the frames by Muntz metal screw bolts, and the outer thickness to the inner by copper bolts, and the whole is then sheathed with copper. There is an internal keel, stern, and stern post of iron and also external ones of wood. This system of construction is but little used in merchant ships.

Some of the ships having iron skins are simply protected by a paint composition. That of Dr. Sim, among others, gives excellent results as to preservation of the iron, and very good results as to cleanliness.

In order to preserve a clean skin and maintain speed longer that can be expected with paint upon the iron, it is the practice in some cases to sheathe the outside either with copper or zinc. If copper is used, two thicknesses of wood sheathing are wanted between the iron and the copper, the outer one fastened with brass, and the stem and stern-post need to be also of brass. There is still some risk, without careful workmanship, of injury to the iron by galvanic action. To avoid this, and to simplify and cheapen the work, zinc has been employed with only one thickness of wood between, and iron fastening. It appears, so far as experiments have gone, to preserve the iron perfectly, to keep as free from weed and shell-fish as copper; to be as readily cleansed or washed down as copper, but to oxidate with a somewhat rougher surface and therefore with more friction.

I would call attention to the life-saving apparatus exhibited, especially to that of the National Life Boat Institution, and Mr. White's boat bridge, as fitted in the troopship "Orontes."

In conclusion, I would refer to what is being done in training naval architects and marine engineers in England.

In 1864 the Education Department, at the instance of the Institution of Naval Architects, established the Royal School of Naval Architects at South Kensington. This school existed nearly nine years and entered 119 naval architects and marine engineers, of whom thirty-eight were private and eighty-one Admiralty students. In 1873 it transferred twenty-four of its students to the Royal Naval College at Greenwich, and ceased itself to exist. The entries at the Royal Naval College between October, 1873, and the present date have been, including those transferred from South Kensington, twenty-four naval architects and 113 marine engineers.

I regret to have to say that of these there is not a single English *private* student in naval architecture, all have been sent by the Admiralty, and there is but one private student in marine engineering. Of foreign students there are at present ten at the College, viz.: three Russians, two Italians, two Danes, one Spaniard, one Norwegian, and one Brazilian. The total number of all classes trained between 1864 and 1876 has been 232, of which twenty-four have been foreign. These foreign students have the full advantages of the Royal Naval College in every particular.

The PRESIDENT: In rising to propose a vote of thanks to Mr. Barnaby for his most important and valuable communication on Naval Architecture, I wish only to point out that it is perhaps the most important branch of engineering. If you take it in a national point of view, it may be said that the safety of the island depends upon it; in a commercial point of view it carries the treasures of the world to and from these shores; and in a humanitarian point of view it involves the safety of thousands of men who work hard for the interests of the country. It is encouraging to find that in naval architecture perhaps the most rapid progress has been made that could well be imagined, in fact a progress which leaves us every year more in advance of what we understood was the newest and the best the year previously. I am sure you will all agree with me that our best thanks are due to Mr. Barnaby, the Chief Director of Naval Construction, who by his earnestness, his impartiality, and devotion, has himself done so much to advance that branch of engineering.

I will now call on Mr. Froude, who it is well known has worked hand to hand with the authorities of the Admiralty in determining the exact form of ships which gives the least resistance, and the greatest form of safety and convenience. Mr. Froude's name has already been mentioned by Mr. Barnaby very prominently, and his work is so well known that I need make no further introduction, but will call upon him to give us his communication.

Mr. FROUDE, M.A., F.R.S.: Mr. Barnaby's allusion to the experiments which I am conducting for the Admiralty on this subject has made this an appropriate occasion for my telling you something of the nature of the work. I regret that I have not had time to put on paper in a carefully prepared and more intelligible form that which

I have to say, but I will endeavour to make it as clear as I can, and I will only ask you to bear in mind that the attempt to be brief generally leads to some obscurity.

I will divide what I have to say into three portions. I will tell you, first, what it is I am doing; secondly, as well as I can I will explain what is the justification of the mode of procedure in using models to test the forms of ships; and, thirdly, I will give you some slight sketch of the result that we have obtained and are obtaining. The thing to be done is, to determine by measuring the resistances of a model at various speeds, what will be the resistances of a ship similar to the model at various speeds, and for convenience I will in the first instance give you a term which is very serviceable in reference to this subject, the term, namely, which expresses the form in which we embody the ascertained merits of the model or ship as a body moving with greater or less resistance. We reduce the results obtained from each form, to what we call a "curve of resistance." We introduced that term because it is a short expression and means a good deal. On a straight base line are set off a series of speeds beginning with zero, and going on to one, two, three, &c., in units of speed, whatever they may be. For every speed for which the resistance is ascertained, the resistance is set off a to scale, and is plotted as an ordinate on the base line at the appropriate speed. Having thus obtained a series of ordinates representing the resistances, a curve drawn through their summits constitutes the curve of resistance. That curve expresses the resistance which the model or ship will experience at any intermediate speed. In speaking of the resistance of a model, I shall generally speak of its curve of resistance as expressing this.

The object, then, is by the use of models to determine the curves of resistance for ships, and in order to do that with effect, it is necessary to be able to produce models, rapidly, economically, and exactly. For this purpose it was necessary to find some more suitable material than wood. It is well known that to make a large exact model in wood, occupies considerable time, requires skilled labour, and costs a great deal; but happily a rather new material in organic chemistry, paraffin, —so called because it has little affinity for anything else—opportunely presented itself, and by the use of this material we are able if we please to complete a model sixteen feet long, of the general dimensions of which this specimen is a *frustum*, in twenty-four hours. We can produce that model, and in eight hours more we can test all its properties of resistance. That is a rate of operation which would soon outrun our powers of rationalised analysis, but we can attain it whenever an occasion requires it. Without describing in detail the *modus operandi* by which the model is produced, I may say that this material lends itself very easily to all the usual foundry processes. We cast the models in a mould made of clay, shaped approximately to the figure required, by means of cross sections planted in due order in the clay, so that we need not produce a pattern model in the first instance. The interior of the casting is kept hollow by a core, just as in the common foundry process, but the core is made of a light frame work, built in cross sections, covered with lath, and with a skin of calico, which is covered with a thin coating of clay and plaister of Paris, and thus rendered impermeable to the melted paraffin. The core, slightly loaded with ballast, is placed in the mould, and rests in it like a ship in a dock, leaving an intermediate space for the paraffin between the core and the mould, and the paraffin is run into this space. As the space is filled, water is poured into the interior of the core, and thus prevents its floating up, and, at the same time, assists to cool the paraffin ; and by the next morning the cold model may be taken out, washed, and placed on the shaping machine.

I will not describe that machine in full detail. It is simply a machine in which the model travelling on a bed, like that of a planing machine, between a pair of horizontal, highly speeded revolving-cutters which possess a horizontal travel, transverse to that of the model, is operated on by them at successive levels, their position being so governed that they cut on the two sides of the model, a succession of true water lines in pairs copied from the drawings of the model by an apparatus under the control of an operator. Originally to guide the apparatus, we used, instead of the water lines on the drawing, a succession of templates of the kind I have here, a very nice form, in which by the help of adjustible steel ordinates, spring-steel ribbons are arranged in curves to correspond with the waterlines in the drawing ; thus the series of templates when fastened together in due superposition, would constitute a skeleton model to the scale of the drawing ; but we have lately found that, working with a magnifying glass, the operator can as correctly follow the line on the drawing as it passes under the indicating point: thus the cutters cut a pair of water lines, simultaneously on the two sides of the model. The model is then changed in level and the successive water lines are cut; you see them here in a form analogous to that which the ship builder adopts when he sets about making a model. He gets his form by cutting out a series of pieces of board to form the several water lines, the thickness of the boards representing the interval between the lines, builds them together, and shapes off the intermediate matter.

Our particular material, paraffin, is most delightful to operate upon. It is quite strong enough for all the purposes of the model, but it is very easily cut with a knife, and may be operated upon with ordinary carpenter's tools, with the advantage that from being without grain it does not misguide the cutting edge. It has the pleasant property that it does not tend to choke the tools, there being no stickiness about it, in fact, with a smoothing plane a shaving of about one thousandth part of an inch in thickness can be taken off quite cleanly. The model, after being cut to the water lines, is shaped by suitable tools to the finished form. That form is represented by the bottoms of these cuts; but, as the bottom of a cut when once it is reached no longer exists as a guiding line, the operator when he had reached it would be "out of soundings," and would not know whether he had gone too deep or not; to get over that difficulty, as soon as the water lines are cut, a spur with short points, the alternate points being very short, is run along each line, leaving a series of corresponding punctures, and a little black powder is run into these punctures, so that when the work is finished it can easily be seen whether the operator has gone deep enough and not too deep. We have another test of the model's correctness. Its displacement is carefully calculated beforehand, and when it is finished its weight is taken. It is then loaded with enough ballast to give it exactly the displacement that it ought to have. It is placed in the water, and, by very accurate means for testing the immersion, we see whether it comes to its proper plane of flotation. With a model weighing 400 or 500 lbs. we generally get the displacement right within about two pounds, and if we were three or four pounds out we should think there was some error, and should probably cut the model afresh, or go over the calculations to see if there were any mistake.

Having told you how I make the models, I must next tell you how I test them. That is as important a part of the business as making them. I will ask you for the present to take for granted, what I will endeavour to explain presently in detail, that there is a true scale of comparison between the curve of resistance for a model, and that for a ship similar to the model. Meantime, I will describe how we ascertain the curve of resistance for the model. Our models vary from six to sixteen feet in length, from eighteen inches to two feet six inches in breadth, and weigh from 200 pounds up to 600 pounds, 700 pounds, and 800 pounds. They are not puny little models. I believe even much smaller models, if correctly interpreted would give valuable results, but these are irreproachably large models. The apparatus for testing consists of the place in which the testing is done, and the testing apparatus itself. The place in which it is done consists of a tank or water space, 280 feet in length, ten feet deep, and thirty-six feet wide, roofed over throughout. Compared with the length of the models we use, the run we are able to give them in testing, is nearly equivalent to the measured mile for the largest ships in the navy, and the depth of the water is about equivalent to the depth of the British Channel, so that we proceed on a large scale of operations, and one by which we have no fear of encountering an adventitious resistance due to the formation of shallow-water waves. The apparatus is arranged as follows:

Just above the surface of the water, throughout the length of the water space, is laid a railway, perfectly level and firmly suspended from the roof of the building without the interposition of any intervening columns near the track of the model, so that there is nothing to obstruct the motion of the water; and a dynamometrical carriage runs on this railway. In the dynamometric carriage there is a form of apparatus, well understood by many of you, a revolving cylinder, the circumferential motion of which represents, to scale, the progressive motion of the railway carriage. The scale is exactly one-fifth of an inch to a foot, so that every foot the railway carriage moves, is represented by one-fifth of an inch of circumferential motion of the cylinder. There is under the railway carriage, a spring balance, and to that the model is hooked or harnessed. The only force which is applied to the model is through this spring balance, which serves as a tow rope, and the extension of the spring balance will, in such a case, correctly represent the resistance experienced by the model at whatever speed it may be moving for the moment, which must be exactly equalled by the whole towing force which is being administered to the model. The extension of the spring is recorded, on an enlarged scale, on the revolving sheet of paper, that is to say, an arm connected with and representing on a large scale the extensions of the spring, makes a trace on the paper as it revolves, and, at the same time, a piece of clockwork connected with an electric circuit makes a contact, which inscribes a mark on the cylinder at each consecutive half second, and by the intervals between these marks, measured by the distance scale, we see how much distance is moved in each half second, that is to say, the speed at which the operation is conducted; we thus obtain a graphic record of the exact resistance experienced by the model and of the speed at which she is moving. I should mention that it is necessary, besides pulling the model by this spring balance, to guide her very inexorably, so that she may not swerve from her course. Because it is a curious circumstance connected with the theory of the motion of such a body through the water, that if you start a perfectly symmetrical model or ship freely on a rectilinear course, though it will perhaps run straight for a little distance, it most likely will suddenly begin to swerve, and will then very quickly turn broadside on. It is not sufficient to lead the head end inexorably in a straight line. If you tow a sixteen foot model by her nose inexorably in a straight line, you would, probably, expect that the stern would follow in the straight line, as a flag will follow the mast. But instead of that being the case, after the model has gone a little distance you will see its stern swerve to one side or the other indifferently, and there remain at an angle of 5°, 6°, or even 10° out of the straight line, acccording to the form, with a curious set of eddies on the two sides of it. To resist this tendency, the only constraint to which the model is subject, is applied at each end by a delicate knee-jointed nicely counterbalanced frame, which allows its head and stern to rise or fall and move backwards or forwards with perfect freedom, so that it does not experience the slightest pressure, except that which restrains its desire to swerve from its course, and this force being transverse and not in the line of motion does not affect the model's resistance.

The diagram we obtain then is an exact record of the resistance of the model experimented on, when moving at the recorded speed. The dynamometrical carriage is drawn by a wire rope, moved by a steam engine, fitted with a delicate adjustible governor, which enables us to administer such speed as is required. We can send the truck at any required speed from 50 to 400 feet per minute, nay, even to 1200 were we to desire it. But besides that, we have the definite record of the speed, and any small inexactness in the speed supplied by the governor, is corrected by the record which I have described. When we have thus got an exact transcript of the resistance of the model at each speed, the analysed results constitute that curve of resistance which I have described to you.

I should have mentioned that there is another phenomenon which we also record in connection with the motion of every model, namely, its rise or fall, that is to say the change of level it exhibits at either end. It is a new fact, which, I believe, we were the first to show, that every form moving along the surface of the water like a ship at reasonable speed, subsides in the water; some forms subside more at the head, some more at the stern, but there is always a bodily subsidence produced at the ordinary speeds of propulson. To record this, a little apparatus connected with the ends of the model makes a second diagram on a revolving piece of paper, one line for each end, showing any change of level exhibited by the model in the course of the run. The model immediately settles into condition, and the condition is maintained throughout, and we have a diagram of it.

I have now described the manufacture of the models, the method of testing them, and the result, viz., that for each model we obtain a curve of resistance such as I have sketched. I will next endeavour to explain the comparison between the curve of resistance of a ship and that of her model, and in order to do that, I must travel a little into theoretical view of the matter. The old notion of a ship's resistance was, that you began with a midship-section, and that as you gave the ship finer and finer ends you reduced the resistance to some fractional multiple of the resistance which would be experienced on moving the midship-section itself through the water, as a plane moving at right angles to itself; and various formulæ, mathematical or emperical, were supposed to define the appropiate multiplier. Now, mathematicians were not unaware of the defects of that mode of viewing the question, although in text books on mathematics and hydraulics, it was always entered as a provisional way of looking at things; but in recent years the higher mathematicians have worked out, in a very correct and conclusive form, what is called the theory of stream lines, a theory which represents correctly all the motions that the particles of the water would undergo in encountering any body which is passing through them, and the force which they would exert on each other and in the body if water were a frictionless fluid; and in order to explain that theory tolerably clearly I must first ask you to go with me beneath the water. We will consider first what happens to a fish. The theory of stream lines tells us that were water a frictionless fluid, the fish would swim absolutely without resistance when once put in motion. The proposition sounds extremely paradoxical, but I do not mind stating it in the most paradoxical form. The theory of stream lines has demonstrated most conclusively that if water were a perfectly frictionless fluid, even a plane moving at right angles to itself would not experience any resistance when once put in motion. It would experience resistance while being put in motion, because of the dynamical conditions of the surrounding fluid; all the particles around it must have some motion duly related to the motion of the plane or other body, and that companion motion has to be established at starting, and the force experienced by the plane or other body while the motion is being established will be felt initially as a resistance; but when once the motion had become steady, no farther resistance would be experienced. The converging stream lines behind the plane or body would exercise just as great pressure on its back surface, as would be experienced by its front surface, while forcing the particles from their natural position into the divergent stream lines in front. This seems paradoxical, but we can arrive at the conclusion by rational steps by changing our mode of approaching the question.

If we imagine that the fish is a fixed body, and the ocean is moving past it, we shall get a conception which will render it more easy to follow the law that I am endeavouring to explain. If we imagine the body, say this egg-shaped body, moving straight through the fluid, in one direction, it is obviously the same thing to imagine the body at rest and the water moving in straight lines past it. Let us trace out this supposition. If the body were out of the way, we could imagine the whole ocean to be cut up into filaments or minute streams of fluid, each following a straight course *ad infinitum.* But when the fixed body intervenes those streams must behave differently in two separate respects. Each stream as it approaches the body, in the first place, is deflected from its course in order to get past the body, and back into its former course ; again when it has passed the body in the next place it has to undergo changes of velocity, chiefly a temporary increase of speed in passing the local contraction of the waterway, which the presence of the body has created, so that the same ocean of water may pass the body as would have passed through the whole of the unoccupied space. These streams as they pass the body have an increased velocity and pass on and lose their increased velocity. Now what the theory of stream lines shows indisputably is, in the first place, that the deflection of a stream from its course if it returns to its former course, causes no total force in the line of motion. That is easy enough to see if you imagine a pipe of parallel diameter all through, bent into an easy curve, like the waterline of a ship, and terminating in the same direction which it had at its entrance. The stream while being first deflected no doubt exercises centrifugal force which tends to push the pipe forward, but when the deflection is reversed the centrifugal force tends to push the pipe in the opposite direction. If that operation is traced through exactly, it is easy to see that the total result is that there is no aggregate pressure in the line of motion. It is the ordinary problem of a marble running round a frictionless curve ; the marble loses no total velocity, the curve it passes round does not obstruct its run, nor does it push the curve in the aggregate in any direction, if it leaves the curve in the same direction, and with the same velocity it had on entering the curve. By the time it has traversed the whole curve it has pushed it just as much backward as forward. The next proposition is, that a stream flowing through a pipe of which one portion is contracted or enlarged, behaves in a very different manner from what one would expect. If we imagine a level parallel pipe of any length, and a stream of water flowing through it with a steady speed, if the flow be frictionless the pressure will be uniform throughout the length of the pipe. Now let a portion of the pipe be tapered to a smaller diameter, and again enlarged to its original diameter, the water must flow through the narrow portion with increased velocity—that is a mere arithmetical consequence. *Prima facie*, it seems certain that the water must be very much squeezed in going through the narrow part, and must there tend to push that part of the pipe forwards; but on the contrary, the moment you look into it, you will see that instead of the water being more pressed at this point than anywhere else, is a great deal less pressed, there is much less pressure there than anywhere else, and if you have a series of vertical pipes connected with the interior of this pipe and the water is allowed to rise in them until it stands at a height representing the pressure, you will see that at the part before the contraction has been created, the water will rise its natural level, but in the contracted part it will have been greatly lowered, and indeed if the contraction is sufficient there may be produced a partial vacuum. There is a little instrument which Lord Rayleigh showed me the other day, in which by connecting a pipe of that sort with a vessel of mercury, and blowing through the pipe with the mouth, you are enabled to raise eight, nine, or ten inches of mercury by the partial vacuum that forms itself in the contracted portion. The rationale is clear. The particles of water at the commencement of the contraction are going at a certain speed, and in the narrow part the same particles are inevitably going with a greater speed. How can particles have possibly acquired increased speed in that direction except by the circumstance that the pressure in front of them is less than the pressure behind them? The very fact of this acceleration is a proof, in terms, that there is less pressure in front than behind, and for that reason this contraction is a region of small pressure, and the original pipe a region of comparatively large pressure. *Vice versa*, it you take a pipe and put an enlargement into it, we find the opposite thing happens. There, the particles which were going with a certain speed at the commencement, when they arrive at the enlarged portion are going with a proportionately reduced speed, and that can only have happened by there being more pressure on each particle in front than behind it. Each particle as it enters this region is being pushed backwards. As the pipe gradually resumes its natural diameter after passing either a contraction or an enlargement, the original pressure is also gradually resumed, and regarded as endways forces, the alternations of pressure exactly balance each other, and the pipe experiences, on the whole, no endway push. The exact *modus operandi* by which the particles adjust themselves mutually to the change of speed is somewhat complicated, but that the fact must be as described is I think obvious, and the fact is that every contracted stream in the flow of an infinite ocean of fluid is a region of increased speed and decreased pressure, and every enlarged current is a region of decreased speed and increased pressure, and these changes of pressure, if the fluid be frictionless, can produce, on the whole, no endways push on the body whose presence causes the changes of speed and pressure.

Now I think I may proceed to treat of the case of a fish moving under water, or, rather, of the streams, so to speak, in the infinite ocean flowing past a stationary fish, and show how the alterations of speed and pressure which they experience would, but for friction, result in an equilibrium of endways force on the fish. At the nose of the fish, when the streams are being deflected outwards to get past him, forming lines convex towards his nose, their aggregate centrifugal force piles up an excess of pressure, constituting what I have called a region of increased pressure, which must also be a region of diminished speed and of enlargement of sectional area in the streams contiguous to the nose, and there is thus established an increase of sternward pressure on the nose, which however we shall presently see is counteracted by an equivalent increased pressure on the tail. Along the fish's middle body, where the streams, which have been put into outward motion by the nose, are losing that motion, and by a reversal of curvature are being deflected inwards into convergence, the modification of pressure just described is reversed, establishing in the contiguous streams a region of diminished sectional area, of increased speed, and of diminished pressure, which acting partly on the widening forward half of the middle body, partly on its narrowing after end, results in an equilibrium of endways force: while at the tail where the converging streams are again being deflected into purely parallel sternward lines, an excess of pressure, depending on the same conditions as that experienced by the nose, and exactly equivalent to it, becomes established. Thus, in the frictionless fluid, equilibrium of endway pressure on the fish would on the whole subsist, and when once put in motion with a steady speed the fish would experience no resistance.

But water is not a frictionless but a frictional fluid. The consequence of that circumstance is that the streams as they flow along the side, do experience a great frictional drag, and pull the fish backwards by dragging along its skin. It may seem absurd to say that so smooth a surface as that of a fish can experience such a frictional drag, but it is the fact; indeed experiment shows that a slippery surface like that of a fish, experiences more, not less, frictional resistance than a hard smooth one; and I may tell you that this friction on the surface not only pulls the fish back, but like the friction of any other body also disturbs that beautiful arrangement which constitutes the system of stream line motions and forces. The mutual drag of the surface and the water, extinguishes in frictional eddies some of the energy of the contiguous streams, and prevents them from re-instating the pressure at the stern of the fish. In the flow of water through pipes a great amount of energy is lost in friction. Even in a straight pipe, as in a contracted pipe or an enlarged pipe, there is a loss of work by friction, and that is exactly what happens in the case of a submerged fish. The fish would experience no resistance in frictionless water, but it does experience resistance from friction in actual water.

Now to go to the case of a ship moving on the surface. First I will suppose the ship to be cut off at the level of the water, and I will suppose the surface of the water to be held down by a sheet of rigid frictionless ice which the ship will touch, but with no pressure; that ice-sheet, offering a reaction to the surface of the water equivalent to that of an infinitely extended ocean, would enable the streams I have just described in the ocean of a frictionless fluid, to flow past the ship just as they would past a fish in a frictionless ocean at a great depth beneath, that is to say, without tending to push her stern-ways.

The ice surface would in effect constitute the remainder of the ocean; but there would be impressed upon the ice considerable force by the differentiated pressures induced in the stream lines as they flow past. At the head end of the ship, where there is an excess of pressure from the streams being retarded, the ice would be pushed upwards; along the sides, where the water is accelerated and there is a diminution of pressure, the ice would be pulled downwards, and at the stern again the ice would be pushed upwards; there would be a stress put upon it. Now, if we were to remove the ice, the water which ha been held by it would arrange itself into hills and hollows. These hills and hollows represent, not precisely the stream line pressures, but the effect of those pressures. The pressures are in fact modified by the deformations of surface, and those hills and hollows, when once formed, assume the form of and behave as, waves, and become thus to a certain extent independent of the parent that created them. Thus, they travel off into the surrounding ocean, taking with them more or less, indeed at high speed very much, of the energy which was put into them in their creation, and this is one great cause of resistance to a ship moving at high speed at the surface. Thus when the ship is moving only at moderate speed—so moderate that the waves cannot raise themselves but immediately subside in their track, they do not travel away into the surrounding water, and then the ship's resistance consists, I may say solely, of surface friction. There is indeed, besides, a small amount of head resistance, due to the distortion of the stream lines, which results from the friction of the streams against each other, in addition to that due to their friction against the sides of the ship, and to the slight loss of absolute pressure which thus ensues; but if the ship has fine lines this is inconsiderable, and at moderate speed the ship's resistance is, practically, simply that of surface friction. With our models, for instance, whenever we go at very low speed—say fifty feet per minute, with a model ten or twelve feet long—we find by actual experiment the resistance is just exactly that due to surface friction. I state that as a fact, but I must tell you that in order to be able to state it with certainty, it was necessary to ascertain the measures of the surface frictional forces, and we instituted a very extensive series of experiments to determine that. We tried plane surfaces nineteen inches wide across the line of motion, varying in length in the line of motion, from three and four inches up to fifty feet, and scarcely more than one-eighth of an inch in thickness, and by a dynamometer we ascertained exactly the resistance at all speeds of all portions of these planes. Several curious facts are connected with the law of such resistance. One is this. The resistance is more ardent, if I may use the expression, at the anterior end of the plane than at the posterior. When one begins to think of the surrounding conditions, it seems wonderful that the difference is not greater than it actually is. Inch by inch along the length of the plane there must be growing into existence, in virtue of the frictional force exerted on the contiguous particles by the friction of the plane, a pair of forward streams, alongside, and if the plane is fifty feet long, they are seen to have, at the tail end of the plane, a breadth of, say, seven or eight inches on each of its sides, and to have a mean velocity, something like half that of the plane itself. It is extraordinary how it happens that particles possessed of such large concurrent speed as those which form these streams, can still exert a not very greatly diminished backward drag on the plane; it is true that the front edge of the plane the intensity is greatly increased, and is quite incommensurable with what it is at the after end. If the anterior edge is of polished metal quite keen, and a very thin coating of tallow is put on it, the friction will strip this back, and arrange it in a pair of thread-like ridges immediately behind the edge; but after this very narrow strip of maximum intensity is past there is for the first few feet a moderately graduated diminution of intensity; and after ten feet of length the diminution of intensity is relatively insensible. However, without affecting to have solved the question theoretically, we have gone through it practically with sufficient completeness to know what mean resistance per square foot is exerted by any given lengths of the different qualities of surface, and we know that surface friction is a very sensitive function, and that until we learn the exact quality of the surface of the ship or body, it is impossible to predicate what its resistance will be. I may tell you that a plane, coated with tin foil six or eight inches long, makes only half the resistance of a plane coated with clean paint or varnish, but when you get to fifty feet long the tin foil makes almost exactly the same resistance as the clean paint; and again a slimy surface, like that of a fish, makes more resistance than a perfectly hard smooth surface. By the help of these experiments I am able to say approximately what the surface friction of a ship of given quality of surface should be, and in virtue of that determination I am able to tell you that with a model of say twelve feet in length, at a speed of fifty or sixty feet per minute, which is a moderate speed for the model, and is equivalent to a speed of six knots for a ship 300 feet long similar to the model, the resistance is nothing more than that due to surface friction: we find, in fact, by the dynamometer that there is not any more than that. I have, therefore, no hesitation in saying that the fundamental proposition I have asserted is borne out by facts.

But when we come to higher speeds we immediately see that the other principal cause of resistance begins to operate ; the waves begin to form themselves. The whole theory of an investigation of merit in form is complicated by that condition, and you have to arrange your form so as to produce the smallest amount of wave motion. I ought to say that the problem to be solved in finding the best form for a ship is one rather difficult even to state satisfactorily. When we say we want to make a ship of a given length, breadth, and depth, which shall go with the least resistance, or a ship of a given displacement, there are, besides, a multitude of collateral conditions to be taken account of, such as comparative first cost, comparative liabiliiy to wear and tear, &c. ; but these considerations can perhaps be more properly introduced as make-weights when an approximate solution has been supplied on broader grounds. Viewing the matter thus broadly, the clearest statement of the proposition I can frame to myself is, that we wish to carry an amount of *useful displacement* at a given speed with the least expenditure of power. By useful displacement, I mean displacement available for the purposes for which the ship is primarily intented, *e.g.*, the carriage whether of goods or passengers, and exclusive of that devoted to weight of hull, of engines, coals, rigging, crew, and stores ; not that these things are, as a matter of fact otherwise than useful, and indeed necessary, but that any reduction in their amount, compatible with the primary purpose of the ship, is a gain, not a loss. Now as any variation in the ship's form must affect the outcome of each of these conditions separately, it is plain that the solution is a very complex one, and in point of fact we cannot arrive at anything like a general determination of what is the best form for a ship. Another circumstance that makes that difficult, is that the relative merits of different forms will vary with the speeds at which they are compared. The form which is best for a ship to go seven, eight, or nine knots, if it is a long ship, would be very sensibly different from what it would be if she is to go sixteen or eighteen knots, again you should adopt a much shorter ship for moderate speeds, because at such speeds the surface friction is the greatest element of resistance, and is relatively greater in the longer ship.

What I have now to explain to you is how to frame the comparison between a ship and her model. When we have got the curve of resistance for the model, we draw on it a second curve, the ordinates of which are lines representing the surface friction of the model. The differences between the ordinates of the two curves are the forces solely due to the formation of the models waves. Now it is a property of waves that similar waves move with speeds proportionate to the square roots of their dimensions, and from that, if it is fairly looked into, it follows that comparing a ship and a model of given lengths, the waves made by the ship, and the waves made by the model will be exactly similar if the speeds at which the ship and the model respectively move, are proportional to the square roots of their respective dimensions. Bear in mind that the resistance consists in effect of two elements, one surface friction and eddy-making, the other wave-making; I deal now with the latter in its comparative effect with ship and model: If the ship of which we make the model is sixteen times as long as the model, the waves that the ship makes at a speed of ten knots, will be similar to those which the model makes at two and-a-half knots, since four is the square root of sixteen, and two and-a-half is a quarter of ten; and on that basis we calculate the resistance that the ship will experience in wave-making, compared with that which the model experiences in wave-making. It comes out that if the speeds are what I call the "corresponding speeds," that is are proportionate to the square roots of the dimensions, at such speeds the resistances due to wave-making are proportioned to the cubes of the dimensions, that is to say, in the instance just given, the ship going ten knots will have a wave-making resistance 4096 times that which the model has at two and a-half knots, 4096 being the cube of sixteen, and, as before, two and a-half being ten divided by the square root of sixteen. In that way we can calculate the ship's resistance due to wave-making from the model's resistance due to the same cause. Then again in calculating the ship's surface friction, knowing her surface, and the speed at which she is moving, and the quality of the surface, we can calculate what will be the resistance due to it at each speed, and by adding this to the resistance due to wave-making we can construct a complete curve of resistance for the ship from the resistance of a model similar to the ship. If the resistance due to surface friction were simply as the area, and as the square of the speed, irrespective of the question whether the given area were a long one or a short one, it would follow that in comparing ship and model the relation between their entire resistances would follow the law just enunciated as governing the relation between their wave-making resistances; but as has been explained the mean resistance per square foot of a long area is less than that of a short area, and thus the ship's surface friction is relatively a little less than the models, and her surface friction must be calculated separately; but when similar ships are compared the differences in length of area will be unimportant, and the law may be treated as exactly true, namely, that *the entire resistances of similar ships at corresponding speeds are as the cubes of their respective dimensions.* The principles on which this law depends render it equally true, however abnormally the wave-making resistance may grow, in terms of the speed.

I am afraid it would take me too long if I were to attempt to tell you in detail all the results of this investigation, but speaking generally, we must for high speeds have long ships, because short ships at high speeds make very big waves; but at moderate speeds short ones do best, because there is less surface friction, and at such speed, that is the chief element of resistance. Practically, what is meant by moderate speed is a speed which, in knots, is quite short of the square root of the ship's length in feet. If you make a well-shaped ship 100 feet long to go at a speed of ten knots she will begin to make a rapidly growing resistance, and at such a speed with a ship of ordinary proportions the total resistance will be about one two-hundredth part of the ship's entire weight. At a speed in knots about 1·35 times the square root of the length in feet, she would be making a very large resistance indeed, probably there increasing nearly as the sixth power of the speed. When that is the case, then a ship of larger dimensions and with the same lines would make absolutely less resistance at that speed. I may say the "Shah" in her recent measured mile trial, going at seventeen knots was propelled by a nett force which was equal to one two-hundreth part of her whole weight, while Mr. Thorneycroft's swift launch on the contrary, which goes at such an extraordinary speed, requires, at a speed of a little over eighteen knots. a propulsive force of one-fifteenth part of her weight. So that you see at what an enormous rate the resistance grows even with the finest forms when driven at extravagant speeds. I think, therefore, the speed at which you may rationally aim is in knots about equal to the square root of the ship's length in feet. At that speed with the best form the resistance is about, as I said, one two-hundredth part of the ship's weight.

The PRESIDENT: Ladies and gentlemen—Every one who has listened to Mr. Froude's lucid explanation must have been struck by the important bearing which his experiments must exercise upon naval architecture. Not long ago it was supposed by every naval architect that the chief element of the resistance of a ship going through the water was its mid-ships section, and that if only the mid-ships section could be cut down the total resistance of the ship would be reduced and its speed increased.

Hence the tendency to add to the length of the ship, by which we reach the proportion of one in ten. Mr. Froude's experiments prove the fallacy of that train of reasoning, and show that the midships section has really nothing to do with the resistance of the water. That resistance is made up by the skin resistance and by the waves engendered by the rapid motion of the ship through the water. Hence this new principle will give rise to new results, and the ship of the future will no doubt differ very materially from the ship of the past. I wish now to propose a vote of thanks to Mr. Froude for his very interesting communication.

The Conference then adjourned for luncheon.

On re-assembling the President called upon Mr. Thomas Stevenson to read his communication

ON LIGHTHOUSE APPARATUS.

Mr. THOMAS STEVENSON then made the following extemporaneous statement: Perhaps I ought in the first instance to explain that the object of my being asked here is not to give an address upon the subject of light-houses generally, but merely to describe certain of the more important improvements which originated in Scotland, and which have been illustrated by models in the museum, some of which are now on the table.

There are many interesting subjects connected with lighthouses, to which, therefore, I shall not refer; for example, I shall not need to refer to the electric light which emanated from the researches of our illustrious countryman Faraday, and which in the hands of Professor Holmes and the Trinity House of London, has been carried out so successfully. Nor shall I need to refer to the very interesting experiments on sound now being carried on by Professor Tyndall, also under the auspices of the Trinity House. All that I intend to do is to describe some of the improvements which have originated in Scotland since 1822.

Perhaps I shall consult your comfort as well as my own convenience best by adopting a somewhat historical form. Prior to the year 1822, the best form of lighthouse apparatus consisted of a silver plated parabola. The optical principle of the parabola is perfectly well known. It is simply this: that all rays emanating from the focus and incident on its surface are rendered parallel to each other, and also parallel to the axis of the apparatus. Of course, if the radiant were a mathematical point, the rays which emanated from the reflector would be strictly parallel, but, inasmuch as instead of employing a mathematical point, a bulky flame is used, the rays which proceed exfocally have a certain amount of divergence, and without that the instrument would be practically useless for lighthouse illumination. Owing to this divergence you are enabled to place reflectors round about a frame, so as practically to light up the whole 360 degrees of the horizon. In like manner, if you had three or four parabolas, each having a flame in its focus, placed with their axes parallel upon a frame which was made to rotate, then whenever the common axis of the parabolas was pointed to a distant observer, that distant observer would receive a powerful flash of light, and as the frame moved round, and the axis was turned away from his eye, it would gradually die out and produce darkness.

In the year 1822—a year which will ever be memorable in the history of lighthouse optics, the distinguished philosopher Augustin Fresnel introduced the dioptric system of lighthouses. For this purpose he used a plano convex lens of the same form as had been proposed, but for burning purposes only, by Buffon in 1748, and in an improved form by Condorcet in 1788. A specimen, the finest I ever saw, is in the museum from the workshop of Messrs. Barbier & Fenestre, of Paris. The principle of the lens was simply this, that instead of having a continuously spherical surface, Buffon proposed to cut out the lens at the back so as to reduce the thickness of the glass, and to save the light lost by absorption in passing through the glass. Condorcet not only did this, but assuming different radii for curvatures of the different faces, was enabled to correct to a large extent the spherical aberration, and not only so, but to make the lenses in separate pieces.

Fresnel further improved the lens by grinding the rings to different centres, and so practically got rid of spherical aberration altogether. Then, Fresnel, instead of using a number of lenses, and with separate flames, used one central flame and surrounded it by a number of those thin annular lenses, so that whenever the axis of one of those large lenses is pointed to a distant observer he gets a flash of light, and when it passes away he loses that light. But in order to save the light which passes over the top of this apparatus, Fresnel had recourse to two agents, namely, smaller lenses inclined to the horizon, and plane silvered reflectors also inclined. By this compound arrangement the light was thrown out above the main lenses in horizontal beams parallel to those coming from the large lenses below. Such was Fresnel's revolving light, figs. 1 & 2, but he did not stop there. He introduced

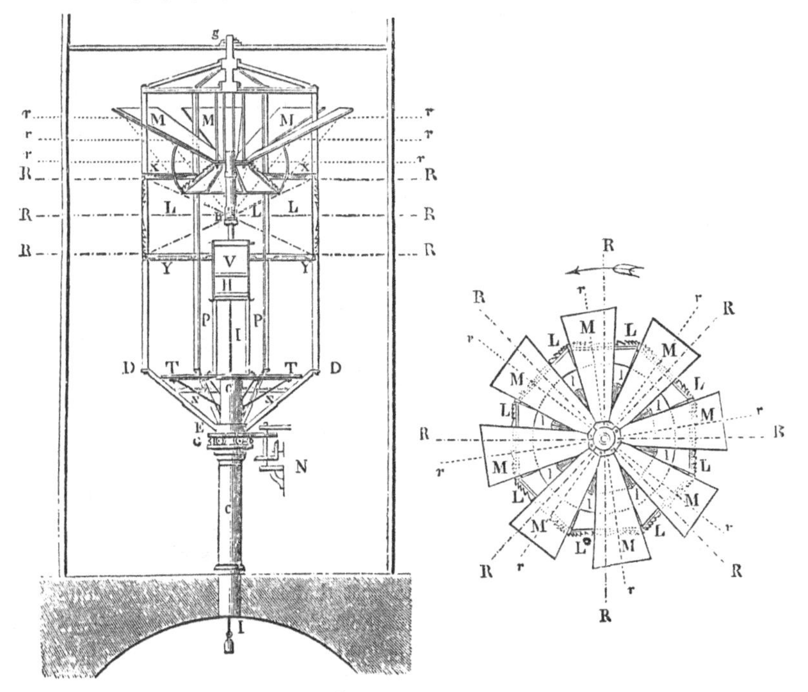

Figure 1 and 2.

L principal lenses, *l* inclined lenses, M plane mirrors.

very important improvements on the fixed light where he employed two totally new instruments. One of these was called a cylindric refractor. This consisted of the solid, which is generated by the revolution round a vertical axis of the middle section of the great annular lens. This cylindric refractor formed a hoop encircling the light, having lenticular action only in the vertical plane. These refractors, when of a large size, were made in the form of polygons, but in 1836, Mr. Alan Stevenson, who was the first to introduce the dioptric system into this country, made the central belt of the first order lights truly cylindrical, and adopted inclined instead of vertical joints for the glass sectors by which an obscuration of the light by the brass settings at any part of the horizon was rendered impossible. He afterwards applied the same principle to the astragals of the outer lanterns of the largest class of lights. The refractor does not in the least degree interfere with the spread of the light by natural divergence all round the horizon, but it parallelizes all the light that falls upon it in a vertical plane. In order to utilize the light which passed over the cylindric refractor, Fresnel afterwards introduced a totally new instrument, and not only so, but for the first time introduced the principle of total reflection into lighthouse illumination. The way in which he did this was by means of a series of prisms of a triangular section. The rays striking on the first face of the prism were somewhat refracted, then on the second face they were reflected, and passing to the last surface, emerged, after another refraction, parallel to the light coming from the cylindric refractor. Those rings are arranged horizontally round the vertical axis, and, in connection with the refracting hoop, enclose the flame in a complete cage.

I have now come to the year 1849, when certain improvements were made in Scotland. It is quite obvious even to a superficial observer that the parabolic reflector is a very imperfect instrument, inasmuch as only a portion of the rays are incident on its surface. What becomes of the rays which escape past that surface? A large cone of rays does escape past the lips of the reflector, and all of that cone is utterly lost. In order to save this light, what is required is to place a lens inside of the reflector (shown in fig. 3) and at such a distance in front of the focus as to give parallel rays, and not only so, but this

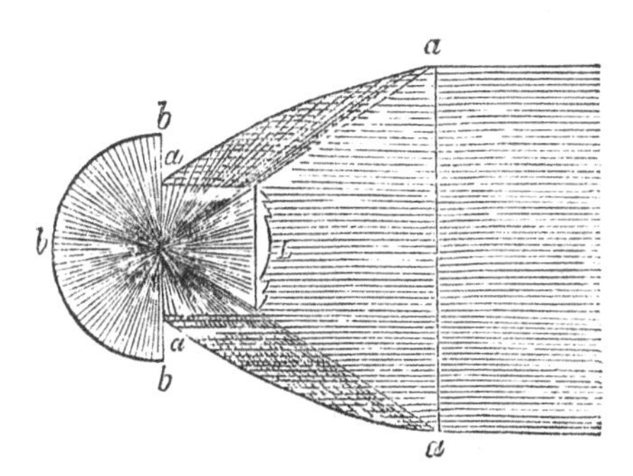

Figure 3.

a paraboloid, L lens, *b* spherical mirror.

lens must subtend the same angle at the flame as the cone of rays which escapes past the lips of the reflector. If this lens be put in the position I have spoken of, it is quite obvious that all the rays are rendered parallel, and there is no loss of light. But I have only spoken of one half of the light, namely, the front half, I have said nothing about the back half. Now the rays falling on the back portion of a parabolic reflector are parallelized by it. The object of the lens is also to parallelize the rays, and therefore, if the rays being already parallelized by the reflector were to fall on the lens they would be caused to converge, and afterwards to diverge, and thus be lost as before. In order therefore to utilise the light from behind, it is necessary to cut off the back of the parabolic reflector, and to beat it into a spherical form, so that the rays falling normally on this surface will be returned back again through the flame, and proceeding in a diverging direction will be intercepted either by the paraboloid or by the lens in front. In this way, the whole of the light will be utilised, and this therefore ought to be the light of maximum intensity. So far, it is an obvious improvement upon what Fresnel used for utilising the light which passed above the lens, for you observe he used two agents in his revolving light, whereas, if his plane mirrors had been made portions of parabolas, he could have dispensed with the inclined lenses which are inside the apparatus, and one agent would have done instead of two. But after all, although optically considered, this form of holophote is geometrically quite perfect, it is really not so perfect an instrument as might at first sight be considered, owing to the inevitable loss of light from the metal reflector by which a large portion of the rays is absorbed, and annihilated. What was wanted was to substitute for metallic reflection the principle of total reflection, which Fresnel had already successfully employed in the fixed light. In order to do that it was necessary to surround the lenses by prisms, having the same profile as those of the fixed prisms of Fresnel ; but instead of being generated round a vertical axis, they are generated round a horizontal axis, so as to parallelize the rays not only in one plane as in the fixed light, but in every plane. (Shown in fig. 4). On the table is an example of this

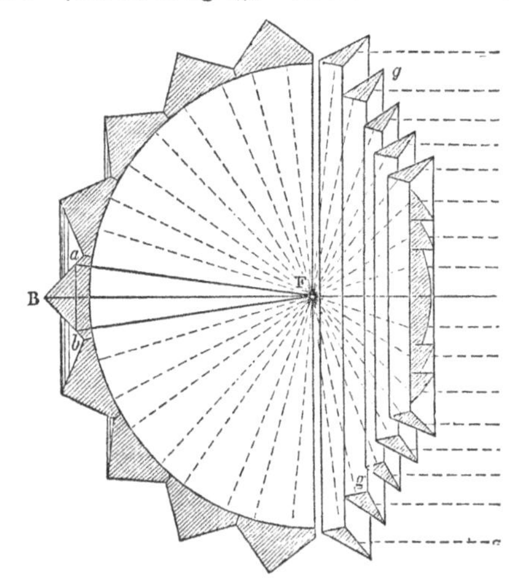

Figure 4.
g holophotal prisms, *a*, B, *b* prisms of dioptric spherical mirror.

light. But the apparatus described deals only with one half of the diverging rays. The other half falls upon the dioptric mirror which is composed of prisms which have their inner faces ground spherically to the flame as a centre. Entering those surfaces, normally, the rays suffer no refraction, and pass onwards to the reflecting surfaces of the prisms where being twice totally reflected, they return back again through their first surfaces to the flame. In order to do this it is quite obvious that the two back faces of the prism must be portions of parabolas, having their foci in the common focus of the apparatus, and their axes of generation coincident with the horizontal axis. This is represented in the model on the table. (Shown in fig. 5). It is quite plain that if this

Figure 5.

Section of one of the prisms of dioptric spherical mirror showing path of ray of light.

instrument be properly constructed—and for this purpose I may mention that I was indebted to Professor Swan, of St. Andrews, for the formula—a person standing behind the apparatus will see no light, although there is nothing between his eye and the intense flame which is burning in the focus but a screen of transparent glass. You will observe that in the focus of the apparatus on the table there is placed a large red ball which you can see through the front half, but when I turn the back of the apparatus to you, you will not be able to see it. You will see part of the stick that supports the ball because a large portion of it is exfocal, but the focus, and whatever is near the focus, will be totally reflected by this screen of transparent glass, and therefore you are not able to see it at the back of the apparatus.

I have to explain that since I designed this apparatus, which was in 1850, my friend, Mr. James Chance, of Birmingham, suggested some valuable improvements both as to the way of setting, and the mode of generating the prisms. These totally reflecting spherical mirrors are now very generally used in lighthouses, and there is an example of it down stairs to which I shall refer again presently.

I have now described the means of producing the most intense single beam of light from a single flame, but for the purpose of producing the revolving light it is better to adopt the plan which Fresnel employed of having one large flame surrounded by a framework containing large lenses. This is the holophotal revolving light (fig. 6) with the great annular lens of Fresnel, and holophotal prisms above and below, which intercept the light in the way I have spoken of. There is a beautiful example of this apparatus made by Messrs. Chance, exhibited by the Trinity House in the yard outside. The frame revolves round about the light, and whenever the lenses and the holophotal reflecting prisms come opposite the eye, you get a powerful flash which dies out as the axis is turned away. You will observe the distinction between this apparatus and the one I formerly described as constructed by Fresnel. He required two agents for the purpose of parallelizing the light, whereas it is done here by a single agent which produces a great saving of light. (Fresnel's revolving light is shown in fig. 1 & 2, and the holophotal in fig. 6).

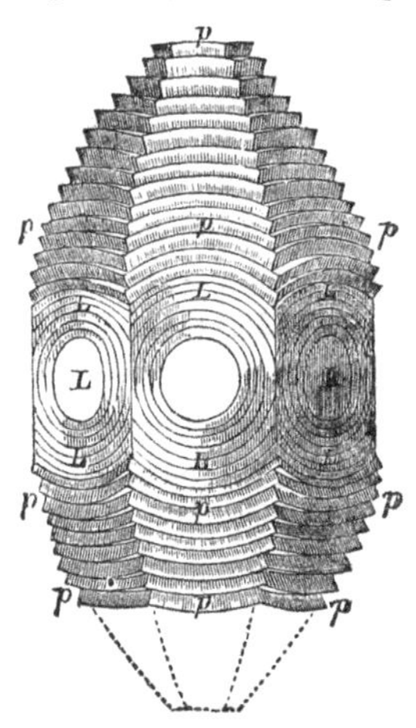

Figure 6.

L lenses, *p* the holophotal prisms.

Hitherto I have restricted myself rigidly to matters which were published, but I must now refer to an apparatus which I saw for the first time last Saturday in the museum down stairs. It is a small apparatus which is represented to have been made in 1825 by M. Tabouret for Augustin Fresnel, and it contains prisms which though not arranged for a revolving but for a fixed light, are of the same forms as those I have described, which were first published in 1849. I know nothing of the history of that apparatus, but I can state positively that no description of it was ever published until 1852, and that holophotal prisms were not introduced into lighthouses until 1850 when I introduced them, and not only so, but the fact is that during no less than a quarter of a century after 1825, apparatus of even the smallest size, where no difficulty of construction could possibly arise, continued to be made at Paris and elsewhere in which two agents were employed. Nor did Mr. A. Stevenson, who spent about a month at the Lighthouse Works in Paris, in 1834, see or hear anything of those prisms. Again, in Leonor Fresnel's description of apparatus which he published in 1844, he represents some improvements made in his brother's light, but he still employs two agents. I may also state that in 1850 I went over to Paris with my late brother, Mr. Alan Stevenson, in order to explain by means of a drawing and model this, among other inventions, but at that time M. Fresnel did not say anything about this old model; and still more striking is the fact that M. Degrand, who succeeded Fresnel as one of the engineers of the Lighthouse Board in Paris, claimed these prisms as an invention of his own in 1851. And equally inexplicable is the fact that M. Tabouret, who is said to have made the model of 1825, exhibited a new revolving light at the Great Exhibition of 1851, in which no such prisms were used. The only apparatus at the Exhibition in which they were, was sent by the Scotch Board. Lastly, so late as 1851 a first order revolving light was constructed for Cape L'Ailly in France. It was originally ordered on the 7th of May, 1851, by the French Administration of the old construction in which there were no prisms of the kind to which I have referred, but subsequently that order was recalled and the light was made holophotal, after the French engineers had seen the holophotal apparatus which was then being made for the Scotch Lighthouse Board, by M. Letourneau.

I shall now pass on to the azimuthal condensing principle for lighthouses. There are many peculiar situations which the lighthouse

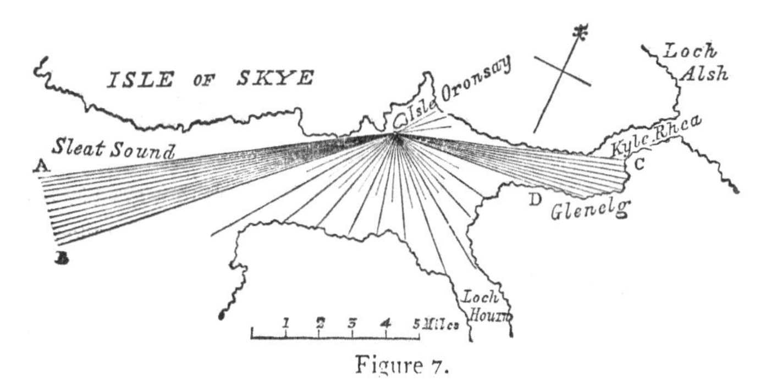

Figure 7.

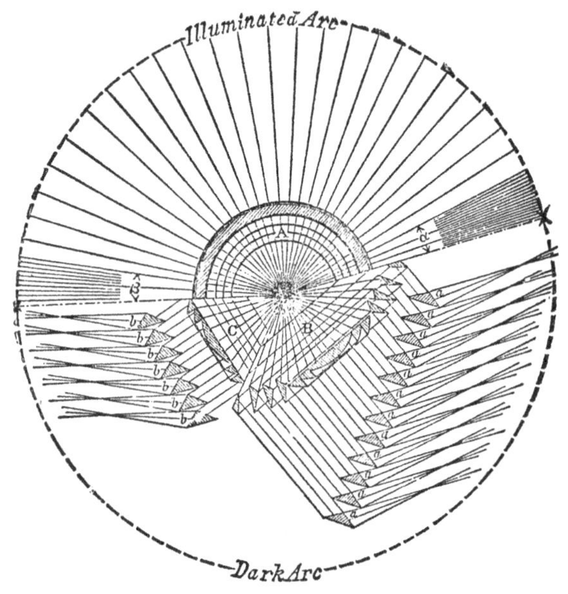

Figure 8.

engineer has to meet in a particular way. One of these is the case of a narrow sound. Suppose, for instance, it were required, as in the case of a place in Scotland, to light up a narrow sound. (Shown in figs. 7 & 8.) It is quite obvious that if you make the light sufficiently powerful to show across the sound, which is perhaps two or three miles wide, it would be very insufficient for showing up the sound, where the mariner had to see it at seven miles, and still more insufficient for showing down the sound where the light is required to be seen for fifteen miles. What is required, then, is a system by which the whole light from the lamp is spread horizontally with strict equality over any given arc in azimuth. Or in a light of equal range, where it must be seen at different distances in different azimuths, the light must be allocated to each of such arcs in the compound ratio of the number of degrees, and the distance from which it is required to be seen. I might describe this by a very simple case, but I prefer to allude to the case of Buddon Ness condensing light, the apparatus of which may be seen down stairs. (Shown in fig. 9 in section, and in fig. 10 in plan). The object there, was to condense the whole of the light

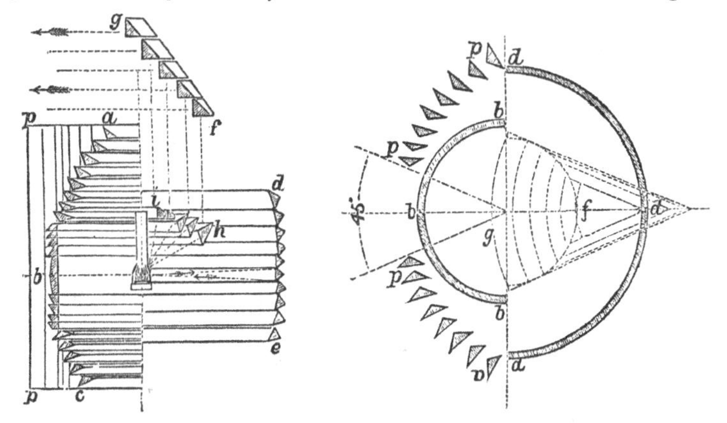

Vertical Section. Horizontal Section.

Figs. 9 and 10.

a, b, c Fresnel's fixed light, *p* straight condensing prisms, *i. h* suspended portion of holophote, *g, f* conoidal prisms.

coming from the flame into an arc of forty-five degrees in azimuth. The first thing to be done was to intercept the front half of the light by 108° of Fresnel's fixed light apparatus. But all that is in this case required to be illuminated is forty-five degrees. Fresnel's apparatus answers the purpose perfectly for the forty-five degrees, but what becomes of the light which comes from the other part of his fixed apparatus? In order to condense that light into the required directions there are straight prisms which were first used at Isle Oronsay in 1857, which straight prisms stand in front of the apparatus, as will be seen down stairs, and all the light (less forty-five degrees) which passes out of Fresnel's fixed light is received upon those straight condensing prisms, and they spread the light over the forty-five degrees, so that the whole of one half of the light is condensed into an arc of forty-five degrees; that is to say, the light which passes through the sides is so bent by refraction and total reflection as to reinforce that which comes from the forty-five degrees in front. Of course those straight prisms do not reflect the light in the vertical plane, but only in the horizontal. With regard to the back part of the light, we can at once avail ourselves of the spherical mirror; but the light that is passing above that would still be lost. In order to intercept that, there is suspended above the flame a half holophote, which has the property of combining the rays which fall upon it in a parallel beam. Then, above this vertical beam of parallel rays there are prisms which were first used at Buddon Ness—right angled conoidal prisms, which also have the property of spreading the light which falls upon them over forty-five degrees. In this way, by the union of the following instruments, the whole light coming from the flame is compressed into the arc of forty-five degress, viz., Fresnel's lens, Fresnel's fixed light prisms, the azimuthal condensing prisms, the cylindric refractor, the dioptric spherical mirror prisms, the holophote, and the conoidal prisms above.

The next thing I shall allude to is a simple apparatus for illuminating sunken rocks at sea. At Stornoway, in the Island of Lewis, in 1851, it was proposed to place a lighthouse upon a sunken reef, which would have been a very serious matter. Instead of that, a main light was placed upon shore, and in order to point out the position of the reef at night, the following plan was adopted. A perch or beacon, which is comparatively an inexpensive structure, was erected upon this sunken rock and on the top of the perch there was a lantern containing certain optical prisms. Exactly on a level with that lantern there was a window cut in the tower of the lighthouse, and in

Figure 11.

that a holophote was placed, and a powerful beam of parallel rays was projected upon the lantern in the perch. The optical prisms redistributed the rays which fell upon them, and to a seaman coming into the harbour the light appears to come from the perch, whereas in reality it comes from the shore about 530 feet distant. This kind of light is used in several other places. (Shown in fig. 11). The only other matter which I think it is necessary to mention is some new forms of prisms which have been recently introduced at Islay. (Shown in fig. 12). They have the remark-

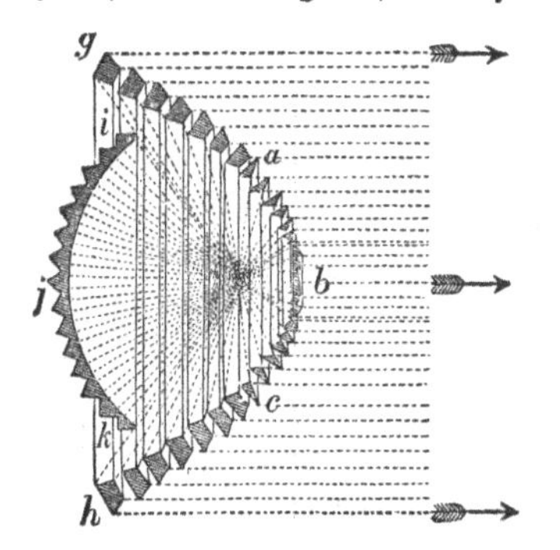

Figure 12.

a, b, c, lens and holophote ; *a, g, c, h,* new back prisms ; *i, j, k,* dioptric spherical mirror.

able property, of parallelizing rays, for an incidence as low as about 180 degrees. I think it right to state that these prisms were invented independently by Mr. Brebner and myself, and by my friend Professor Swan of St. Andrews, who gave the formula for constructing them. Another new form of prism, (fig. 13), which was described in *Nature* some months ago, is called a twin prism. This prism has its apex cut out, the object being not only to carry out an ingenious suggestion of Professor Swan of passing the light from a set of prisms through the chinks in another set placed before them, but to reduce the thickness of the glass through which the light has to pass. In this way you save much light, which would otherwise be lost by absorption. The first example of this kind of apparatus is now being made by Barbier and Fonestre of Paris, for the Scotch Lighthouse Board.

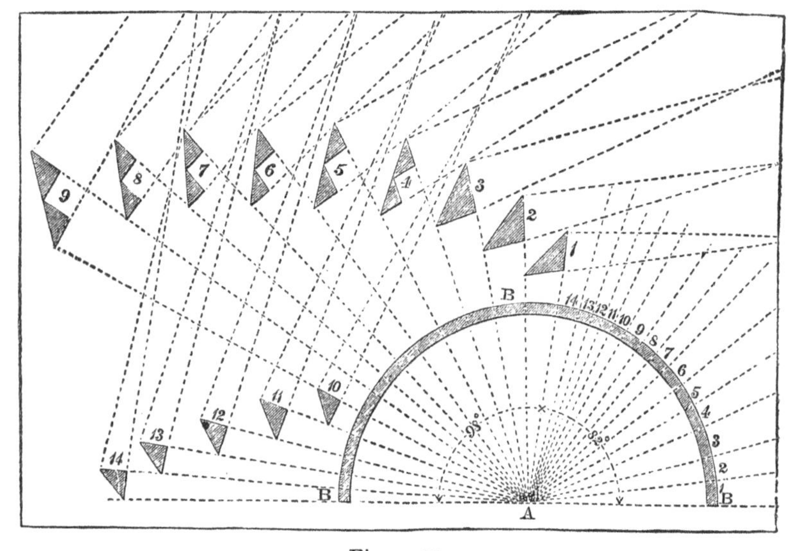

Figure 13.

B, Fresnel's fixed light; 1, 2, 3, 10, 11, 12, 13, 14, straight condensing prisms; 4, 5, 6, 7, 8, 9, new straight twin prisms.

I have nothing more to add, except with regard to the characteristics of different lights, which I may mention in two words. Formerly there used to be the fixed light and the revolving and coloured lights. But we have now two other characteristics, namely, the flashing light, which is a light producing extremely quick flashes, very distinctive from the revolving light, and the intermittent light which consists of occultations produced by frames coming down in front of the apparatus, and so cutting off the light suddenly, as opposed to the gradual waxing and waning which you have with a revolving light. These lights were introduced by the late Mr. R. Stevenson in 1827 and 1830. In 1871, Mr. R. L. Stevenson proposed an optically perfect intermittent light in which the periods could be made of unequal duration by means of eclipsing mirrors moving horizontally within the glasswork of a fixed light apparatus.

The PRESIDENT: I will now call upon you to pass a vote of thanks to Mr. Stevenson for his very interesting communication. I have little or nothing to add except to say that the art of projecting light by mechanical means seems to have attained a wonderful perfection. There are two branches of this subject which Mr. Stevenson might have mentioned, which he has not done, for want of time. There is the generation of light, either by the combustion of oils, or by electricity, all of which form part of the general subject. There is also the construction of lighthouses themselves, but we must be content with what we can practically attain to in collecting information on the exhibits which are here.

I must therefore ask you to pass a vote of thanks to Mr. Stevenson for his paper.

The vote of thanks having been carried,

The PRESIDENT: I will now call on General Morin for his communication on ventilation, a subject which no one has investigated more deeply than he has.

NOTES ON WARMING AND VENTILATION.

ON VENTILATION.

GENERAL MORIN, Director of the Conservatoire des Arts et Métiérs at Paris : The subject which I am about to bring before you, cannot be compared—with regard to scientific interest—to those questions which have already been discussed; but it affects so closely the well-being of humanity, that it has, for a long time past, attracted public attention.

Indeed the principles and the results of which I intend to-day giving you some examples, are by no means new. I will do no more than briefly call your attention to them.

In a work entitled *Illustrations of the Theory and Practice of Ventilation*, by Dr. Reid, of Edinburgh, and which was published in 1844, the principle is already expressed that *the introduction of fresh air must take place from the ceiling, and its escape from the floor.* The author had applied this system to the Club Room of the Royal Society of Edinburgh. Mr. Goldworthy Gurney, a skilful engineer, had also expressed this principle in these words: "It is desirable that fresh air should enter from above, and escape from below. (Committee of Enquiry of the House of Lords, page 41.)

The advice of these gentlemen was disregarded, and about the year 1845 there was still to be seen in the subterranean galleries of the Houses of Parliament two ventilators, 6 feet in diameter. They have since been removed.

I will but say a few words upon heating apparatus. Ordinary chimneys have the advantage of allowing an agreeable and wholesome view of the fire, whilst at the same time they cause the renewal of the air in the room ; but they are inconvenient inasmuch as they occasion draughts through the doors and windows. Only from 0.10 to 0.12 of the heat given forth by the fuel is made use of.

The ventilating chimneys proposed long ago by Mr. Belmas, colonel in the French Engineers, in the eleventh number of the "Mémorial de l'officier du génie," and for the establishment of which Captain Douglas Galton has made excellent suggestions, have the advantage of bringing into the room as much air at a moderate temperature as they carry out through the smoke-pipe, and of making use of from 0.30 to 0.33 of the heat furnished by the fuel.

Hot air stoves, made of cast iron, have the serious defects of altering the air ; those made of hollow bricks are to be preferred to them. Both kinds must have the necessary means at hand for the introduction of fresh air.

The warming apparatus by means of the circulation of hot water, have the advantage of carrying heat to almost any distance. At the Sydenham Palace twenty-five boilers supply water to pipes, the united length of which is nearly equal to the distance from London to Dover.

VENTILATION.

1. The object of ventilation is to get rid of tainted air, and to replace it by fresh air.
2. The escape must take place near the deteriorating cause and far from persons.
3. Experience shows the superiority of the "Système d'appel" over the "Système d'insufflation."
4. The fresh air can be taken by "appel" at any height. Guy's Hospital in London, and l'hôspital de Lariboisière in Paris are examples of this.

5. The means of causing the draughts is perfectly simple, it is merely necessary to have a grate at the bottom of the chimney for the escape of the air.

The volume of air to be renewed in places requiring to be purified may be fixed as follows :—

HOSPITALS :—	Per hour and per individual. M. C.
Ordinary cases	60 to 70
Wounded persons	100
Women lying-in	100
At times of epidemics	150
PRISONS	50
WORKSHOPS, ordinary	60
„ unhealthy	100
BARRACKS, by day	30
„ by night	40 to 50
THEATRES, MUSIC-HALLS, &c.	40
LONG GATHERINGS, MEETINGS	60
SHORT „ „	30
INFANT SCHOOLS	12 to 15
ADULT „	25 to 30
STABLES, &c.	180 to 200

These figures agree with those of English, American, and German hygienists.

With regard to temperature, the necessary results to be obtained by warming apparatus are the following :—

TEMPERATURE.

DAY-NURSERIES, for babies	15°c
ASYLUMS, SCHOOLS	15°
HOSPITALS, ordinary	16° to 180°c
„ for the wounded	12°
WORKSHOPS, BARRACKS	15°
PRISONS	15°
THEATRES, MUSIC-HALLS, &c.	19° to 20°
GATHERINGS, MEETINGS, &c.	19° to 20°

NECESSARY CUBIC SPACE AND QUANTITY OF AIR.

Important studies were made in 1842 by M. Leblanc, a French physician, under the title of *Investigations upon the composition of air in enclosed places;* the conclusion to which he came may be expressed as follows :—

The proportion of carbonic acid increases according to the degree of unhealthiness (of the locality) and may be taken, so to speak, as its gauge.

In a note entitled: *On the Cubic Space and volume of air, necessary to insure the healthiness of inhabited places*, Dr. Chaumont, an English army doctor, has given a *Table of comparisons between the proportions of carbonic acid and the impressions on the sense of smell.*

He has proved that a bad smell, indicating unhealthiness, manifests itself, as soon as the proportion of carbonic acid in the air is greater than 0·0006.

I found it easy to deduce from it the volume of air which ought to be renewed according to the cubic capacity of the inhabited places. Here are the results:

Cubic space for each person:

M.C.	M.C.	M.C.	M.C.	M.C.	M.C.	M.C.	M.C.
10	12	16	20	30	40	50	60

Volume of air to be introduced per hour and for each person:

M.C.	M.C.	M.C.	M.C.	M.C.	M.C.	M.C.	M.C.
90	88	84	80	70	60	50	40

These figures agree with the practical rules adopted by French and English engineers.

INFLUENCE OF THE EXTERNAL TEMPERATURE.

The fundamental condition of the uniformity of the "appel" is that the temperature in the escape-chimney should always be equally greater than that of the external air. A difference of 20°c is generally sufficient. Natural ventilation has the inconvenience of being subject to accidental distubances in its motions and consequently is inadequate to ensure healthiness.

The application of the preceding principles has given the results summed up below.

Creche de St. Ambroise a Paris. (Day nursery for babies). This building was intended for fifty children, and the volume of air to be drawn out had been fixed at 1800 M.C. per hour. It was equal to 1800 M.C. The consumption of coal for this purpose was 1k40 per hour.

Primary Schools, Rue des Petits-hotels, Paris.

	Volume of air required.	Quantity obtained.
First Floor	4000 M. in an hour.	4030 M.
Second Floor	2000 M.	1989 M.

And this in spite of serious defects in carrying out the process.

Barracks.—The arrangements for barracks are similar to those noticed above.

Ships.—Dr. Reid has, in his work, pointed out very sensible arangements. M. Bertin, an engineer in the French navy, has, on a ship adapted for the transport of 308 horses, made use of the heat escaping from the steam boilers and cooking-fires. The quantity of air to be got rid of, for each horse had been established at 150 M.C. per hour. Total, 33,900 M. in one hour. The result of experiments made in calm weather, with the auxiliary fires alone, gave 143 M.C. per horse, in all 31,934 M.C. per hour.

Hospitals.—At the hospital Ste. Eugénie, Lille, the volume of air, to be renewed per bed and per hour had been fixed at 45 M.C.—too low a figure—corresponding, in a room with twenty-two beds, to a total of 45 × 22 = 990 M.C. It was much exceeded, as may be seen by the following results:

		Total per Room.	Per Bed.
22 beds on each floor.	Ground floor	3,906 M.C. per hr.	178 M.C.
	First do.	3,440 „ „	157 „
	Second do.	4,823 „ „	217 „
		Total: 12.169 „	Average: 184 „

RESULTS.

During the war, the number of sick-beds was augmented from twenty-two to sixty, and for the wounded to fifty. But the ventilation rose to 5060 for sixty beds, or 84 M.C. per bed. Thanks to this thorough ventilation, *no case of hospital gangrene broke out.*

With regard to *Field Ambulances*, the American system might be adopted.

The Lying-in Hospital at St. Petersburg. Here M. le Baron de Derschau, a skilful Russian engineer, has carried out the following arrangements:

Single bedded rooms, for private patients	8
Four-bedded rooms for married women	16
Four-bedded rooms for unmarried women	28

The results obtained were: Volume of air got rid of per bed and per hour = on average, 127 M.C.

Volume of air introduced by special apertures = 92 M.C.

Guy's Hospital in London, built by Mr. Rhode Hawkins. The space per bed alloted is but from 45 to 48 M. The volume of air got rid of has been found to be equal to 109 M.C. per hour and per bed. Fresh air is taken in by two towers, 20 or 25 M. in height. It descends into subterranean vaults, where it is heated by means of circulating hot-water apparatus.

Its escape takes place by means of a central tower 60 M. high.

Corps Legislatif, Paris. The volumes of air got rid of per hour have been:

Salle and Tribunes	28,639 M.C.
Lobbies, etc.	12,744 M.C.
Total	41,417 M.C.
Plus passages	2,500 M.C.

The proposed quantity was but of 30,000 M.C.

The places to be ventilated were as follows:

Halls, galleries, passages, staircases ...	11,354 M.C.
Eight saloons	8,988 M.C.
Total	20,342 M.C.

There is a renewal of air *once and a half* every hour.

Theatres.—The introduction can take place by interjoists at the boxes, by the cupola, and by the tympans; and the escape at the bottom of the boxes, in the passages, stair-cases, and from under the orchestra seats. Use ought to be made of the lights, for the exhaustion.

At the "Théâtre Lyrique," the quantity to be got rid of had been fixed at 51,000 M.C.; the result reached from 55,000 to 60,000 M.C.

Railway stations ought to be ventilated by means of lanterns placed up high.

With regard to *workshops and laboratories*, it is merely necessary to apply the general rules mentioned above.

Glass roofs and ceilings are inconvenient inasmuch as they allow the interior air to become much heated in summer, and very cool in winter. They ought to be double. In the winter season it is necessary to heat a room with a glass roof from 30° to 40°.

In *kitchens* the heat which is wasted by the stoves may be employed to warm baths.

Water-closets can easily be ventilated by means of gas burners.

The lecture-rooms of the "*Conservatoire des Arts et Metiers*" have for eight years been warmed and ventilated according to the principles of the above-named system. Air is introduced by the ceiling at the rate of about 0 M. 50 in 1" and at a temperature of from 18° to 20°.

In the interior, the temperature is kept between 18° and 22° at the most. The volume of air got rid of is 25 M.C. per hour and per individual, when the amphitheatre is perfectly full.

In summer the renewal of the air keeps the temperature 2° or 3° below that of the external air.

As an example of the advantages to be derived from good ventilation, Dr. Reid, quotes, in his work, the agreeable effect produced in the Banquetting Hall of the Royal Society of Edinburgh, by the system which he had introduced.

At one of the entertainments, at which fifty guests were present, the comfort produced by the renewal of air had the effect of keeping up the flow of spirits and the appetite of the assembly, to such an extent that the consumption of liquids was greatly increased and the number of carriages required to carry the members home was doubled. But the good doctor adds that these gentle excesses did harm to no one.

General Morin winds up by invoking the assistance of the ladies—who are as much interested in the matter, as any one—in order to carry out all necessary measures for ensuring the healthiness of all inhabited places.

Drawing-rooms and ball-rooms, magnificently lit up, decorated with flowers and filled with noble youths and lovely maidens, are no less dangerous for the health of the body, than for the peace of the soul.

The air is tainted with a thousand different effluvia, and however little agreeable it may be to hear it spoken, this ill, which all the human race is heir to, cannot be avoided. We may say of it, as the poet says of death :—

> "Le pauvre en sa cabane, ou le chaume le couvre
> Est sujet a ses lois
> Et la garde qui veille aux barriéres du Louvre
> N'en défend pas les roís."

Let us therefore unite our efforts to ward off these evils, and to establish healthy dwellings for the poor as well as for the rich, in order that all alike may enjoy these two great blessings :—

"Mens sana in sano corpore."

The PRESIDENT : I am sure we have all enjoyed the discourse which General Morin has given us upon this very important subject of ventilation. The time has now come when we have all experienced the great advantage of ventilation, and have all come to the conclusion that it is very desirable. But opinions differ very widely on the best modes of carrying into effect under different circumstances. It is a very difficult subject, which General Morin has followed up with great care and perseverance and the results which he has brought before us, are, I think, such as will commend themselves to our most earnest attention. I beg now to pass a most hearty vote of thanks to him for his communication.

The vote of thanks having been passed,

The PRESIDENT : I will now call on Messrs. E. Dent & Co. to give us their communication

ON TIME MEASURERS.

Messrs. E. DENT & CO. : A clock consists of two distinct parts—a pendulum oscillating in a certain small interval of time—and a mechanism whose duty it is to register the number of these oscillations and to maintain them by giving to the pendulum periodically a little push or impulse—for without this it would soon come to rest.

In order that a clock may perform accurately it will be at once seen that the time occupied by each oscillation of the pendulum must be a constant quantity.

For this it is necessary that its theoretical length should be invariable;

that its arc of oscillation should be invariable; and finally that it be unaffected by changes in the density of the atmosphere.

These conditions may at first sight appear very simple, but cannot be so readily fulfilled as might be imagined.

In the first place, of whatever material the pendulum may be composed, some small alteration in its length (which we have mostly to consider) will occur at every change of temperature; and as we cannot ensure a perfectly uniform temperature, a correction or compensation must be applied to counteract the effects of a variable one.

The arc of oscillation must not vary because the pendulum in its swing does not describe a cycloidal curve, as it should do in order that the long and short oscillations might be performed in the same space of time; but almost a segment of a circle which diverges rapidly from the cycloid, consequently a change in the arc described altering the amount of divergence from the true theoretical path, alters the time of oscillation. This variation is known as the circular error, and cannot be corrected. All we can do is to endeavour to make the error a constant quantity by maintaining a constant arc of oscillation.

The variations in the density of the atmosphere affect the duration of the oscillation of the pendulum although to a much less extent than a change of temperature or a change of arc. Still the variation is sufficiently large to make a correction desirable. It amounts to about 0·3 sec. in 24 hours for a change of pressure indicated by one inch fall or rise of the mercury in the barometric tube.

As we propose to invite your attention more particularly to the compensation of the pendulum for changes of temperature, and to Sir Geo. Airy's correction for changes of barometric pressure (as exemplified in the clock downstairs which is almost a counterpart of the Greenwich Sidereal Stand one), we will dismiss the problem of the maintenance of a constant arc of oscillation after stating that for its accomplishment we use an escapement and train of wheels actuated by a weight, the whole being so constructed as to impart to the pendulum at every oscillation or alternate oscillation a certain push or impulse which shall be of uniform force, and just equal to the resistance encountered by the pendulum in its swing from side to side. We will not stop to describe the forms of escapement best fulfilling this condition but will simply mention that the three varieties most in favour are the dead-beat of Graham, invented early last century; the detached escapement invented by Sir Geo. Airy some thirty or forty years ago, and the more modern gravity escapements.

The compensation of the pendulum for changes of temperature is effected in various ways, but the principle is the same in all. Two metals, one expanding more freely than the other, are used, and so applied that the expansion of the one counteracts and nullifies the expansion of the other, so that the centre of oscillation of the pendulum is maintained at the same point.

Originally, brass and steel were the metals employed, but mercury and steel *or* iron, or zinc and steel *or* iron are now generally used—so we will at once pass to them.

A mercurial pendulum consists of a steel rod at the lower end of which hangs a jar or vessel of glass or iron containing the mercury. Now the jar of mercury forms the bob of the pendulum, and is at the same time the compensating element. The action in changes of temperature is as follows. In heat, the steel rod lengthens and the jar descends, but the same cause which produces this, makes the mercury expand and rise in the jar, so that one expansion compensates for the other. In cold, the metals contract similarly, and the same effect is produced by the contrary action.

In a zinc and steel compensation, as now made, the construction is somewhat more complicated, but still is easily explained. The bob or principal mass of the pendulum is of lead, and does not form part of the compensation. It is supported by a steel tube, to the lower end of which it is fastened at its centre. The upper end of this tube is fitted with a collar which rests on the top of a zinc tube, this in its turn is supported by a nut on the lower end of the central steel rod. In heat, the steel tube and rod expand and would allow the bob to descend, but the zinc expanding raises it up an equal amount, so that the same theoretical length is preserved : the length of the zinc tube being such, that its longiditudinal expansion in any given temperature exactly equals the expansion of the steel rod and tube together. A drawing of a zinc and steel compensation pendulum is here.

Now, for astronomical clocks in which the greatest accuracy is required, the mercurial pendulum so long stood unrivalled that it may fairly be asked how it is that it now seems likely to be superseded by the zinc and steel compensation. The reason is this. The mercury in the jar may be considered as a solid mass presenting but a comparatively small surface to the action of the atmosphere by which changes of temperature are communicated to it, whilst the steel rod and suspension spring present a much larger surface in proportion to their bulk, and are acted on much more quickly, so that in a variable temperature you have one part of the pendulum expanding or contracting, as the case may be, before the other, and an unsteady rate is the result. It is obvious that the greater the diameter of the jar the more sluggish will be the compensation, and it has been proposed to put the mercury into several small jars, but there are inconveniences to this.

Now, in a zinc and steel pendulum, those parts on which we depend for the compensation are much more nearly of the same bulk, and present nearly the same surface to the action of the atmosphere, so that it is fair to infer they act simultaneously, or almost simultaneously. In these pendulums you will observe that the bob is supported at its centre, so that its expansion or contraction, let it occur when it may, does not alter the theoretical length or affect the time of oscillation.

Some persons overlook or ignore the importance of suspending the pendulum bob at its centre, and so nullifying the effect of changes of temperature upon it; but we cannot but think that a little careful consideration on their part will lead them to allow that it is not without advantage.

We have been favoured by the Astronomer Royal with a copy of the rate of one of our clocks having a zinc and steel compensation. This clock was going for several months in one of the small observatories used in the Transit of Venus Expedition. It has been printed for the purpose of showing the behaviour of a zinc and steel pendulum under as trying circumstances as can well be conceived, the daily variations of temperature being frequently fifty degrees.

We will now pass on to the barometric connection. The extreme steadiness of rate of the Greenwich Standard Clock at equal pressures, led Sir Geo. Airy to attempt the corrrection by means of the magnetic apparatus you may have seen below, a drawing of which we have here. Two bar magnets, having their poles reversed, are fastened vertically one in front and the other at the back of the pendulum bob, and beneath them is a horse-shoe magnet carried by one arm of a lever, the other end of which is in connection with a float resting on the lower limb of a bent barometric tube. Between the poles of the horse-shoe magnet and those on the pendulum there is constant attraction, so that you have a force acting in the same direction as the force of gravity, but variable with the density of the atmosphere.

It is found that the clock loses as the pressure increases, so with a high barometer the horse-shoe magnet is raised, the attractive force increased, and the clock made to gain a corresponding amount.

We have printed a copy of an instructive diagram prapared at the Royal Observatory under the direction of the Astronomer Royal, which shows in a very marked way the effect of changes of pressure on the rate of the Greenwich standard clock before the magnetic apparatus was applied. The red line shows the calculated rate, allowing for the pressure, and the black line shows the actual observed rate of the clock. It will be seen how nearly the one coincides with the other, the greatest difference between them being less than one-tenth of a second in twenty-four hours. The correction is now effected automatically by the apparatus we have just described, and we have authority for stating that it completely answers the purpose for which it was designed.

The time measuring power of a chronometer is dependent upon its balance spring, which consists generally of a slight ribbon of hardened and tempered steel, coiled round and round with an upward twist somewhat in the form of a cylinder.

The elasticity of this little spring performs just the same duty for the balance which gravity does for the pendulum—it keeps it swinging backwards and forwards—with a degree of uniformity which is truly surprising.

But as a controlling power the elasticity of this little spring labours under one serious disadvantage as compared with gravity: it varies rapidly with any increase or decrease of temperature.

It varies more or less according to the material of which the spring is composed; thus a gold spring suffers a greater change than a steel spring, a steel spring than a palladium spring, and a palladium spring than a glass spring.

It may surprise you to hear us speak of a glass spring, but the thing has been done, and is here.

This glass spring is the invention and handiwork of the late Frederick Dent, and the chronometer before us represents the only perfect specimen now in existence.

The elasticity and strength of this spring are so extraordinary that until you had examined it you could scarcely credit its being glass, but as seeing is believing, and we shall be happy to exhibit a portion of another one to any person who may be interested in it after the demonstrations.

But meanwhile, for illustration, we will tell you two experiments, the second one being quite unintentional. In the first a chronometer was taken having an ordinary steel spring, the balance of which was oscillating 180° from rest—a glass spring was substituted, everything else remaining exactly the same, the oscillation of the balance rose to 200°.

The unintentional experiment was that the chronometer was knocked off a table, and though both points of the staff were bróken the balance spring sustained no injury.

With regard to the specimen before us you will see that with the exception of having rather more coils, it is shaped like a steel spring. The oscillation of the balance is 180° from rest. The glass spring was under trial at the Royal Observatory for a period of nearly three years, and the following results were attained. The figures are the average daily rates calculated from periods of a month, the mean temperature being given opposite each. There is one peculiarity about this performance to which we must briefly call attention.

This rate shows that the chronometer had always a tendency to accelerate or to advance upon its rate, and with regard to this it may be interesting here to state that steel springs exhibit the same tendency just in proportion to their hardness. And that a soft steel or gold spring "accelerates" in the reverse direction, *i.e.*, loses upon its rate.

Now let us return and inquire what is the behaviour of balance springs of different materials under a given change of temperature.

Suppose that we take four chronometers, the first with a balance spring of gold, the second with a balance spring of steel, the third with a balance spring o' palladium, and the fourth with a balance spring of glass ; and in place of the ordinary balance we substitute upon each a balance composed of a glass disc, so as practically to eliminate any error due to the expansion of the balance. Then, if we regulate each chronometer to go right at the temperature of freezing, thirty-two degrees fahr., we shall find that when we raise the temperature to 100 degrees the chronometer with a *gold spring* will lose *eight min. four sec.* a day, the chronometer with a *steel spring six min. twenty-five sec.*, the chronometer with a palladium spring *two min. thirty-one sec.*, and the chronometer with a glass spring *forty seconds ouly.*

It is with the compensation for these errors that we have practically to deal in chronometers, the loss of time due to the expansion of the balance being slight in comparison.

In the case of the glass spring chronometer, the smallness of this error made it necessary to apply a very different to the usual form of compensation balance: compound laminœ formed of silver melted upon platinum weighing about two grains, each are mounted upright upon a disc of glass, and with any increase of temperature the tending of these towards the axis of motion affords a sufficient compensation.

Unfortunately the amount of compensation required for a steel spring is so great that in correcting it we introduce a secondary error, the nature of which I will try to explain to you.

The loss of time due to the loss of elasticity of the spring varies nearly in the same proportion as the temperature, and an ordinary compensation balance advances or withdraws its weights to or from the axis of motion practically in the same proportion.

But *the time in which the balance swings, varies not as the distance of its weights from the axis of motion*, but as the square of their distance, and thus it requires a greater motion of the weights inwards, to produce the same amount of effect as a lesser motion outwards.

Thus, in an ordinary compensation balance, if the weights were adjusted to advance a sufficient distance to compensate the chronometer for a rise in temperature of thirty degrees, a fall of the same amount would make them retire so much too far as to make the chronometer lose four seconds a day. And the best that could be done under the circumstances would be to so adjust the chronometer as to divide the error, and make it lose two seconds a day in the heat, and two seconds a day in the cold.

Many methods have been introduced for correcting this secondary error; some are based upon the principle of increasing the motion of the weights when they move inward, and reducing it when they move outward; others are arranged to move an auxiliary weight towards the axis of motion, both for an increase and decrease of temperature; others check the motion of the main weight away from the axis of motion; and others assist it only when it is going towards it. Illustrations of all these different forms are here exhibited, but not to occupy your time we propose only to describe one of them.

Here is a bar formed of brass, melted upon steel, the brass being underneath. Do not consider the staples at present, but imagine for the present that the weights by which they are crowned are mounted upon little pillars straight from the main bar.

Now let us see what happens with any increase of temperature: the brass expands, so does the steel, but the brass expands most, and the only method by which the two can accomodate matters, is that the bar shall form a curve, the brass being outside. In other words the bar bends upwards and tilts the compensation weights in towards the axis of motion. The reverse action takes place in the cold.

Now see, the weights move inwards and outwards equally for equal rises and falls of the temperature. Now let us consider the action of our staples. They themselves are compound pieces, and when the weights are pointed towards the axis of motion in the heat, they open and advance their progress a little; and in the cold, when the weights are pointed away from the axis of motion, they close and retard it, producing in this way the necessary correction for the secondary error.

But although balance springs as compared with gravity are at this very great disadvantage with regard to changes of temperature, they have one exceedingly important peculiarity which thëir rival has not: that is to say, that with a spring of suitable form and strength, by slightly altering the length, or for very small amounts by arching it, you are able almost completely to isochronise the long and short arcs of oscillation of the balance. You can make the chronometer keep the same rate when its balance is swinging 180° from rest as when it is swinging 225°. The direction you alter the spring in is this:—if the chronometer gains during the short arcs you shorten the spring, and if during the long you lengthen it.

The property of the spring is exceedingly convenient, because by leaving the spring rather fast in the short vibrations you can correct the chronometer for any occasional irregularity owing to an increase of friction. For friction meaning retardation, in other words a checking of the action of the balance. The oscillation falling off at the same time, and the balance moving rather faster in consequence the tendency to lose is corrected.

It would appear that this plan of leaving the chronometer fast in the short vibrations also answers as a barometric compensation. For instance, we took a chronometer which was going exactly right at the normal atmospheric pressure and placing it under the receiver of an air pump reduced the pressure to about 15 in. Before the pressure was reduced the balance was swinging about 214° from rest, but in the thinner atmosphere it rose to 270°, and yet the time of the chronometer did not vary more than three-tenths of a second a day—the tendency to gain by the removal of the friction due to the denser atmosphere being corrected by the tendency to lose during the longer oscillations.

The advantage due to the power of isochronising the long and short arcs of oscillation, and the mechanical superiority of its escapement over any clock escapement ever yet invented, almost place the chronometer alongside of the best astronomical clocks.

In conclusion, allow us to remind you that there is a fundamental distinction between a clock and chronometer—the time-measuring power of the one depending upon gravity, the time-measuring power of the other depending upon the elasticity of its balance spring—that is to say, upon a property of matter. The chronometer is therefore much more absolute in its measurement of time than the clock, and if for no other reason every effort should be made towards bringing it to that perfection which would render it available for the highest astronomical purpose.

The PRESIDENT: Mr. Glasgow, the representative of the British Horological Institute, wishes to add to the information given to us just now, and although the time for adjournment has arrived I am sure you will be glad to hear him.

Mr. GLASGOW: I must apologise for coming before you at so late an hour when I had no previous idea of saying anything, but I thought it would be well as there is such an opportunity, and as the subject is of such general interest to the public, that I should carry the matter a little further than Messrs. Dent have done. The rather restricted paper which they have read on the compensation of pendulums I will not deal with. A few years ago it would have been vastly more interesting to scientific men than it is now, inasmuch as electricity is now the means of taking the exact time to all the large cities, and electric clocks have to a great extent taken the place of the accurate pendulum which we used to have so many experiments upon. The very ingenious contrivance of the Astronomer Royal, shown on the clock of the Messrs. Dent, and referred to in the paper just read, for correcting the influence of the atmospheric pressure on the pendulum, known as the barometric error, is, in the opinion of so high an authority as Sir Edmund Beckett, unnecessary. He tells us that when testing the Westminster clock for the barometric error, he found it had none; which he attributed to the rather unusual length of the arc, or swing of the pendulum, and after careful experiments, he found that this error of the pendulum could always be corrected by regulating the length of arc. But as chronometers and pocket watches have become vastly more important to the general public than they used to be, I will say a few words upon chronometer balances, and those of pocket watches. There are a variety of balances which may be seen down stairs lent by the Horological Institute, and I am sorry they are not properly displayed or catalogued. There are amongst these balances specimens of various inventions, from the first made by Arnold and Earnshaw to the latest sent for trial to Greenwich, and it is curious to note how very little difference there is between the earliest and latest examples, and not encouraging to see how little improvement has been made during the seventy years that have elapsed since the invention of Earnshaw: although there are in this collection alone some forty differently constructed balances, all being the results of attempts to improve upon it.

There is also a collection of balances shown by Messrs. Dent, most of them being a modification of the form, known in the trade as "Dent's Balances," and others exhibited of very complicated construction, but after much time and trouble spent on these ingenious contrivances, it has been found that simplicity of form is indispensable to a reliable timekeeper, and the balance of Earnshaw without modification is what is all but universally used for ship's chronometers at the present time. So also is it with the pocket watch, although the manipulative skill of the workman has greatly advanced, science has not kept pace with it, and pocket chronometers are fast giving place to the best form of lever escapement—that invented by Mudge and strangely enough abandoned by Mudge himself in favour of a much inferior escapement.

The model of the balance shown by Mr. Dent differs in form from that in general use, the ordinary form being that of a ring or rim composed of a lamina of brass and steel melted together and so proportioned to each other that when cut open the rim contracts in heat, thereby decreasing its diameter and expands in cold, which increases its diameter, and thus compensating for the error which is principally in the balance spring, caused by its elongation and loss of elastic force in heat, and its greater resistance when cold ; and it may be of importance to those who wear watches without compensation balance to know on the authority of the Astronomer Royal, that for every difference of ten degrees in temperature there is a difference in twenty-four hours of one minute in the time of these watches.

Unfortunately for the scientific progress of watchmaking the trade spirit so predominates that scientific men have not of late years devoted themselves to its study as might have been expected from the nature of the great interest attached to it, otherwise we might have been further advanced than we are at present.

But what I want to say is this, and I am glad of the opportunity of saying it here, the mode of testing and purchasing chronometers for the navy is not satisfactory, and is not calculated to encourage the men by whose efforts alone we can hope for a higher standard of excellence. In the first place the prices paid by the Government are too small to tempt such men from their ordinary pursuits to make the long and careful experiments necessary for perfecting any new form of balance or even improving upon old forms, and secondly, chronometers are sent to Greenwich by scores from all the little towns in the kingdom, bearing names of men who perhaps have never seen the instruments to which they are affixed, but if purchased they are "chronometer makers to the Admiralty," the chronometers being made in London.

I think this should not be, and that some means should be adopted to insure that "chronometer maker to the Admiralty" be a distinction worth coveting and competing for, and not merely to be used as a trade "puff" by men who know nothing of chronometers.

The PRESIDENT: It only remains for me to move a vote of thanks to Messrs. Dent for their communication, which is very interesting as coming from gentlemen whose names have been connected with true chronometry for so many years.

The conference then adjourned.

SECTION—MECHANICS (including Pure and Applied Mathematics and Mechanical Drawing).

Thursday, May 25th, 1876.

Dr. SIEMENS, President, in the Chair.

F. J. BRAMWELL, M. Inst. C.E., F.R.S. : The subject on which I have now the honour to address you, the subject which is to occupy our attention to-day, is that of "Prime-Movers," that is to say, we are about to consider that class of machines which, to use the words of Tredgold, "enable the engineer to direct the great sources of power in Nature for the use and convenience of man."

Although machines of this kind are in truth mere converters or adapters of extraneous forces into useful and manageable forms, and have not any source of life, power, or motion in themselves, nevertheless they impress us with the notion of vitality; and it is difficult to regard the revolving shaft of a water-wheel or turbine set in motion by some hidden stream, or to gaze upon the steam-engine actuated by an unseen vapour, without, as I have said, the idea being raised in our minds that the machines on which we are looking are really endowed with some kind of life; and thereupon not inaptly, although not quite accurately we speak of them as Prime-Movers.

The invention of Prime-Movers marks a very great step in the progress of mechanical science in the world, as it commences an era distinct from that in which mere machines, to be acted on by human or animal muscular force, were alone in existence. Machines such as these, highly useful as they may be, are, after all, only tools or implements more or less ingenious, and more or less complex.

Mankind could not have been very long upon the earth before they must have found the need and must have discovered the utility of some kind of tool or implement; they must soon have found that the direct action of the power of the arm, which was not enough by itself to break up some obstacle, became sufficient if that action were applied by the wielding of a heavy club, or by the putting into motion of a large stone; and thus the hammer, or its equivalent, must have been among the earliest of inventions. Such an implement would soon teach its users that muscular force when exercised through a considerable space, could be stored up, and could be delivered in a concentrated form by a blow.

Similarly it could not have been long before it must have been found that to raise water in the hollow of the hand in small quantities and by repeated efforts was not so convenient a mode as to raise at one operation all that was required, and to do so by the aid of a bent leaf or by the use of a shell, and in this way another implement would speedily be invented. We might pursue this line of speculation, and doing so we should readily arrive at the conclusion that (without attributing to the early inhabitants of the earth any profound acquaintance with mechanics) the hammer, the lever, the wedge, water vessels, and other simple tools and utensils must soon have come into existence; and we should also be led to believe that when even with the aid of tools such as these a man singly could not accomplish any desired object, the expedient of combining the power of more than one man to attain an end would soon be thought of; and that the requisite appliances, such as large beams, for levers, numerous ropes (which must very early in the history of the world have been twisted from filaments) and matters of that kind would come into use. In corroboration of this view, if a corroboration were wanted, the fact may be cited that on the discovery of any isolated savage community, it is, I think I may say invariably, found to have advanced thus far in mechanical art.

But passing from such machines as these, which are rather of the character of mere tools and implements than of that of machines, as we now popularly use the word, one knows that even complicated mechanism for the purpose of enabling muscular force to be more readily applied, is of very ancient date. On this point I will quote from only one Book, that is the Bible, and from that I will cite but a few instances. At the tenth and eleventh verses of the eleventh chapter of Deuteronomy, a statement is made clearly indicating that in Egypt irrigation was carried on by some kind of machine worked by the foot, whether the tread-wheel, with water buckets round about it, mentioned so many centuries later by Vitruvius, or whether the plank-lever, with a bucket suspended at one end and raised by the labourer running along the top of the lever to the other end (an apparatus even now used in India), we do not know; but that it was some machine worked by the foot is clear, the statement being that when the Israelites had reached the promised land they would find it was one abounding in streams, so as to be naturally watered, and that it would not require to be watered by the foot as in Egypt. Again in Chronicles it is related that King Uzziah loved husbandry, and that he made many engines, unhappily not in connexion with agriculture, but for warlike purposes, "to shoot arrows and great stones withal." Further, in the seventh chapter of the Book of Job, we have the comparison of the life of man passing away swifter than a weaver's shuttle, this points unmistakably to the fact that there must in those days have been in existence a loom capable of weaving fabrics of such widths that the shuttle required to be impelled with a speed equal to a flight from one side of the fabric to the other; and no doubt such a fabric must have been made in a machine competent at least to raise and depress alternately the halves of the warp threads. The potter's wheel also is frequently mentioned in the Bible.

Such instances as these are sufficient to show that considerable progress must have been made in the very earliest days of history in the construction of machines whereby muscular force was conveniently applied to an end; but if we leave out of account, as we fairly may, the action of the wind in propelling a boat by sails and the action of the wind in winnowing grain, I think we shall be right in considering that in the times of which I have been speaking there did not exist any machine in the nature of a power giver or Prime-Mover.

Doubtless the want of a greater force than could be obtained from the muscles of one human being must have soon made itself felt; and intelligent men, conscious of their own ability and of their mental power of directing a large amount of work, must have been grieved when they found the use of that power circumscribed by the limited force of their own bodies ; and therefore, early in the world's history, there must have been the attempt by the offer of some consideration or reward to induce the robust in body but not in mind, to work under the directions of these men of superior intelligence. But when such aid as this became insufficient, the way in which, in all probability, the people of those days endeavoured to satisfy the further demand would be to make captives of their enemies, and to reduce them to a state of bondage—to grind at the mill, to raise water, or, yoked by innumerable cords and beams, to draw along the huge blocks required in the foundations of a temple, or for the building of a pyramid, or to act in concert on the many oars of a galley—although by what means this last-named operation was performed is not very clear. And doubtless, coupled with this condition of bondage, there must have been an amount of human suffering which is too frightful to be contemplated.

Such machines as those to which I have called attention could not have been invented and brought into use without the exercise of much mechanical skill ; but considerable as this skill must have been, it had never originated a Prime-Mover, it had given no source of power to the world, but had left it dependent on the muscular exertions of human beings and of animals.

Great then was the step, and a most distinct era was it in mechanical science, when for the first time a Prime-Mover was invented and a machine was brought into existence which, utilizing some hitherto disregarded natural force, converted it into a convenient form of power by which as great results could be obtained as were obtainable by the aggregation of a large number of human beings, and could be obtained without bondage and without affliction.

There are probably few sights more pleasing to one who has been brought up in factories than to watch a skilful workman engaged in executing a piece of work which requires absolute mastery over the tools that he uses, and demands that they should have the constant guiding of his intelligent mind. Handicraft work of such a kind borders upon the occupation of the artist ; and to see such work in the course of execution is, as I have said, a source of pleasure. But when, descending from this, the work becomes more and more of the character of mere repetition, and when it is accomplished by the aid of implements which from their very perfection, require but little mind to direct them, and demand only the use of muscle—then, although the labour when honestly pursued is still honourable, and therefore to be admired, there comes over one a feeling of fear and of regret that the man is verging towards a mere implement. But when one sees, as I have seen in my time, in England, and as I have seen very recently on the Continent, men earning their living by treading within a cage, to cause it to revolve and thereby to raise weights—an occupation demanding no greater exercise of intelligence than that which is sufficient to start, to stop, and to reverse the wheel at the word of command—one does indeed regret to find human beings employed in so low an occupation; an occupation that places them on the level with the turnspit, and one which is most properly meted out in our prisons as a punishment for crime; accompanied, however, in the instance of the prison, with the degradation that the force exerted shall be entirely wasted in idly turning a fan in the free air; and thus the prisoner, in addition to the fatigue of his body, undergoes the humiliation of, as he expresses it, "grinding the wind."

If they played no other part than that of relieving humanity from such tasks as these, Prime-Movers would be machines to be hailed.

True it is that the labourers who were thus relieved would not thank their benefactors; and indeed, so far as the individuals subjected to the change were concerned, they would have cause not to thank them, because those individuals having been taught no other mode of earning a livelihood, and finding the mode they knew set on one side by the employment of a prime-mover, would be at their wits' end for a means of subsistence; and would be experiencing those miseries which are caused by a state of transition. But in some way the men of the transition state must be relieved; and, in the next generation, it no longer being possible to subsist by such wholly unintelligent labour, the energies of their descendants would be devoted to gaining a livelihood by some occupation more worthy of the mind of man.

Early Prime-Movers, from their comparatively small size, probably did little more than thus relieve humanity, but when we come to consider the prime-movers of the present day, by which we are enabled to contain within a single vessel, and to apply to its propulsion 8000 indicated horse-power, or an equivalent of the labour of nearly 50,000 men working at one time, we find that the prime-mover has another and most important claim upon our interest, namely, that it enables us to attain results it would be absolutely impossible to attain by any aggregation of human or other muscular effort, however brutally indifferent we might be to the misery of those who were engaged in that effort.

Excluding from our consideration light and even electricity as not being up to the present sources of power on which we rely in practice, there remain three principal groups into which our Prime-Movers may be arranged :—viz., those which work by the agency of the wind, those which work by the agency of water, and those which work by the agency of heat.

Of these three great groups two, heat and water, are capable of division, and indeed demand division into various branches.

A water Prime-Mover may be actuated by the impact of water, as in some kinds of water-wheels, turbines, and hydraulic rams, or by water acting as a weight or pressure, as in other kinds of water-wheels, and in water pressure engines, or by streams of water inducing currents, as in the case of the jet pump and of the "trombe d'eau," or by its undulating movements, as in ocean waves. The ability of water to give out motive power may arise from falls, from the currents of rivers, from the tides, or, as has been said, from the oscillation of the waves.

Prime-movers which utilize the force of the wind, are few in number, and in all cases act by impact.

As regards those prime-movers which work by the aid of heat. We may have that heat developed by the combustion of fuel, and being so developed, applied to heating water, raising steam, and working some of the numerous forms of steam-engines, or, as in the case of the Giffard injector, performing work by currents induced by the flow of steam ; or we may have the heat of fuel applied to vary the density of the air, and thus to obtain motion as by the smoke-jack ; or the fuel may be employed to augment the bulk and the pressure of gases, as in the numerous caloric engines ; or we may have heat and power developed in the combustion of gases, as in the forms of gas engines, or in the combustion of explosives, as in gunpowder, dynamite, and other like materials, used not only for the purposes of artillery and of blasting, but for actuating prime-movers in the ordinary sense of the word.

Again, we may have the heat of the sun applied through the agency of the expansion of gases to the production of power as in the sun pumps of Salomon de Caus, and of Belidor, to the production of power through the generation of steam as in the sun engine of Ericsson; or, finally, we may have it applied direct, as in the radiometer of Mr. Crooks.

A consideration of the foregoing heads under which prime-movers range themselves, will speedily bring us to the conclusion that the main source of all mechanical force on this earth is the sun. If the prime-movers be urged by water, that water has attained the elevation from which it falls, and thus has been made competent to give out power, by reason of its having been evaporated and raised by the heat of the sun. If the power of the water be derived from the tidal influence, that influence is due to the joint action of the sun and its assistant the moon. If the prime-mover depend upon the wind for its force, either directly, as in windmills, or indirectly, as in machines worked by the waves, then that wind is caused to blow by variations of temperature due to the action of the sun. If the prime-mover depend upon light or upon solar heat, as in the case of the radiometer and of the Sun engine, then the connexion is obvious; and if the heat be due to combustion, then the fuel which supports that combustion is, after all, but the sun's rays stored up. If the fuel be, as is now sometimes the case, straw or cotton stalks, we feel that there we have the growth of the one season's effect of the sun's rays; if the fuel be wood, it is equally true that the wood is the growth of a few seasons' exercise of the sun's rays; but if it be the more potent and more general fuel—coal—then, although the fact is not an obvious one, we know that coal also is merely the stored-up result of many years of the exercise of the sun's rays.

And even in the case of electrical prime-movers these depend on the slow oxidation—that is, burning of metal which has been brought into the metallic or unburnt state from the burnt condition (or that of ore) by the aid of heat generated by the combustion of fuel.

The interesting lecture-room experiment with glass tubes charged with sulphide of calcium, or other analogous sulphides, makes visible to us the fact that the sun's rays may be stored up as light; but that they are as truly stored-up (although not in the form of light) in the herb, the tree, and the coal, we also now know, and we appreciate the far-seeing mind of George Stephenson, who astonished his friend by announcing that a passing train was being driven by the sun: and we know that the engineer was right and that the satirical author was wrong, when he instanced as a type of folly the people of Laputa engaged in extracting sunbeams from cucumbers. The sunbeams were as surely in the cucumbers as they are in the sulphide of calcium tubes, but in the latter case they can be seen by the bodily eye, while in the former they demand the mind's eye of a Stephenson.

Although the sailing of ships and the winnowing of grain must from very early times have made it clear that the wind was capable of exercising a substantial power, nevertheless being an invisible agent, it is not one likely to strike the mind as being fit to give effect to a prime-mover, and therefore it is not to be wondered at that prime-movers actuated by water are those of which we first have any record; unless indeed the toy steam-engine of Hiero may be looked upon as a prime-mover, anterior to those urged by water. It would appear that in the reign of Augustus water-wheels were well known, for Vitruvius, writing at that time, speaks of them as common implements; not so common, however, as to have replaced the human turnspit, as we gather from his writings that the employment of men within a tread-wheel was still the more ordinary mode of obtaining a rotary force. It would seem, however, that water-wheels driven by the impact of the stream upon pallet boards were employed in the time of Augustus not merely to raise water by buckets placed about the circumference of the wheels, but also to drive millstones for grinding wheat. Strabo states that a mill of this kind was in use at the palace of the King of Pontus.

Having thus mentioned the earliest record of hydraulic (or indeed of any) prime-movers, I will not endeavour to trace their history down to modern times, as it would be impossible to do so usefully within the limits of an address. I will therefore, without further reference to the historical part of the subject, ask you to join me in considering what are the conditions which govern the application of water to hydraulic prime-movers.

After all, water must be looked upon as a convenient form of descending weight. When the fall is not great it is always practicable by means of water-wheels having buckets which retain the water to employ, as I have said, its mere gravity as a motor, and probably it is by this mode that the highest result is procured from any given quantity of water falling through a given height. By the use of a backshot wheel as much as seventy-five per cent. of the total power is rendered available. The twenty-five per cent. of loss arises from the friction of the axle of the wheel and of the gearing transmitting the force to the machine which is to utilize it; from some of the water being discharged out of the buckets before the bottom of the fall is reached; from the necessary clearance between the wheel and the tail-water; from the eddies produced in the water as it enters the buckets; and (to a small extent) from the resistance of the air.

When the difference of level between the source of water and its delivery exceeds, however, forty or fifty feet, the water-wheel becomes very unwieldy and expensive and revolves so slowly that it ceases to be a desirable prime-mover. Then recourse can be had to water-pressure engines, engines wherein pistons move in cylinders, and, being pressed alternately in opposite directions by the head of water, set up rotary motion in the machine, in the same way as if the pistons were acted upon by steam. In the construction of such water-engines great care must be taken to have ample inlets and outlets, in order that the loss incurred either by the power requisite to drive the water through restricted orifices, or by surface resistance caused by a too speedy flow along the various passages, may be a minimum. Care has to be taken also in the arrangements of the valves that the engines, when employed for rotary movement, may be able to turn their centres without producing an injurious pressure upon the water within the cylinders. Water engines employed for pumping, but without rotary movement, are mentioned by Belidor, in his "Architecture Hydraulique," published in 1739, Article 1156. In England Sir William Armstrong has brought these machines to great perfection. The first of his make, erected many years ago, is still working most successfully at the Allan Head Lead Mines. This machine is driven by a natural head of water and not from an accumulator, and is employed in the mine as a winding engine.

An extremely useful feature in engines of this kind, is their adaptability to be driven by the pressure of water derived from an ordinary water works, and in this manner small manufacturers carrying on business in their own houses are enabled to obtain a prime-mover with great ease and all things considered at small cost; and not only is advantage taken of such machines for the purpose of driving manufactories, but water cylinders are now largely used for working the bellows of church organs, for which purpose, however, an overshot water-wheel is shown as being employed as far back as Salomon de Caus's book, date 1615.

Large water-wheels or even water-engines are comparatively costly machines, and as large water-wheels make but few revolutions per minute they require expensive and heavy gearing to get up speed, and thus it is that it frequently becomes a desirable thing to dispense with such machines and to resort to other modes of making available high falls of water. In former times this was done by suffering the impetuous stream of water to beat upon the pallets of water-wheels, but from such machines only a poor effect could be obtained, as a large portion of the energy in the water was devoted to the formation of eddies and the generation of heat and to the production of lateral currents, leaving but a small percentage available as motive power.

Much of the evil effect, however, attendant upon using the impact of water as a means of driving water-wheels is obviated by the construction invented by the distinguished French engineer Poncelet. For high falls, however, the implement now generally employed is the turbine, of which the well-known Barker's Mill may be looked upon as the germ.

I have got before me No. 1983, a model of Fourneyron's turbine.

This is not an apt model for my present purposes, inasmuch as it represents a turbine to be employed with a comparatively low fall of water; but even in such instances the turbine gives most excellent results, and it has the advantage over the water-wheel of being capable of working with great efficiency, although there may be a considerable rise in the "tail-water," a rise which would materially check the action of an ordinary water-wheel. In this turbine every care has been bestowed to give a proper form to the pallets on which the water acts, so as to take up step by step, as it were, the whole of the energy residing in the stream; in order that the water may pass away from the turbine in an inert condition, and that in acting upon the vanes of the turbine eddies may not be formed, and that thus energy may not be wasted.

There are probably few sights more surprising to the old-fashioned millwright who has been used to see water-wheels of fifty or even seventy feet diameter, employed for the utilization of a high fall, than that of a turbine occupying only a few cubic feet of space, but running at such a velocity as to consume the whole of the water of a considerable stream, and so to consume it as to deliver nearly as large a percentage of useful effect as would the cumbrous water-wheel itself.

If a fall of water be employed with the mere object of raising water, this end can be attained without the employment of either water-wheel or turbine by the aid of the Montgolfier ram, a very useful machine for those cases wherein but a small percentage of the whole fall is required to be raised, but is required to be raised to a considerable height. No. 1996, which I have before me, is a glass model of such a ram, but I fear it is too small to be visible, except to those who are very near to the table. You are, however, all aware that the principle of action consists in the sudden arrestation of a column of water flowing with a velocity due to the head. When the water is arrested, a small portion of it raises an outlet valve, and thereby passes into an air vessel against a pressure competent to drive the water up to the desired height, while the main body recoils along the supply pipe; then an escape valve having fallen open the water that has recoiled returns, a large portion passes out by this valve, and then the velocity being once more fully established, the escape valve shuts and causes another arrestation and a repetition of the working. This is an implement by which a large volume of water, and having but a low fall, can be made to raise a portion of itself to a great height; but there is a converse use of water, wherein the employment of a small stream moving rapidly owing to its having descended from a considerable height, is caused to induce a current in other water, and to draw it along with itself with a diminished velocity, but with a velocity still competent to raise the united stream to the insignificant height which suffices for delivering the water from swamps and marshy land.

This employment of the induced current as a prime-mover is described by Venturi in the record of his experiments made at the latter end of the eighteenth century, and within the last few years Professor James Thomson has applied the same principle with great success in his jet pump.

The next mode I shall notice of obtaining motive power from water, is also one where it operates by an induced current—this is, the trombe d'eau, an apparatus wherein water, falling down a vertical pipe, induces a current of air to descend with it, and the lower end of the vertical pipe being connected with the top of an inverted vessel, the bottom of the sides of which vessel is sealed by a water joint, the water dashing upon a block placed below, the mouth of the pipe, is separated from the air, so that while the water descends and escapes from under the sides of the vessel the air rises and is accumulated in the upper part, and can be led away to blow a forge fire. These machines are described in Belidor's work.

The utilization of the rise and fall of the tide is also fully described by Belidor, who gives drawings of channels so arranged that during both the rise and fall of the tide the water-wheel, notwithstanding the reversal of the current, revolves in one and the same direction. The tide is a source of power which it is highly desirable should be utilized to a greater extent than it is; if we consider the enormous energy daily ebbing and flowing round our shores, it does seem to be a matter of great regret that this energy should be wasted and that coal should be burnt as a substitute.

The last mode in which power may be obtained from water, to which I have to allude, is that of the employment of the waves.

Earl Dundonald, better known as Lord Cochrane, proposed by his patent of 1833, to utilize this power for propelling a vessel. This he hoped to accomplish by means of cylinders containing mercury, the oscillations of which were to cause a vacuous condition in the cylinders, and thereby give motion to an air-pressure engine; and lately we have had produced before the Institution of Naval Architects, and also before the British Association at Bristol, the apparatus of Mr. Tower, by which the motion of the waves is to be utilized. A model constructed on this principle has driven, it is said, a boat against the wind at some two or three miles an hour.

The next kind of prime-movers in order of date to be considered are those that are driven by the wind.

Although, undoubtedly, the propelling of a ship by sails, and even the winnowing of grain, must have long preceded the invention of a prime-mover driven by water, yet the employment of the wind as a source of motive power for driving machinery appears to be but of comparatively recent date. It is said that the knowledge of this kind of prime-mover was communicated to Europe by the Crusaders on their return from the East, but it is difficult to see what foundation there is for this statement. It appears to be certain, however, that wind motors were commonly employed in France, Germany, and Holland in the thirteenth century.

We can easily understand that in countries where waterfalls and rapid streams are abundant, the windmill would not (owing to its uncertainty) be resorted to; on the other hand, in arid countries, or even in countries like Holland, where the streams are sluggish and where there is a large amount of land to be drained, the wind, although still uncertain, would nevertheless be a valuable power, and therefore would be utilized.

Prime-movers to be worked by the wind appear to have been made practically in only two forms—viz., the common one wherein a nearly horizontal axle carries four or more twisted radial sails, and that one wherein the axle is vertical and the arms project from it laterally, either as radial fixed arms, as curved fixed arms, or as arms having a feathering motion similar to that of paddle wheels. Where the arms are straight and fixed, some contrivance must be resorted to, to obtain a greater pressure of wind on one side than on the other.

Bessoni in his work, "The Theatre of Instruments and Machines," published at Lyons in 1582, describes a windmill with vertical spindle and curved horizontal arms placed in a tower with a wind-guard, and by the drawing shows it working a chain pump. Belidor also says, in article 852, that windmills with vertical axles were well known in Portugal and in Poland, and he describes how "that they work within a tower," the upper part of which was fitted with a moveable portion to act as a screen to one side of the mill.

I will not detain you by an allusion to the ancient sailing chariot mentioned by Uncle Toby in "Tristram Shandy," nor will I pause to describe one of which, about thirty years since, was employed upon Herne Bay Pier ; in fact, this exhibition in the midst of which we are assembled, gives but little encouragement to pursue the subject of prime-movers worked by wind, as I have not as yet come across in the catalogue a reference to any apparatus illustrative of this branch of mechanics.

It is to be regretted that the use of this kind of prime-mover, the Windmill, is on the decline. It is a power that costs nothing; the machinery can be erected in almost any situation, and although such power cannot be depended on, being of necessity as uncertain as the wind, it nevertheless might be commonly employed as an auxiliary to steam, diminishing the load upon the engine in exact proportion as the windmill was urged by any wind which might happen to blow.

I may say to the credit of our American brethren that they employ on their sailing ships a windmill, known by seamen as "The Sailors' Friend," to pump, to work windlasses, and to do all those matters which in a steamship fall to the lot of the donkey-engine and steam winch, unless, as in a recent voyage in which all Englishmen have been so much interested, these duties are imposed upon a baby elephant.

There is one motor which may be put either into this class or into the next where we consider the application of heat. I allude to the smoke-jack; but beyond recognising its existence as a prime-mover, and a very early one indeed (it is to be found in Zonca's work published in 1621), attention need not be bestowed upon it.

We now come to consider those prime-movers which are worked by heat, and we will commence with those which are worked by its immediate and not by its secondary action.

The direct rays of the sun have for a very long time past been suggested as a means of obtaining motive power. Salomon de Caus, in his work published in 1615, describes a fountain which is caused to operate by the heat of the sun's rays expanding the air in a box and expelling thereby through a delivery valve the water from the lower part of the box. When the sun's rays ceased to impinge upon the box the air cooling, contracted a suction valve opened and admitted more water into the box, to be again displaced on the following day. De Caus also gives a drawing of an apparatus where the effect of the sun's rays is to be intensified by a number of lenses in a frame. He proposes these apparatus as mere toys to work ornamental fountains, but Belidor, by article 827, describes and shows a sun pump consisting of a large metallic sphere fitted with a suction pipe and valve and a delivery pipe and valve and occupied partly by water and partly by air, the suggestion being, as in the case of Salomon de Caus, that the heat of the sun in the day time expanding the air should drive up the water into a reservoir, while the contraction of the air in the night time should elevate the water by the suction pipe and thus re-charge the sphere for the next day's work. In modern times, as we know, some attempts to obtain practical motive power from the direct action of the sun have been made, and notably by Mr. Ericsson.*

The temptation to endeavour to bring into ordinary and commercial use a machine of this character is very great. We were told by our President in a lecture delivered by him to the British Association at Bradford, that the solar heat if fully exercised all over the globe, and supposing that globe to be entirely covered with water, would be sufficient to evaporate per annum a layer fourteen feet deep. Now assuming ten pounds of water evaporated from the temperature of the air into steam by the combustion of one pound of coal (a much larger result than unhappily we get in regular work), this would represent an effect obtained from the sun's rays on each acre of water equal to the combustion of 1680 tons of coals per annum, or about ninety-two hundredweight of coal per acre per twenty-four hours, that is to say, enough to maintain an engine of two hundred gross indicated horse-power day and night all the year round. When, however, we consider the effect of the sun, not upon the surface of water but upon the earth, and deal with its power of producing heat-giving material, that power compares very unfavourably with the results above stated, and this no doubt arises, first, from the fact that the sun is

* In the number of the "Revue Industrielle," of the 24th November last, page 455, are an engraving and a description of the solar boiler of M. Mouchot, Professor at Tours. The apparatus employed by M. Mouchot consisted of a framework shaped like the frustrum of a cone, and mounted so that the axis could always be moved to point to the sun. The interior of the framework was lined with mirrors, the diameter of the large mouth of the cone was 2·6 metres, and of the base 1 metre, giving an effective area of sun's rays intercepted of 45 feet. The mirrors were placed at an angle of 45°, and reflected the rays of the sun upon a blackened copper boiler, placed within a glass envelope in the axis of the cone. It is stated that on a very hot day as much as 11 lbs. of water were evaporated per hour by this apparatus.

frequently obscured, and secondly, from the fact that a large portion of the energy of the sun is spent in evaporating moisture from the ground and not in the direct production of combustible material. I have found it extremely difficult to obtain any trustworthy data as to the weight of fuel grown per acre per annum. If we take the sugar cane we find that in extremely favourable cases as much megass and sugar are produced per acre as together would equal in calorific effect about five tons of good Welsh coal. Coming to our own country and dealing with a field of wheat, the wheat and straw together may be taken as being equal probably to about two tons of coal as a maximum. The statements made to me with regard to the production of timber per acre per annum, when grown for the purpose of burning, are very various, but the best average I can make from them is that in this country there is produced as much wood as is equal in calorific effect to about one and a half tons of good coal per acre. Comparing these productions of heat-giving material with the energy of the sun, as shown in the evaporation of water, one sees how tempting a field is that of the direct employment of the solar rays as a source of power, more especially when it is remembered that those rays are obtained from week to week and year to year without having to wait the tardy growth of the fuel-destined tree.

I will now ask you to consider with me the prime-movers that owe their energy to the heat developed by the combustion of some ordinary kind of fuel, coal or wood. Passing by as a mere toy and as being not an actual prime-mover, the reactionary steam sphere, the æolipile of Hiero, I will come at once to those simple forms of heat engine, intended, whether worked by steam or by the expansion of air, for the raising of water. Salomon de Caus, in his work of 1615, already mentioned, says that if a globe be filled with water and have in its upper part a pipe dipping nearly to the bottom, and if the globe be put upon the fire the heat will cause the expansion of the contents, and the water will be delivered in a jet out of the tube.

The Marquis of Worcester in his "Century of Inventions," published in 1659, makes, as is well known, a similar proposition, but it does not appear that these machines were seriously contemplated for practical use. Papin (I take Belidor's article, No. 1276, as my authority), in 1698 (as appears in his pamphlet of 1707), experimented by order of Charles the Landgrave, of Hesse-Cassel, with the view o ascertaining how to raise water by the aid of fire. But his experiments were interrupted, and he did not resume them until Leibnitz, by a letter of 6th January, 1705, called his attention to what Savery was doing in England, sending him a copy of a London print of a description of Savery's engine. This engine, which of course is well known to you, is illustrated by a model in this collection and now on the table before me. Savery employed a boiler, the steam from which was admitted into a vessel furnished like the sun pump of Belidor with a suction pipe and clack, and a delivery pipe and clack. The steam being shut off, cold water was suffered to flow over the vessel, a partial vacuum was made, water was driven up into the vessel, and was expelled through the delivery pipe upon the next admission of steam, the cocks being worked by hand. This machine came into very considerable use, and was undoubtedly the first practical working steam-engine. It had, however, the defect of consuming a large quantity of steam, as the steam not only came into contact with the cold vessel, but also with the surface of the water in that vessel. Papin, as we know, obviated a portion of this loss by the employment of a floating piston placed so as to keep the steam from actual contact with the surface of the water. I have put a rough diagram of Papin's water raiser on the wall.

We have in the collection, No. 2007, a cylinder from Hesse-Cassel, said to be of the date of 1699, and to have been intended for employment in Papin's machine; but it is difficult to say for what part of the apparatus it could have been designed, inasmuch as the cylinder is provided with a flange at one end only, and no means, so far as I can ascertain, exist for closing the other end. You will see from the diagram (as no doubt is already well known to you), that Papin did not propose to condense the steam and by its condensation to "draw up" the water (to use a familiar expression), but intended that the vessel should be charged by a supply from above, and that the steam should be employed only to press on the floating piston and to drive the water out. Papin, however, hoped to use his engine not merely as a water raiser, but as a source of rotary power by allowing the water to issue under pressure from the air vessel, and so as to impinge upon the pallets of a water-wheel and thus produce the required revolution.

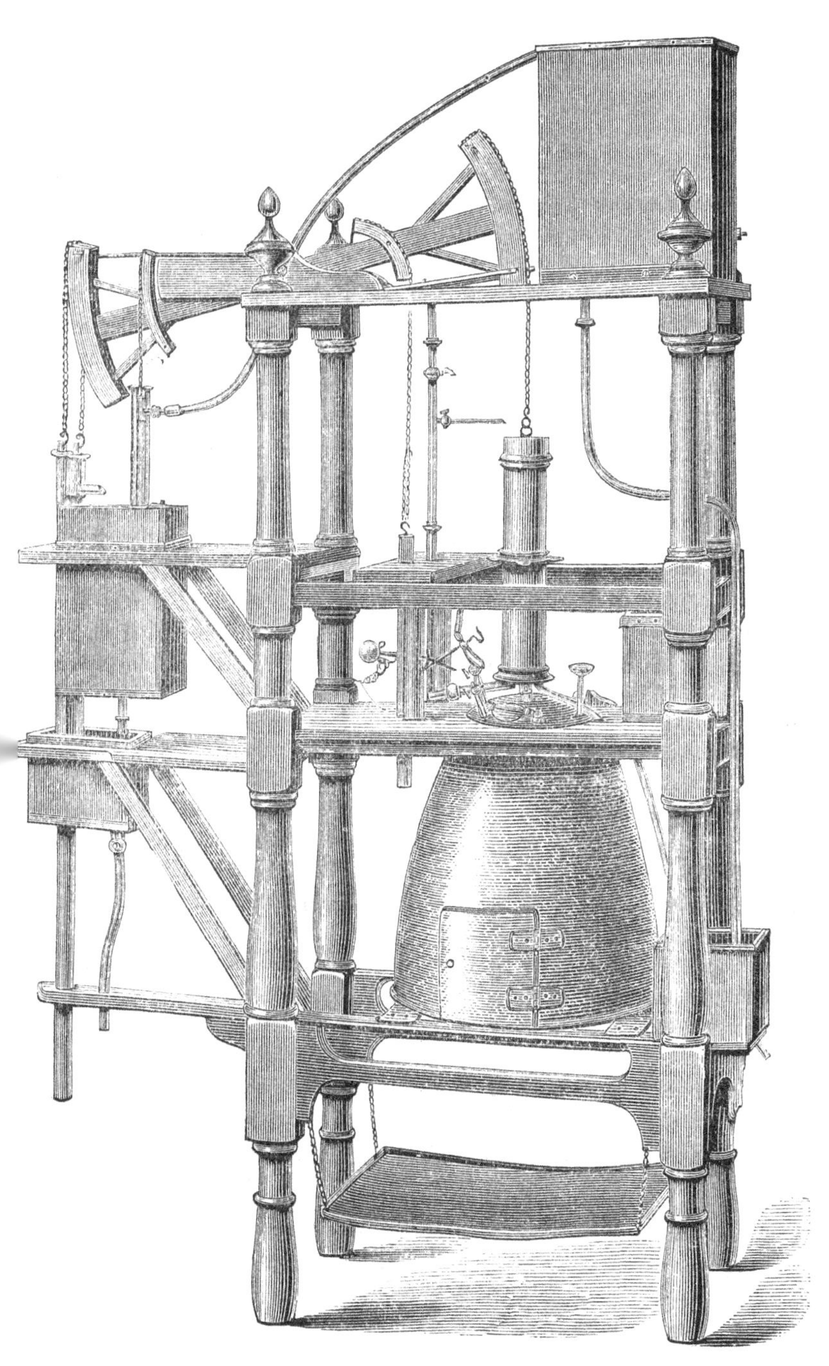

Original model of Newcomen's Steam Engine.
Made about 1705. Now in the Museum of King's College, London.

We now come to Newcomen, who, I think, may fairly be looked upon as the father of the steam-engine in its present form. No. 1942 is a model of his engine, which is further illustrated by a rare engraving of 1712, the property of Mr. Bennet Woodcroft.

Here we have the steam boiler, the cylinder, the piston and rod, the beam working the pumps in the pit, the injection into the cylinder, and the self-acting gear making altogether a powerful and an automatic prime-mover.

That conscientious writer, Belidor, to whom I have already frequently referred, says that he hears of one of these machines having been set up in the water-works on the banks of the Thames at York Buildings. (I may say to those who are not aware of it, that those works were situated where the Charing Cross Station now stands). He is much interested in the accounts he receives, and on a Newcomen engine being erected in France at a colliery at Fresnes near Condé, Belidor paid several visits to it in order that he might understand its construction thoroughly and be thereby enabled to explain it to his readers. He has done so with a minuteness and faithfulness of detail in description and in drawings that would enable any mechanic to reproduce the very machine. This engine had a thirty-inch cylinder with a six-foot stroke of the piston and of the pumps; the boiler was nine feet in diameter and three and a half feet deep in the body; it had a dome which was covered with masonry two feet six inches thick to hold it down against the pressure of the steam. It had a safety valve (the Papin valve) which Belidor calls a "ventouse," and says that its object was to give air to the boiler when the vapour was too strong. It had double vertical gauge cocks, the function of which Belidor explains; it made fifteen strokes in a minute, and he says that, being once started, it required no attention beyond keeping up the fire, and that it worked continuously for forty-eight hours, and in the forty-eight hours unwatered the mine for the week; whereas, previous to the erection of the engine, the mine was drained by a horse-power machine working day and night throughout the whole week, and demanding the labour of fifty horses and the attendance of twenty men. I should have said that the pumps worked by the steam-engine were seven inches bore, and were placed twenty-four feet apart vertically in

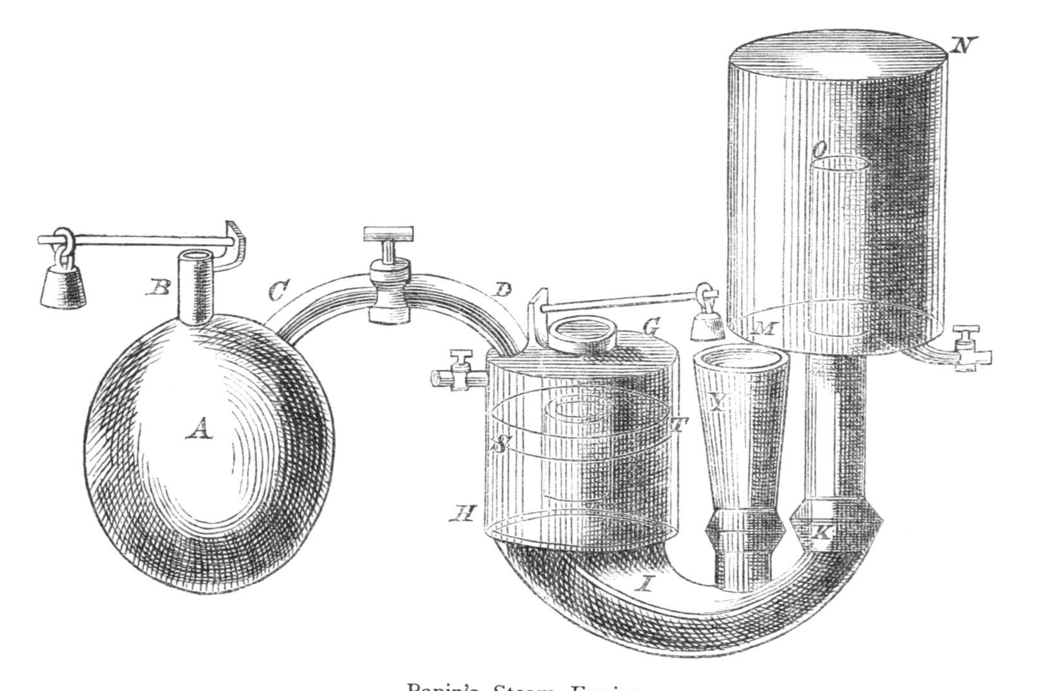

Papin's Steam Engine.

Copied from Plate iv., *Chap.* iii., *Book* iv., *Vol.* ii., *Belidor's* "*Architecture Hydraulique.*"

A, the "Alambic." *B*, the "Tuyau." *C D*, the "Syphon." *G H*, "Cylindre" acting as a "Corps de pompe." *S T*, "Piston," copper and hollow, so as to float. *I K O*, "Tuyau recourbé." *M N*, "Cylindre," 3 feet high and 23 inches diameter. *Y*, Trumpet-mouthed vessel.

the pit, which was 276 feet deep, and that each pump delivered into a leaden cistern from which the pump above it drew.

After having given a most accurate description of the engine, Belidor breaks out into a rhapsody and says (I will give you a free translation): "It must be acknowledged that here we have the most marvellous of all machines, and that there is none other of which the mechanism has so close a relation to that of animals. Heat is the principle of its movements; in its various tubes a circulation like that of the blood in the veins is set up; there are valves which open and shut; it feeds itself and it performs all other functions which are necessary to enable it to exist."

Smeaton employed himself in perfecting and in properly proportioning the Newcomen engine, but it was not until James Watt that the next great step was made. That step was, as we all know, the doing away with condensation in the cylinder, the effecting it in a separate vessel, and the exclusion of the atmosphere from the cylinder. These alterations made a most important improvement in the efficiency of the engine in relation to the fuel consumed; but they were so simple that I doubt not if examiners into the merits of patents had existed in those days, Mr. Watt would have had his application rejected as being "frivolous." We have here from case No. 1928, a model made by Watt, which appears to be that of the separate condenser and air pump. We have also 8*b*, which is a wooden model made by Watt of a single acting inverted engine having the top side of the cylinder always open to the condenser, and a pair of valves by which the bottom side of the piston can be put into alternate connexion with the boiler and with the condenser, the contents of which are withdrawn by the air pump, 3*b*. From the same case is a model of a direct acting inverted pumping engine made in accordance with the diagram; 8*b*, 1*b*, is a model of Watt's single acting beam pumping engine, while 2*b* is a model of Watt's double acting beam rotary engine. 10*b*, from the same case, is Watt's model of a surface condenser. To Watt we owe condensation in a separate vessel, exclusion of the air from the cylinder, making the engine double acting, employment of the steam jacket, employment of the steam expansively; the parallel motion, the governor, and, in fact, all which made Newcomen's single acting reciprocating pumping engine into that machine of universal utility that

Jonathan Hull's Steamboat.

Copied from the original engraving published in his pamphlet in 1737.
Now in the Library of the Institution of Civil Engineers.

A A

the steam-engine now is, and not only so, but Watt invented the steam-engine indicator, which enables us to ascertain that which is taking place within the cylinder, and to detect whether or not the steam is being economically employed. I have on the table before me a very excellent model illustrating an inverted direct acting pumping engine in its complete form, and I have also a model, of French manufacture, the cylinder and other working parts of which are in glass, which shows a form of beam condensing engine at one time common. I do not say, however, that Watt was the first to make the suggestion of attaining rotary motion from the power of steam. Leaving out of consideration Hiero's toy, Papin, as I have remarked, hoped to get rotary movement second hand by working a water-wheel with the water that had been raised by his steam engine; moreover, as early as 1737, Jonathan Hulls proposed to obtain rotary motion from a Newcomen engine, and to employ that motion in turning a paddle wheel to propel a tug boat which should tow ships out of harbour, in a calm, or even against an adverse wind. I have here one of the prints of his pamphlet, and in order that you may better appreciate Hull's invention I have put an enlarged diagram upon the wall, and I think I may take this as the starting point for saying a few words about the steam-engine as a prime-mover in steam vessels.

We have in the collection, No. 2150, Symington's engine tried upon the lake at Dalswinton in 1788. Here a pair of single acting vertical cylinders give by the up and down motion of their pistons reciprocating movement to an overhead wheel; this wheel gives similar motion to an endless chain, which chain is led away so as to pass round two pairs of ratchet wheels loose upon two paddle shafts. By the use of a pair of ratchets the reciprocations of the chain are converted into rotary motion in one direction only, and that the driving direction, of the two paddle-wheels, placed one behind the other. Symington's arrangement for obtaining the rotary motion always in one direction of his two paddle-wheels is very similar to that proposed by Jonathan Hulls for his single stern-wheel.

Want of time forbids me to do more than to just allude to the names of Hornblower and Wolff in connexion with double cylinder engines—engines wherein the expansion of steam is commenced in one cylinder and continued in another and a larger one.

I wish to say a few words which will bring before you the changes that have been made within a very few years in the construction of steam-engines: when I was an apprentice the ordinary working pressure of steam, except in the double cylinder engine, was only 3 lbs. above atmosphere, and in those days there was in a marine boiler more pressure on its bottom when the steam was down due to the mere head of water in the boiler than there was pressure on the top when the steam was up due to the force of the steam, whereas now condensing marine engines work commonly at 70 lbs.; and there is a boat under trial where the steam is, I believe, as high as 400 lbs.

To those who are curious on the subject I would recommend a perusal of two Blue Books, one being the evidence taken before a Parliamentary Commission in 1817, and the other before a Parliamentary Committee in 1839, they will find there the weight of evidence to be that the only use of high pressure steam is to dispense with condensing water, and that as a steamboat must always have plenty of condensing water in its neighbourhood, no engineer, knowing his business, would suggest high pressure for a marine engine.

I have before me a model of a pair of engines which, although they were made not so very long ago (for I saw them put into the ship), have nevertheless an historical interest. This model shows Maudslay's engines of the *Great Western*, the first steamer built for the purpose of crossing the Atlantic; I think I am right in saying that 7 lbs. steam was the pressure employed in that vessel, and that, in order to extract the brine from the boiler, it was necessary to use pumps, as the pressure of the steam was not sufficient to expel the brine and to deliver it against the pressure of the sea.

Time does not permit of my touching upon the various improvements in boilers, condensers, expansive arrangements, and other matters which have gradually been introduced into our best engines for land and for ocean purposes. I have hung upon the wall a rough diagram, showing a pair of oscillating engines as applied to driving a paddle-steamer, and another showing a pair of inverted compound cylinder engines to drive a screw propeller, a model of such a pair of engines, with surface condensers and all modern appliances, (being Messrs. Rennie's engines of the P. and O. Company's S.S. *Pera*, by which I have had the pleasure of travelling), is now before me.

I will conclude this part of the subject by saying, that to the combination of science and sound practice is due the fact of the consumption of coal having within the last ten years been reduced in the marine engine from 5 lbs. per gross indicated horse-power per hour to an average of 2¼ lbs., and in exceptional instances to as small a quantity as 1½ lbs. per horse per hour.

Let us now devote a little of the time that is left to the consideration of the locomotive on the common road as well as on the railway. I have before me, No. 2145, a model of the actual Cugnot engine in the Conservatoire des Arts et Métiers, which, in 1769, journeyed, slowly it is true, but did journey, and did carry passengers along the roads in Paris.

It is a most ingenious machine; it has three wheels, and the motive power is applied to the front—the castor or steering wheel—so that engine and boiler turn with the wheel, precisely as within the last few years Mr. Perkins has caused the engine and boiler to turn with the steering wheel of his three-wheeled common road locomotive. The steam makes the pistons in a pair of inverted single acting cylinders to reciprocate, and their rods, by means of ratchet wheels, give rotary motion to the castor wheel, and thus propel the carriage. I think there is no doubt but what we must look upon this engine of Cugnot as the father of steam locomotion in the same way as we must regard Symington's engine as the parent of marine propulsion. I have before me, No. 1926, Trevethick's engine of 1802. I have also before me a Blenkinsop rail, one that has been in actual use for many years, provided, as you will see, with teeth, into which a cogged flange on the side of the driving wheel, geared, to insure that adequate traction should be obtained. This plan has been revived within the last few years, to enable the steam locomotive to climb the Righi. A sketch of the Righi engine and rail is on the wall. It will be seen that in the Righi instance the teeth, instead of projecting, as in the Blenkinsop plan from the side of the rail, are ranged between two parallel bars, like the rungs of a ladder.

On the ground floor of the Exhibition we have the veritable

Locomotive Engine, known as "Puffing Billy."

Constructed in 1813 by Jonathan Foster, under William Hedley's Patent. Now in the Patent Office Museum.

"Puffing Billy," an engine which began work in 1813, and got along without the aid of cogs by mere adhesion upon plain rails; it is a rude-looking machine, but it laboured up till the date of the last Exhibition (1862), doing its work for forty-nine years on the railway belonging to the Wylams Colliery, and as tradition says, interesting George Stephenson, who as a boy saw it in daily operation.

On the ground floor also we have, 1954, the "Rocket," with which seventeen years after the starting of "Puffing Billy," George Stephenson carried off the prize in the Manchester and Liverpool Railway competition. The leading particulars of this engine are as follows: A pair of $7\frac{1}{2}$ inch cylinders, 1′ 5″ stroke, placed at an inclination, driving 4′ 6″ wheels, the boiler multi-tubular, having twenty-four three and a half inch tubes, while the fire is urged by the waste blast. Before alluding to this I ought to have mentioned that in one of the Blue Books to which I have called your attention, that which gives the evidence before the Commission in the year 1817, there is a statement by a witness that in "those parts there are machines called locomotives," &c. &c.

Once more I am compelled to say that time will not admit of my entering into any detail in respect of the modern locomotive, except to remark that by the aid of excellent boilers, of high pressure (steam 140 lbs. to the inch), of considerable, although rather imperfect expansion, effected by the link motion, there is provided for the use of our railways a machine which in the "Passenger" form is competent to travel with ease and safety sixty miles an hour; and in the "Goods" form is competent to draw a load of 800 to 1000 tons, and to attain these results with a very commendable economy in fuel. I have put on the wall two diagrams of Locomotives of the convenient form for local traffic that we call Tank engines; and I have before me No. 1957*a*, a most beautifully made sectional working model of a Russian six-wheeled "Goods" engine.

Within the last twenty years another description of steam-engine has acquired a prominent and important place among our Prime-movers; I allude to the Portable engine, or to the Portable engine in its more complete form of a self-propelling or Traction engine. The general construction of these machines borders closely upon that of the locomotive. Very great attention has been paid to all their details, and the Royal Agricultural Society of England, by their excellent

The Locomotive Engine "Rocket."

Constructed by Messrs. Stephenson & Co. in 1829.

Now in the Patent Office Museum.

arrangements for periodical trials, have stimulated engineers to devote their best energies to the subject. No. 1942 is a model of one of Aveling and Porter's common road traction engines, capable also of acting as a source of power for driving farmyard machinery, or for effecting steam ploughing. Upon the wall I have placed rough diagrams of another kind of traction engine—a kind wherein india-rubber tires are used. This is manufactured by Messrs. Ransome, Sims, and Head; and I have also placed diagrams of the ordinary Portable engine and of another most useful kind of Portable engine—the Steam Fire engine. I have there likewise a sketch of Hancock's common road steam coach, which some thirty years ago regularly plied for hire from the Bank to Paddington in opposition to the ordinary horse omnibus. Hancock's carriage was a vehicle which in my judgment has never since been surpassed, and I am sorry to say never to my knowledge equalled, as regards the various points which should be attended to in making a steam carriage to circulate safely among horse traffic.

There is another way in which steam may be employed as a prime-mover. We saw that water in the form of the trombe-d'eau could induce a current in air, and thereby blow a forge fire, and that a rapid stream could be caused to induce a current in other water, and thus drain marshy lands. Similarly steam can be caused to induce a current in water, and thereby impel the water so as to raise it to a height, or to force it as feed water into a boiler against a heavy pressure. When used for a mere pumping apparatus such a mode of employing steam is very wasteful, because the steam is condensed in large quantities by the water, and the water is needlessly heated at the expense of the steam; but when used in feeding a boiler the whole of the heat is taken into that boiler, and thus this objection does not apply. By means of that most elegant and scientific apparatus, the Giffard injector, it is possible by a jet of steam to induce a current in surrounding water powerful enough to take the water and the condensed steam into the boiler from which the steam had previously issued. No. 1976, which I have before me, is a sectional model of a Giffard injector.

I believe it was I who first gave a popular explanation of the principle of action of the Giffard injector; and although a Scientific Congress is probably not the place for a popular explanation, I will venture to repeat it. The principle may be summed up in one word, "concentration." The steam that issues from an orifice of an area of 1, when condensed, has a sectional area (according to the original pressure of the steam) of only $\frac{1}{200}$th, or $\frac{1}{400}$th, or $\frac{1}{800}$th, as the case may be. Thus the velocity remaining the same, and the weight the same, the energy of the steam issuing from an area of 1 is concentrated 200, 400, or 800 times upon the area, due to the smaller transverse section of the liquid stream.

This concentration of energy is far more than sufficient to enable the fluid stream to re-enter the boiler from which the vaporous stream started; and so much more than sufficient, that it may be diluted by taking with it a certain quantity of water which was employed in the condensation of the steam, and is required for the feeding of the boiler.

With a view to obtaining economy in fuel many attempts have been made to employ some other agent than steam as the means of developing the power latent in fuel, but it is imperative that I should dismiss these with a mere enumeration. A very interesting engine of this kind (because, excluding Hiero's toy and smoke jacks, it is, so far as I know, the first proposition for obtaining rotary motion by the aid of heat), was the fire wheel of M. Amonton, of which an account is to be found in the first volume of the French Academy of Sciences, date 1699. On referring to that volume I do not see it is stated in terms the machine was ever put to work, although it is said that M. Amonton made many experiments to convince the Academy of the practicability of his invention. M. Amonton proposed to have a metallic wheel revolving on a horizontal axis; the outer rim of the wheel was to be divided into a number of separate air cells, each of which had a channel so as to communicate with other cells, arranged round the wheel nearer to the centre than the air cells; the air cells as they passed over a fire were to be heated, and the air was to drive the water up to one side of the wheel, so as to keep that side always loaded, and thus give the wheel a tendency to revolve. The cells, after leaving the neighbourhood of the fire, were to be cooled by passing through water, to re-contract the air ready for the next operation.

No. 1940, which is before me, is a model of Stirling's hot-air engine, but time does not remain to describe it.

Besides hot-air engines, we have had engines working by the explosion of gunpowder, and others working by the explosion of gases. No. 1945 is Langen and Crossley's gas engine, from which, I believe, extremely excellent results have been obtained.

I will now ask you to look at a tabular statement which shows the consumption of fuel in agricultural engines when under trial, expressed in pounds per horse-power per hour, and also in millions of pounds raised one foot high by the consumption of one hundredweight of coal. I told you how excellent were the results at which our agricultural engineers had arrived. You will see that one of those machines working with 80 lbs. steam, and of course without condensation, has developed not a gross indicated horse-power, but an actual dynamometrical horse-power for 2·79 lbs. of coal per horse per hour, giving a duty of as much as seventy-nine millions. This high result was obtained by the excellence of the boiler and of the combustion, as well as by that of the engine. If you look at the column of evaporation you will find that as much as 11·83 lbs. of water were converted from the temperature of the boiling-point into steam by the combustion of 1 lb. of coal. This was due not to the excellence of the boiler alone, but to the extraordinary ability of the stoker, and to the care and labour bestowed—a care and labour far too expensive to be employed in practice. But should not we engineers endeavour to ascertain whether we cannot by mechanical means practically, with certainty and cheapness, procure an accuracy of combustion as great or even greater than that which can be got by the almost superhuman attention of a highly-trained man, who at the end of four hours of such work is utterly exhausted? Many forms of fire-feeders have been attempted and used with more or less success; but I cannot help thinking that in order to obtain the accurate proportioning of air and fuel by which alone can we get efficient and economical combustion, we shall have to turn our attention in the direction of dealing with the fuel in a comminuted state, either by converting it into gas, as is done by our President, Dr. Siemens, by availing ourselves of liquid fuel, or by employing the process of Mr. Crampton, the making of the fuel into an impalpable powder, which can be driven into the furnace by the air which is to consume it there.

By these and by other means we may hope to improve combustion; by strict attention to the proportioning of the parts of the boiler we may hope to make the best use of this improved combustion; by higher initial pressure, by greater expansion, and by the general employment of condensation wherever practicable (and by the use of the evaporative condenser there are very few cases in which it is not practicable), we may trust that the steam-engine, even on its present principle, may be rendered more economical than it has ever yet been; and may give us more than that one-eighth or one-ninth of the total energy residing in the fuel which now alone we get under the very best and most exceptional conditions. A large loss, however, must with steam-engines as we now know them always be incurred. We cannot hope to deal with initial pressures and temperatures corresponding with steam of a density equal to that of water; nor to carry expansion down to the point where ice would be formed in the condenser. But wonderful as the steam-engine is, worthy as it was and is of Belidor's eulogium (which I read to you), we know it is not the only heat motor; and we are aware that there are other forms of such motors which theoretically at all events promise higher results.

By improvements in the existing steam-engine, by the invention and development of other heat motors, by the employment of the power of water and of wind, either as principal motors or as auxiliaries, we may hope for further progress in the machines the subject of my address, "Prime-Movers."

I have brought before you, of necessity hastily, and therefore (and also on account of my own incapacity for the task) imperfectly, the leading improvements which have been made in Prime-Movers, from the date of the water-wheels of Vitruvius to the best devised steam-engines of our own day. These improvements have been effected by men like Papin, Savery, Newcomen, Watt, Symington, Fulton, Stephenson, and others, who were not mere makers of engines, but were men full of an ardent love of their noble profession, who followed it because of the irresistible attraction it had for them; followed it from their boyhood to their grave, and in that very following found their great reward.

These men undoubtedly possessed that combination of science and practice, which combination Dr. Tyndall has told us is necessary if either science or practice is to continue to live, for to use his expressive language, without this combination both die of atrophy—the one becomes a ghost, the other a corpse.

We have every reason to believe that this combination will become more intimate, not only in the engineers of the present day, but in those of the next and of succeeding generations; and to men thus endowed with science and practice we may trustfully leave the continued improvement of Prime-Movers, and may rest assured that as a more general application of these machines must of necessity follow such improvements, the day will soon dawn when in no civilized country will there continue to be, from the hope of gain, the temptation to employ intelligent humanity in the brutal labour of the turnspit or of the criminal on the treadwheel.

The PRESIDENT: Ladies and Gentlemen,—In coming here on this wet morning to listen to Mr. Bramwell on Prime-Movers we all expected to hear something worth listening to, but I think our expectations have been surpassed by the discourse to which we have actually listened. Mr. Bramwell, after dealing with the theoretical principles upon which prime motion must depend, has brought before us in review all those early attempts by which the labour of man has been gradually supplanted by machine labour, and he has been aided in a way he could not have been aided in any other place by the models which he has had at his disposal. These models show us not only the principles upon which those early pioneers acted, but they bring home to our minds the difficulties—the mechanical difficulties—with which they must have had to contend. Altogether I look upon the communication to which we have listened as a standard treatise on this subject: and I hope that the department will do it ample justice in publishing it in the Proceedings with such illustrations as will make it valuable. I call upon you to accord your thanks to Mr. Bramwell for his communication.

Mr. BRAMWELL: Ladies and Gentlemen,—I have to thank you most sincerely for having responded in the way that you have to the kind expressions of our President, Dr. Siemens. I regret the great length to which my Address has extended, but I could not deal with it more concisely; in fact, I had not time to write a short one.

The PRESIDENT: The time is rather short before we adjourn, and

I propose that Mr. Hackney should deliver his discourse on furnaces, as far as he can before we adjourn, and that we should after the adjournment view those models which there is not time at present to display.

ON FURNACES.

Mr. HACKNEY: Furnaces, as used in the arts, may be defined as the arrangements under which fuel is burned to produce heat, for the purpose either of inducing permanent changes in the substances heated, or of preparing them, by softening or fusion, for subsequent treatment.

This excludes two of the applications of fuel which together take up the larger proportion of that consumed. The domestic use—namely, for cooking, and for warming, lighting, and ventilating inhabited places; and that for the generation of motive power.

Even thus limited, the amount of fuel burned in furnaces is very great, and the possibility of effecting further economies in its use is a problem of vast importance. The quantity of coal burned, in this kingdom alone, for smelting ores and for refining and manufacturing metals, was, according to the statistics published by the Mining Record Office, 35,802,000 tons in 1873, and 34,400,000 tons in 1874. Of these amounts the manufacture and working of iron, in its different forms, took nearly the whole :—35,039,000 tons in 1873, and 33,562,000 tons in 1874; or more than 100,000 tons on every working day.*

The subject of the construction and mode of working of the several forms of furnaces in use is one of great extent, and can only be dealt with briefly in such a paper as the present. They may be divided into those in which solid fuel is intermixed with or directly surrounds the matters to be heated, and those in which the heating is done, in one way or another, by flame, without direct contact between the solid fuel and the work.

The characters of the fuel best fitted for these two kinds of furnace are essentially different. Where the matters to be acted on, or the vessels that contain them, are in direct contact with the fuel, as in a

* Mineral Statistics, 1874, p. xiv.

smithy fire, a cupola, or an ordinary coke furnace for melting steel or brass in crucibles, an intense local heat is required in the mass of the fuel itself, and any heat developed above its surface is useless. In flame furnaces, on the other hand, such as those for glass melting, or for puddling or heating iron, in which the materials to be heated are not embedded in the fuel, but placed in a chamber above or at the side of it, the heat made use of is that of the flame; the heat that is carried into the working chamber by the current of gases rising from the fire, together with that due to the further combustion of these gases, on admixture with an additional amount of air.

Thus, for furnaces of the first class, the most suitable fuel is one, such as charcoal, coke, or anthracite, consisting of nearly pure carbon, free from volatile matter, as this is useless in them as a source of heat, and the driving of it off renders latent a certain amount of that generated by the combustion of the carbon, and so lowers the temperature of the fire.

In flame furnaces, a lowering of the temperature at the fire-grate, where the air and the solid fuel meet, is immaterial, or may be even advantageous, as tending to diminish loss by radiation and to preserve the furnace from injury by excessive heat. The only use of the heat at the grate is to generate a full supply of combustible or partly burned gases, at a high temperature, which in completing their combustion, as they pass over the working bed, shall heat as strongly as possible the matters placed there. The fuel preferred for use in such furnaces, is thus either a combustible gas, or a solid fuel containing hydrogen as well as carbon, such as coal or dried wood, that will produce on burning a long and powerful flame. A flame, it is true, may also be obtained from fuels that contain little else than carbon and mineral matter, by burning them in a thick bed, so that the greater part or nearly the whole of the CO_2 formed in the first instance by the combustion of the carbon, is transformed into CO as it passes up through the mass; and by introducing with the air as large a proportion of steam as can be used without lowering too much the temperature of the fire. The steam is decomposed by the hot carbon, producing, according to the temperature and thickness of the fire, a mixture of either H and CO_2 or H and CO. The gases thus generated, together with the mixture of CO and N, produced by the passage of the air itself through the mass of fuel, flow forward into the working chamber, and there burn, on mixing with a further supply of air, introduced above the fire.

Examples of the simplest form of the class of furnace in which solid fuel is mixed directly with the matters to be heated are the heaps in which brick clay is burned to make ballast, and in which iron and other ores are often calcined. In these, the ore or dried clay, in pieces of convenient size, is thrown into a heap, together with a little coal; and the mass, being lighted at one end, burns through to the other. In calcining in such heaps, there is a considerable waste of heat, as a great proportion of the burned gases from the fire pass off at a high temperature; and when the calcination is completed, all the heat that the red-hot mass contains is lost.

In an ordinary limekiln, and in kilns for calcining iron ore, such as are used in Wales and Cleveland, a much larger proportion of the heat produced by the combustion of the fuel is utilized, and the amount of this required is proportionately reduced. The process in such a kiln is carried on continuously; the coal and raw stone being filled in, together, at the top, and the calcined material drawn at intervals from the bottom. By this mode of working, the air that maintains the combustion of the fuel is drawn up through the material already burned; cooling this down from a red heat to a temperature not much above that of the atmosphere, and being itself considerably heated: while the burned gases and excess of air, passing away, rise through the mass of cold freshly charged material, and leave in it the greater part of their available heat, escaping nearly cool.

In such a kiln, according to Gjers, calcining 800 tons of Cleveland iron ore per week, the consumption of small coal is about one ton to twenty-four or twenty-five tons of ore.* In kilns of somewhat smaller size, both in Wales and in Cleveland, in which the regenerative action is less complete, and in which the loss of heat also, by radiation and conduction, is greater, the consumption is stated by Phillips to be one

* Journal of the Iron and Steel Institute, 1871, p. 213.

ton of small coal to twenty tons of ore; and in calcining in open heaps, in South Wales and Staffordshire, it is given as one ton of small coal, and five hundredweight of large, to ten tons of ore.*

A similar economy in fuel to that which characterizes such continuous kilns is attained, even more perfectly, in calcining limestone or other materials, and more especially in burning bricks, by the system of annular kilns patented some years ago by F. Hoffmann of Berlin, and now very generally used. In this form of kiln, which is continuous in its working, the air to supply the fire is drawn through the bricks already burned, cooling them down to the temperature of the air, and carrying forward their heat to the part where it is required; and the burned gases, mixed with excess of air, pass, on their way to the chimney, through the stacks of damp air-dried bricks, that have not yet been fired; and escape, finally, at a comparatively low temperature, and saturated with moisture.

The kiln is built in the form of an arched passage, eight or nine feet high, and of the same width. This tunnel or passage is bent on itself, in the form of a ring of any convenient diameter, and provided with twelve equidistant doorways for putting in and removing the bricks, and twelve corresponding flues, leading to a central chimney, and each provided with a damper. The entire kiln thus consists of twelve or more chambers or compartments, arranged in a circle and communicating freely with each other; each compartment having an independent entrance door and chimney flue.

The successive compartments being numbered one to twelve, the working of the kiln is as follows:—Supposing No. 2 to be the compartment from which the burned bricks are being removed, and No. 1 to be the adjoining compartment, last emptied, and now being filled with new unburned bricks; then No. 12, on the other side of No. 1, will be the compartment last filled with new bricks, which are now being dried preparatory to burning. When the work is in this position, the damper in the flue leading from compartment No. 12 to the chimney, and the entrance doorways of Nos. 1 and 2 are open, all the other chimney flues and the other door openings being closed; and

* Metallurgy, J. A. Phillips, pp. 181, 189.

a large damper or partition is fixed between the compartments No. 1 and No. 12. Thus the air entering through the open doorways of No. 1 and No. 2 has to make the entire circuit of the kiln, before it escapes, through the flue leading from No. 12, to the common chimney. In the course of this circuit, it passes first among bricks almost cold, and takes up their heat, and then goes forward to warmer bricks, and then to hotter and hotter, carrying the heat of the cooling bricks forward with it, until it reaches the part of the ring diametrically opposite to the two open and cold compartments. At this place it gets a final accession of heat, from the burning of a small quantity of coal dust, which is dropped in among the bricks, from time to time, through numerous small openings furnished with air-tight moveable lids. Thus, at this part of the kiln, there is generated the full intensity of heat which is required for burning the bricks. The products of combustion then pass forward to the bricks not yet burned, which are thus heated by their continuous current, and so from the hottest bricks to those that are less and less hot, heating them as they go, and thence to those that are still damp, drying them as they go ; and thence they pass finally to the chimney, in a state almost cold, and loaded with moisture from the damp bricks.

On the following day, the partition is removed from between the compartments Nos. 12 and 1, and placed between Nos. 1 and 2 ; the compartment No. 1 having been by this time filled with fresh damp bricks, and its doorway built up. The damper in the flue leading from No. 12 compartment to the chimney is shut, and that from No. 1 is opened. No. 2 compartment being now empty is refilled with unburned bricks ; and the removal of cold burned bricks from No. 3 is commenced. The place where the small coal for fuel is thrown in is also advanced round the circle by one compartment ; and the products of combustion, at the end of their circuit in the annular chamber, and just before their escape to the chimney, now pass among the fresh bricks that were built in on the day before ; and so the process goes on, just as on the previous day ; and the fire makes a complete circuit of the kiln in twelve working days.

The saving in fuel effected by thus utilizing the heat of the burned gases, and of the red-hot bricks, as well as the greater part of that taken up by the walls of the kiln, is very great: amounting to from two-thirds to three-fourths, in quantity, and frequently to much more in value, as cheap dust coal does the work for which in the older kilns a better coal is required.

The large amount of water to be evaporated from the unburned bricks can only be carried off, without risk of condensing a portion on the cold bricks next to the chimney, and so wetting and softening them instead of drying them, by drawing through the kiln a very large excess of air, even more than the rude mode of firing adopted renders inevitable; and as this makes it difficult to maintain a temperature so high as is required for burning fire-bricks, particularly for Welsh Dinas bricks, Mr. Ensor, a fire-brick maker near Burton, adds a by-pass flue to kilns for work of this kind, by which part of the warm air, from the bricks already burned, is led round to the unburned bricks that are being dried, without passing through those compartments in which the highest heat is required; no more being passed through these than is needed for the combustion of the fuel. In a Hoffmann kiln, thus modified, sufficient heat may be maintained to burn Dinas bricks very satisfactorily.

Since the continuous Hoffmann kiln has been in use, several forms of what may be termed semi-continuous kilns have been brought forward, in which, as in it, the heat of the burned gases, and of the hot burned bricks, is more or less perfectly utilized. Tunnel-kilns have also been put up, with the same object: in these the position of the fire is fixed, and the bricks or other articles to be burned are drawn past it, stacked on suitable trucks or waggons.

The small furnaces, fired with coke, that are commonly used for melting steel or brass in crucibles, require no detailed notice. In these, the crucible is imbedded in the fuel, and a rapid combustion and high temperature are maintained round it, by closing the upper part of the furnace and connecting it to a high chimney.

Where, as in the case of a smithy fire, the top of the furnace cannot be conveniently closed in, or where a keener combustion is required than can be obtained by chimney draught, the plan is adopted of forcing air into the fire by mechanical means.

Blast furnaces and cupolas, so arranged, are used largely in smelting the ores of iron, lead, and copper, and in fusing cast iron and other substances. In all these, the fuel and the materials to be melted or otherwise acted on are charged, together, into the upper end of a vertical shaft; and the combustion is maintained by air forced in through one or more openings or tuyères near the bottom. Of such furnaces, those employed for the manufacture of cast iron are by far the largest and most important, and may be taken as the type of the class.

The blast furnace for iron making has attained to its present construction, and to its colossal size, by what may be described as a process of natural selection; a gradual advance and improvement, and the survival of the fittest. Representatives of its earliest stages are to be found in the small and simple furnaces that are described as still in use, for the direct production of malleable iron, in the more remote parts of Africa, India, and Borneo. Some of these are little shafts of clay, hardly larger than a chimney-pot, and charged with a mixture of rich iron ore and charcoal. A draught, to maintain the combustion, is obtained either by placing the furnaces so that they are exposed freely to the prevailing winds, or by forcing air in at the bottom of the shaft by some simple form of hand-bellows. The Osmund furnace of Sweden, with a shaft six feet high, and making one and a half to two tons of malleable iron per week, and the Stückofen, ten to sixteen feet high, and yielding blooms of iron weighing four to six hundredweight, were merely enlargements of these primitive furnaces, and like them produced, not cast iron, but malleable masses resembling puddled balls. The blast furnace, yielding liquid cast iron, grew however directly from the Stückofen; as, in a furnace of this height, cast iron was obtained at will, by increasing the proportion of fuel and keeping the metal in the hearth covered with slag; and when the height was still further increased, nothing but liquid metal was produced.* The possibility of obtaining iron from the ore in a liquid form, that could be tapped out, greatly cheapened its production, and permitted the use of still larger furnaces and the smelting of poorer ores; and it was found that the cheaper

---

* Metallurgy, Iron and S J. Percy, 1864, pp. 325 et seqq. Ibid., J. A. Phillips, p. 170.

way to make malleable iron, with the appliances then available, was not, except in rare cases, to produce it directly from the ore, but to reduce the metal first to the state of cast iron, and to make malleable metal from that by a second process.

Among the further steps that have led up to the modern form of blast furnace are the very general substitution of coke or coal for charcoal as fuel, the use of hot blast, a very great increase in the size of the furnaces, and the collection and use of the waste gases.

The most recently erected furnaces in the Cleveland district are eighty to ninety feet high, and from 26,000 to 30,000 cubic feet in capacity.* The blast supplied to them is at a temperature of 500° C. to 750° C., and the waste gases are collected, and so efficiently employed, to heat the blast and to raise steam for working the blowing engines, that in many works no fuel, except the coke charged into the furnaces, is used.

The economies effected by the late increases in the height of blast furnaces and in the temperature of the blast have been very great. Thus, with blast at a temperature in each case of about 540° C., and with other conditions the same, the consumption of coke which in furnaces forty-eight feet high and of 6000 cubic feet capacity, built at the Clarence Works in 1853, was twenty-nine hundredweight per ton of iron made, is reduced, in more recently erected furnaces, by increasing the height to eighty feet and the capacity to 12,000 cubic feet or more, to twenty-two and a half hundredweight.†

Again, in furnaces of the same size, each increase in the temperature of the blast has been attended with a marked economy; the extent of which, however, diminishes, for equal increments of temperature, the higher this is raised. When blast, for instance, heated from the temperature of the air, which may be taken at 10° C., to 150° C., was first introduced by Nielson, at the Clyde ironworks in 1830, the saving in fuel effected by raising the temperature by 140° C. was equal to about two tons of coke per ton of pig iron made, or thirty-six per cent. of the whole consumption;‡ whereas, in modern practice, the further economy

* J. Gjers, On Cleveland Blast Furnaces, Journal of the Iron and Steel Institute, 1871, p. 202.

† Isaac Lowthian Bell, Blast Furnace Phenomena. Ibid., Proceedings of the Institution of Mechanical Engineers, 1875, p. 364.

‡ J. Percy, op. cit., p. 394.

obtained by an equal rise in temperature, from 620° C. to 760° C., is not more, even according to the strongest advocates of highly-heated blast, than two hundredweight per ton, or nine per cent.; a reduction, that is, in the amount of coke used to smelt ordinary Cleveland ores, and in furnaces eighty feet high, from rather less than twenty-two hundredweight per ton to twenty hundredweight.*

Mr. Bell's investigations† show that the remarkable saving in fuel, obtained by the use of highly heated blast, is due to the lower temperature at which the gases escape from the furnace top, and the smaller proportion of CO that they contain; the ratio of the amount of CO_2 to that of CO, in the gases, being the index of the more or less advantageous manner in which the fuel is burned; and he has pointed out that the economy of large and high furnaces over smaller furnaces is to be traced to the same causes;—the gases from a high furnace carry off less sensible heat; and up to a certain limit, which he fixes, for the materials in use in the Cleveland district, at a height of eighty or ninety feet, they are poorer in CO than those from lower furnaces.

Thus height of furnace and temperature of blast appear to be capable in a great measure of replacing each other in their effect on economy of working; and the benefit to be derived from working with very hot blast is likely to be greater in the case of furnaces which, from the characters of the fuel and ore used, cannot be made very high, than it is in the high furnaces in use in Cleveland. As an instance of the similar effect, on the consumption of fuel, of the use of hot blast and of an increase in height, Mr. Bell quotes the working of the furnaces at Lillieshall. There, in furnaces fifty-three feet high, the consumption of coke, with cold blast, was forty hundredweight per ton of iron made; and this consumption was equally reduced, to twenty-eight hundredweight per ton, in the case of one furnace, by increasing the height to seventy-one feet; and in another, with the old height of fifty-three feet, by heating the blast.

In the Glasgow district, the fuel generally used in the blast furnaces is uncoked free burning coal; and with this comparatively soft fuel it

* C. Cochrane, Proceedings of the Institution of Mechanical Engineers, 1869, p. 21, and 1870, p. 62.

† Blast Furnace Phenomena.

is not found practicable to work furnaces, of the ordinary form, much exceeding fifty-two feet in height. In the self-coking furnace, however, as it is termed, of Mr. Ferrie, the barrier thus imposed to increased economy appears to be got over successfully, and in a somewhat remarkable way. Mr. Ferrie's furnace is eighty-three feet high, with a closed top, the charging being effected by a bell and cone; and for a height of thirty feet, near the top, the shaft is divided by cross walls into four sections. These cross walls and the corresponding portions of the side walls are built hollow, with flues in their thickness, in which a portion of the gas is burned, so as to assist the ascending current, in the furnace, in coking the coal and heating up the charge. The amount of raw coal that this furnace burns is thirty-four hundredweight per ton of iron made, against nearly fifty-three hundredweight in ordinary furnaces, fifty-two feet high, and working under the same conditions: a difference of about nineteen hundredweight. Mr. Bell, who has studied the working of the Ferrie furnace, attributes this saving, half to its increased height, which is rendered practicable by the additional support given to the charge by its friction against the cross walls, and half to the effect of the heat communicated, through the walls, by the burning of a part of the gases in their flues.*

The older, and still the more common, method of heating the air for blast furnaces is by passing it through a series of cast iron pipes, heated to redness, by the waste furnace gases or otherwise, in suitable hot blast stoves. The greatest temperature of blast that can be maintained, in this way, without causing the rapid destruction of the pipes, hardly exceeds, however, 550° C. to 600° C.; and where higher heats are desired, recourse must be had to the system of regenerative fire-brick stoves, first introduced by Dr. Siemens and Mr. Cowper.†

A model of one of Mr. Cowper's stoves is on the table. In principle it s very simple. Each stove consists of a large cylinder of boiler plate, lined with fire-brick work, and filled with loose fire bricks, stacked together so as to form a series of vertical slightly zigzagged flues, and to expose the greatest possible extent of surface. To heat the stove,

---

* Journal of the Iron and Steel Institute, March 30th, and August 29th, 1871.

† E. A. Cowper, Proceedings of the Institution of Civil Engineers, vol. xxx. p. 309. C. Cochrane, Proceedings of the Institution of Mechanical Engineers, 1870, p. 62.

gas from the furnace is burned, with a suitable admixture of air, in a brick shaft in its interior; and the products of combustion are drawn down, through the mass of bricks, by the draught of a high chimney. When this operation has been continued for three or four hours, the greater part of the mass of brickwork, filling the body of the stove, is at a uniform red heat; and the heat decreases, thence, towards the bottom, so that the gases passing off to the chimney are at a temperature not exceeding 150° C.; no more than is sufficient to maintain a draught. When the stove has thus been heated, the air, gas, and chimney valves are closed; and by opening the cold and hot blast valves, blast is sent through it, in the reverse direction, passing first among the nearly cool bricks in the lower part, next to the chimney, and then over hotter and hotter surfaces, until before it reaches the top of the stove, it has attained nearly to the temperature of the bricks themselves. The hot blast thence rises through the remainder of the column of chequer work, and passes down the brick shaft or combustion chamber, and through the hot blast valve, to the furnace. By the time that the current of blast, flowing through the stove, has begun to cool down, sensibly, the uppermost courses of chequer work and the walls of the combustion chamber, a second stove, which has meantime been heating up, is put "on blast," and the first stove is heated again in turn. By working a set of either two or three stoves, in this way, a constant supply of hot blast is maintained; the temperature of the blast, between the beginning and the end of a shift, not varying more than 50° or 100° C. When three stoves are worked together, two are generally kept at a time "on gas," or heating up, and one "on blast."

The difficulty that was experienced in keeping the earlier forms of the Cowper stove free from deposited dust, carried over from the furnace, led to the introduction of a modification of it, by Mr. Whitwell, that has also been extensively adopted. In this, the brickwork provided to take up the heat of the burning gas, and give it out again to the blast, is arranged in the form of a series of parallel vertical walls. These present less heating surface than bricks arranged on the plan adopted in the Cowper stove; and the stoves, being of less height, take up also more floor space. The advantage claimed them is that they are more easily cleaned.

In cupolas, for the simple melting of cast iron, less alteration has been made, during recent years, than in blast furnaces.

Considerable savings in fuel have, however, been effected, by melting more rapidly, so as to diminish the loss of heat by radiation from the outer surface, and by making the cupola considerably smaller in area at the tuyères than in the rest of the shaft, in order to obtain a more concentrated heat at the point of fusion. Larger cupolas have also been employed, as, for instance, to melt the metal for the Bessemer process, than were generally in use before; and the plan has become common of providing these and other large cupolas with fore hearths, or receptacles for the melted metal, of a capacity of five or ten tons, which are kept warm by allowing a part of the flame to escape through them, and into which the metal flows at once, as it melts, instead of accumulating in the bottom of the cupola in contact with the coke.

An arrangement that presents some points of interest is the "flameless" cupola of M. Voisin. The principle of this is to admit a second and carefully moderated supply of air to the upper part of the shaft, sufficient to burn the carbonic oxide contained in the escaping gases, without wasting any sensible quantity of coke, or producing sufficient heat to transform the CO_2, that has been produced by the burning of the gas, again at the expense of the coke to CO.

It is claimed that in this way the CO in the gases may be burned, without wasting coke, and that the top of the cupola is thus maintained cool and flameless; while as the metal and fuel are heated to redness, by the combustion of the waste gas, a smaller proportion of coke is required. Thus, in foundry cupolas, to which the plan has been chiefly applied, a saving is effected by it, according to some statements, equal to 40 lbs. of coke per ton of metal melted; the consumption per ton having been reduced from 200 to 160 lbs.; and in other cases the coke consumption of foundry cupolas on Voisin's plan, exclusive of the first charge, is stated at as little as 140 or even 134 lbs. per ton.

That when the coke used in a cupola is fairly hard, the materials of the charge may be thus raised to a red heat, by the combustion of the gases, with little or no loss of coke, is shown by one of the experiments made by Mr. Bell, in the course of his investigations of the working of the Cleveland blast furnace. Mr. Bell found that on passing a slow current of CO_2 over ordinary Durham blast furnace coke, contained in a glass tube, no CO was formed when the heat of the coke did not exceed 550° to 650° C., and only traces even when the temperature was raised to such a point that the combustion tube, of hard German glass, began to soften.*

In blast furnaces, no such saving as that above quoted is to be looked for, by burning the gases in the upper part of the shaft ; as in these the proportion that the fuel bears to the other materials of the charge is so much greater than in a cupola, that the sensible heat, alone, of the gases is more than sufficient to heat up the whole charge to their own temperature ; and in addition to this there is, as Mr. Bell has shown, an actual evolution of heat, in the upper part of the furnace, due to the dissociation of a portion of the CO. In the case of the furnace, for instance, eighty feet high, erected at the Clarence Works in 1866, the total capacity for heat of the ore, coke, and limestone charged, is less than half that of the gases ; so that, neglecting the absorption of heat by chemical action, and its evolution by dissociation, the sensible heat alone of these would be sufficient to heat up twice as much solid material as is charged.

The furnaces of the second group, or flame furnaces, are very varied in form and character. In these, the useful effect is obtained by bringing a flame, or current of highly heated and burning gas, into contact with the matters to be acted on, instead of imbedding these in or mixing them with the solid fuel.

The ordinary reverberatory or flame furnace, with a fire grate, a flame chamber or working chamber, and beyond that again a flue leading to the chimney, is well known, and there is no need to go over in detail the variations from the general type by which it is adapted to different uses.

The most important recent modification of this form of furnace is the regenerative gas furnace of Messrs. Siemens, of which models and a diagram are in the room. In this, which dates in its present form from 1861, the fuel is transformed, in a separate gas producer, into a

* Proceedings of the Institution of Mechanical Engineers, 1869, p. 55.

combustible gas, consisting chiefly of carbonic oxide and nitrogen, mixed with hydrogen and hydrocarbon gases and vapours, distilled from the fuel, some vapour of water, and more or less carbonic acid. The gas is led, through flues or pipes, to the furnace, which may be at any distance from the gas producers, and is there burned. Gas, capable of producing, in such furnaces, the highest temperatures ordinarily used in the arts, may be made from any description of carbonaceous fuel, from anything in fact that will burn, however much mineral matter it may contain, and whether it is wet or dry. In Sweden, for instance, damp sawdust is used as the fuel to furnish gas for welding and other high heat furnaces; the large amount of water that the gas from such a material contains being first removed by cooling it, either by sprays of water or by passing it through a surface condenser.

The furnace consists essentially of a heating chamber, of any convenient shape, below which are placed four regenerator chambers, for taking up the waste heat from the flame, on its way to the chimney, and giving it out again to the entering air and gas. These chambers are filled with loosely stacked fire bricks, and each of them is precisely analogous in its action to a Cowper hot blast stove on a smaller scale. The air and gas, entering the furnace, pass up through two of the chambers, and are thus highly heated, before they are brought together, and burn, at the entrance to the working chamber, or furnace proper, in which the matters to be heated are placed; and the spent flame, from this, is at the same time drawn down to the chimney through the other two chambers; and, leaving the greater part of its available heat in the brickwork filling these, escapes to the chimney nearly cool. At intervals of half an hour to an hour the direction of the draught is reversed; the air and gas being introduced, in the opposite direction, through the two chambers that have been heated by the waste flame, and the current passing to the chimney being turned through the first pair of chambers, to reheat them in turn.

As the heated gases are made to pass downwards, through the regenerators, and the cool currents of air and combustible gas ascend, the heating and cooling of the masses of brickwork take place very uniformly; the hot current descending always most freely through the coolest channels, and the ascending current rising chiefly through the hottest. The position of the regenerators, below the level of the working chamber, gives also the advantage of an absence of indraughts of cold air into this; as owing to the ascending draught of the hot passages, through which the gas and air are introduced, a balance of pressure, or even an outward pressure, may be maintained in it, while the furnace is in full work.

The saving of fuel in the regenerative gas furnace amounts in average practice, when the furnaces are well managed, to fully fifty per cent. on the quantity used in an ordinary furnace doing the same work. The saving is greater, the higher the heat that is required in the working chamber; and where the most intense heats are needed, as in making or melting mild steel on the open hearth of a reverberatory furnace, no other than a regenerative furnace can be used; in no other is a sufficiently steady and intense heat maintained, without cutting draughts.

Other advantages of the system are the freedom of the flame from dust; its diminished oxidizing or cutting action, (the waste on iron piles, heated in a well-designed and well-managed gas furnace, being only one-half or one-third as great as in an ordinary coal furnace); and thirdly, the facility with which, whatever fuel is used, a uniform, living flame, of any required length, may be obtained, by making the mixture of the gas and air more or less rapid and intimate; from two or three feet only, as in the furnaces for melting steel in crucibles, to thirty or forty feet, in large plate-glass furnaces; the hottest part of the flame, in the latter case, being not where the gas and air meet, but some five-and-twenty feet away, where they begin to be thoroughly mixed.

Since the Siemens furnace has been in use, several other gas furnaces, with continuous tube regenerators, have been brought forward. Furnaces, that is, in which the air, and in some also the gas, are heated, not by being introduced through masses of brickwork, previously raised to a high temperature, but by being passed, in continuous currents, without reversing, through fireclay tubes or hollow bricks, round which the burned gases are drawn away to the chimney. The Ponsard furnace, which has been recently brought into use, to some extent, in France and Belgium, appears to be the best designed of these modifications; and the quoted results of its working, when in good order, are very satisfactory. In this arrangement, the air only is heated by the regenerator; the gas producer being placed close to the furnace, and the gas from it taken direetly, without further heating, to the point where it is burned. In the case of such a furnace, there are several advantages in thus heating the air, only, by the waste heat; the tubes of the regenerator are not choked up by deposits from the hot gas, nor is there the risk of loss of gas by teakage; and as the volume of burned gases is sufficient to heat nearly twice as much air as is required for combustion in the furnace, a moderate leakage of air, in the regenerator, into the current passing to the chimney, does no harm; since, if the regenerator exposes sufficient surface, as much air, heated to nearly the full temperature of the waste gases, may still be drawn into the working chamber as is required there. The ordinary Siemens furnace, however, though apparently more complicated, is probably a stronger and more durable arrangement than any furnace working with continuous tube regenerators can be made, and is better fitted to bear the rough treatment that is generally the fate of such appliances in actual work.

A proposed modification of the Ponsard system of furnace, that presents considerable theoretical advantages, is to supply the gas producer, as well as the working chamber of the furnace, with highly heated air from the regenerator; the hot gas being taken, as in the ordinary Ponsard furnace, direct, without further heating, from the gas producer to the working chamber in which it is burned. The carrying out of such an arrangement, in the case either of the Siemens furnace, or of a furnace with tube regenerators, appears likely to present great practical difficulty, but if it can be successfully worked out, the increased economy will be great; as, in such a furnace, the whole of the heat evolved from the fuel, except that still inevitably lost by external cooling, and that carried off by the gases passing to the chimney, at a temperature that need not exceed 100° or 150° C. would be available, in the working chamber, as heat of high temperature.

In the Siemens furnace, in which the gas producer is supplied with cold air, the sensible heat of the gases, as they leave the producer, does not affect the temperature of the flame in the furnace; for the amount of heat contained in the products of combustion is sufficient to heat up the entering gas and air, from the temperature of the atmosphere, to a temperature nearly equal to that at which the spent flame leaves the working chamber; and the effect of sending in hot gas instead of cold gas, is not that the gas is much hotter when it reaches the top of the regenerator, but simply that the bottom of the regenerator, where the gas enters, is less cooled down; and, on reversing, the burned gases, after traversing it, escape at a higher temperature to the chimney. In practice, the heat of the gas, as it leaves the producer, is in most cases purposely thrown away, by leading it through overhead sheet iron "cooling tubes," in order to obtain a better pressure, at the furnace, from the syphon action between the ascending hot column and the descending heavier cool column, and to remove the greater part of the vapour of water that it generally contains.

Again, in the Ponsard furnace, as has been pointed out, when only the air, to burn the gas in the working chamber, is heated by the waste flame, the quantity of heat carried into the regenerator is about twice as great as is required to heat this amount of air; and however perfect the action of the regenerator may be, however great may be its extent of surface, the burned gases necessarily escape to the chimney at a high temperature.

If, on the contrary, the air to supply the gas producer, as well as that to burn the gas, were heated by the waste heat; and the hot gas from the producer were led direct, without passing through a regenerator, to the working chamber, and burned there; the volume of air to be heated would be sufficient to take up all the available heat of the waste flame, and the heat that is otherwise inevitably lost, either by the chimney in the one case, or from the cooling tube in the other, would be rendered available.

The system of burning powdered fuel, that has been worked out by Mr. Crampton, is another remarkable deviation from the ordinary form of flame furnace.

A model of such a furnace, arranged for mechanical puddling, is on the table. The coal, burned, is first ground between ordinary millstones to such fineness that it will pass through a sieve with thirty holes to the linear inch (which Mr. Crampton estimates may be done, on the large scale, at a total cost of less than a shilling a ton); and the powdered coal is supplied, at any required rate, by a mechanical feeding arrangement connected with each furnace, and is led into a jet of air, from a fan, at a pressure equal to three or four inches of water column, by which it is carried forward into the furnace. In this, the jet of mixed coal dust and air takes fire, and burns like a jet of combustible gas, except that the flame is solid, not hollow like a gas flame burning in air. The revolving puddling furnace is in form a short cylinder, closed at one end and slightly narrowed at the other, and consists of a double casing, of thin boiler plate, kept cool by a current of water flowing through it, and lined with a "fettling," of oxide of iron, five or six inches thick. The air and coal dust are blown in through the flue-piece, that fits against the open end of the revolving puddling chamber, and the waste flame escapes to the chimney by the same opening, round the entering jet. The combustion of the fuel is thus effected in the working chamber, itself, and as the coal and air are mixed intimately throughout the flame, it is very complete. The consumption of coal, in puddling ten-hundredweight charges of pig iron, run liquid into the furnace, is stated to be nine hundredweight per ton of puddled bar produced.

This system of furnace, though in most respects remarkably perfect, and giving every promise of success as applied to revolving puddling furnaces, is not found to be suited for use in those more ordinary arrangements, in which the working chamber is built of brickwork or other siliceous material, as the fluxing action of the ash of the coal destroys, very rapidly, any such work that is in contact with the flame and exposed to its full heat.

Among the directions in which improvements have been effected in flame furnaces of the ordinary type, are the use of the waste heat to raise steam, a system now carried out, to a greater or less extent, in all iron works where such furnaces are used; the employment of a blast or forced draught under the fire-grate, (the air forced in being frequently more or less heated,) in order to allow of burning cheaper small coal, and to give a command, such as that possessed by the regenerative gas furnace, over the pressure in the working chamber; and, lastly, arrangements for preventing the cooling down of the fire, each time that fresh fuel is put on, and the rush of cold air into the furnace, through the opened fire-door, when a pressure is not maintained in it by blast.

The Newport furnace of Mr. Jeremiah Head* may be taken as an example of those furnaces in which blast, heated by the waste flame, is introduced under the fire-grate. In this, the blast pressure is obtained by a steam jet, and the resulting damp air is heated to about 290° C. by passing it through a cast iron heating stove, round which the waste flame is led on its way to the chimney. The hot blast is conducted partly under the fire-grate, and partly to a row of holes in the furnace roof, immediately over the fire, through which a supply of air is thus introduced, sufficient to complete the combustion of the gases rising from the fire. Mr. Head states that in a hand puddling furnace of this construction, working four-hundredweight charges, the temperature at which the burned gases finally pass to the chimney is reduced from 1112° C. (the average temperature found in the chimney of a similar furnace, in which the air was not heated) to 860° C.; the heat at the same time at the fire bridge being estimated at 1370° C. The consumption of coal is reduced from twenty-four and a half hundred-weight, per ton of puddled bar produced, to sixteen and a half hundred-weight, and the yield is stated to be also from one to three per cent. greater than in the ordinary furnace.

In Price's retort furnace the fire is supplied with air forced in by a fan, and heated by the waste heat to between 200° and 260° C., and the coal used is also heated, before it is pushed forward on to the grate, by charging it through a high vertical retort, round which the spent flame passes to the chimney. The saving in fuel effected, in furnaces for puddling and heating iron, by thus making use of the waste heat, is stated to amount to between thirty and forty-five per cent.

An American arrangement, known as Frisbie's feeder, that has been recently introduced into this country, avoids in a different way the cooling of the surface of the fire each time that fresh coal is put on; and the burst of smoke, after firing, from the evolution for a short time, from the suddenly heated fuel, of hydrocarbon gases and vapours that pass away only partly burned; as well as the necessity for frequently opening the fire door. In this mode of firing, each charge of coal is filled into a moveable charging box, and pushed up, from below,

* Journal of the Iron and Steel Institute, 1872, pp. 220 et seq.

into the middle of the fire-grate, so that the surface of the fire remains always hot, and as the distillation of the gases from the raw coal goes on continuously, the fire remains uniform in character and may be readily kept smokeless. The arrangement has been applied, in this country, to boilers, flint glass furnaces, and puddling furnaces, and is said to effect a decided economy in fuel and give altogether satisfactory results.

The use of hydrogen and hydrocarbon gases, as furnace fuel, is chiefly limited to the heating of small laboratory furnaces, or to work on a scale but little greater; as the cost of such gases, when artificially produced, is too great to admit of their competing on a larger scale with coal. In the comparatively few localities, however, in which such gases have been found to flow naturally from bore holes, penetrating to beds of coal or shale, they form a valuable fuel, that is made use of to a considerable extent. In some parts of Pennsylvania, from bore holes put down for petroleum, a supply of gas, consisting chiefly of marsh gas (CH_4), mixed with other hydrocarbons and with hydrogen, has been found to rush steadily, for years, at a pressure estimated at nearly one hundred pounds per square inch;* and in the case of at least one such "gas well," that at Leechburg, the gas has been very effectively turned to account. There, according to a recent number of the *Enginnering and Mining Journal*,† the gas is conveyed from the well, through a distance of between one hundred and two hundred yards, to adjoining sheet iron works, where it heats five puddling furnaces, six heating furnaces, two annealing furnaces, and so on, and furnishes in fact all the fuel required for turning out nearly thirteen tons of sheet iron a day. Before the gas was brought to the works, the daily consumption of coal had been seventy-five tons, and the make per day of sheet iron did not then exceed ten tons; so that the gas does now as much work as would have required, on the former plan, ninety-seven and a half tons of coal per day.

Petroleum and other liquid fuels have also been used for heating furnaces, both in England and in America, but on a scale only experimental rather than really industrial. The results obtained are, how-

---

* Engineering and Mining Journal (New York), March 18th, 1876, p. 269.

† Engineering and Mining Journal (New York), May 22nd, 1875, p. 367; and June 26th, 1875, p. 476.

ever, very remarkable. In 1869 an extended trial was made, at Chatham Dockyard, of the system of Messrs. Dorsett and Blyth, for heating with creosote or other heavy oils. The furnaces to which the plan was applied were those used for heating ship plates and armour plates for bending. The oil was vaporized in a small boiler or generator, and supplied to the furnaces through jets one-eighth of an inch in diameter, at a pressure of thirty pounds per square inch; and the air, not previously heated, was simply drawn in by the chimney draught. The heating, both of the armour plates and of thin ship plates, was done in one-third the time taken in heating in the ordinary way, with coal; and the weight of oil burned, per ton of plates heated, was between one-sixth and one-eighth of the former consumption of coal. More recently, in the course of last year, a report by Professor H. Wurtz has been published,* of the working, in America, of Eames's system of firing furnaces with petroleum, which quite corresponds with the results obtained at Chatham. On Eames's plan, the crude petroleum used is evaporated and carried into the furnace by a current of highly superheated steam, at a pressure of ten pounds per square inch. The furnace to which it was applied was one for heating scrap iron piles, for rolling into plates; and the air supply, as at Chatham, was cold. The furnace was heated up, ready for charging, in forty-five minutes, and with a consumption of only twenty-two and a half gallons of oil. The output, per shift of ten hours, was eight tons of rolled plates, with a consumption of 300 gallons=2000 pounds of oil: the same furnace, when fired with coal, having heated the piles for only six tons of plates per shift, and burned five and a half tons of coal. The effective heating value of equal weights of oil and coal was thus in the proportion of eight to one.

The much less weight of mineral oil than of coal that is required, according to these two statements, to do an equal amount of work, is probably due in part to the combustion of the oil-vapour being effected with the admission of a smaller excess of air than is found in the products of combustion from a furnace fired with coal; to the combustion taking place wholly in or at the entrance to the working chamber; and to the work being done in less time, so that there is less loss of heat by outside

* Engineering and Mining Journal (New York), Aug. 7th, 1875, p. 122.

cooling; but it is chiefly to be attributed to the greater heating power of the hydrogen, of which such oils largely consist, than that of carbon. As Dr. Percy has pointed out, though the great weight, and the high specific heat, of the gases resulting from the combustion of hydrogen in air, lower the theoretical maximum temperature produced to a little under that due to the combustion of carbon, yet if the products of combustion pass off at any ordinary furnace temperature, considerably below the theoretical maximum, the number of available units of heat produced by the combustion of hydrogen is much greater than that obtained from an equal weight of carbon. Thus, if the burned gases pass off at 1500° C., the available heating effect of hydrogen is nearly four times as great as that of carbon.

It does not appear that any trials have been made of working regenerative furnaces either with mineral oil or with natural gas; nor, indeed, even of burning these fuels with air moderately heated, such as is used in Head's or Price's furnace; though the advantage of thus making use of the waste heat, to increase still further the temperature of the flame, would evidently be great. It is not too much to say, that a regenerative furnace, with regenerators, of course, only for heating the air, and fired either with hydrocarbon gas, or with liquid fuel, on such a plan as either of those above referred to, would far surpass any furnace now in use, (except those, such as Deville's, in which the fuel is burned by a current of oxygen,) in the weight of material that it would heat with a given consumption of fuel.

In nearly all furnaces, the amount of heat that is utilized is an extremely small proportion of the total heat due to the combustion of the fuel; the greater part being carried off by the burned gases, or lost by conduction and radiation. M. Gruner calculates, in a recent paper published in the "Annales des Mines," on the proportion of heat that is utilized in furnaces of different kinds, that in the fusion of steel, in crucibles, in ordinary coke furnaces, the heat utilized does not exceed 1 7 per cent. of the total amount that the fuel would be capable of giving out, if perfectly burned; and that even on the extreme supposition that half of the fuel is burned only to CO, the heat utilized amounts to only 2·6 per cent. of that evolved. In flame furnaces, the proportion of heat utilized is higher; reaching, as a maximum, fifteen to twenty per cent. of the total heat due to the amount of coal burned, in well-arranged regenerative gas furnaces for reheating iron. In those arrangements in which there is little heat lost by external cooling, and in which the heat of the products of combustion is most fully utilized, the useful effect is much higher: thus, in large blast furnaces, M. Gruner estimates that it is as much as seventy to eighty per cent. of the heat actually developed in the furnace and introduced into it by the blast, or between forty and fifty per cent. of the total heat that the fuel could evolve if completely burned; and in the annular Hoffmann brick kiln, it is estimated to amount also to between seventy and eighty per cent. of that given out by the fuel.

A greater proportion of the heat evolved is lost, in the burned gases, the less the difference is between the temperature of the flame and that required to be maintained in the working chamber; for as soon as any portion of the flame is cooled down to the temperature of the matters to be heated, however high this may be, it can impart no more heat to them, and must be drawn away and replaced by hotter flame from the fire. Hence, a small increase in the initial temperature of the flame, such as that obtained by effecting the combustion of the fuel by means of moderately heated blast, or a small diminution in the proportion of heat lost from the working chamber by external cooling, effects a great saving in the consumption of fuel that is required to do a given amount of work.

The effect of a high flame temperature, on the proportion of heat utilized, is strikingly shown by the very economical working of furnaces on Deville's system, that of burning coal gas with oxygen, instead of with air. The theoretical temperature of such a flame, if not limited by dissociation, would probably amount to 7000° or 8000° C., and it is in any case far above the fusing point of platinum, which is estimated at about 1900° C. In an example, of which M. Gruner gives particulars, in the paper above referred to, of the fusion of a charge of 250 kilogrammes of platinum by this method, the cold furnace was heated up, and the metal melted in it, in one and a quarter hours, and with a consumption of only twenty-four cubic metres (848 cubic feet) of gas; the proportion of heat actually utilized, in the fusion of the metal, being fourteen per cent. of that due to the combustion of the gas. Thus, on account of the intense heat of the oxyhydrogen flame, as good an economical result was obtained, in this little furnace, as in the best of the flame furnaces used in ordinary metallurgical work; though the proportionate loss of heat from the surface cooling of so small a furnace, (a little trough not more than thirty inches long,) must have been enormous, the products of combustion cannot have escaped at a lower temperature than 2000° C., and neither the coal gas nor the oxygen was heated.

The system of working furnaces under high pressure, on which some experiments were made a few years ago by Mr. Bessemer, offers a possible method of increasing, considerably, the temperature of combustion of ordinary fuels, and with it the proportion of available heat; but since the first experiments were made, of which an account was published in the technical journals in 1869, nothing further has been done to test the practical value of the scheme.

The diminished proportion of heat that is lost by surface cooling, in the case of large furnaces, and the consequent higher temperature of their flame, render them in all cases much more economical in fuel than furnaces of smaller size. In the case of ordinary puddling furnaces, for instance, when the coal consumption in those working five-hundredweight charges is about twenty-three and a half hundredweight per ton of bar produced, the consumption is reduced to eighteen hundredweight per ton in working ten-hundredweight charges, and to fifteen hundredweight per ton in still larger furnaces working charges of fifteen hundredweight; and in the welding furnaces, in use at Woolwich Arsenal, the larger the furnace is, the higher by actual experiment is the temperature of the flame as it passes over the bridge, and the smaller is the amount of coal required per ton of metal heated. A furnace of ordinary size, heating six tons at a charge, consumes about eight hundredweight of coal per ton; a larger furnace, heating a charge of thirteen tons, does the work with only seven hundredweight per ton; and the largest of all, capable of heating at once a mass of iron weighing sixty-five tons, gets this up to a full welding heat with a coal consumption of only five and a half hundredweight per ton. In copper smelting, in glass making, and in other work, large furnaces are similarly found to use less fuel, in proportion to the work done in them, than furnaces of smaller size.

The Conference then adjourned.

The PRESIDENT: As Mr. Hackney was rather limited in time before luncheon, I will now call upon him to add anything he may desire, or to explain any of the models which were not up here before.

Mr. HACKNEY: I have nothing to add formally to the paper which was read before the adjournment, but I will say a word or two on some of the models which have been brought up since. There is, for instance, a model somewhat old-fashioned of a blast furnace, showing the hearth and the form of the stack. Here, also, is an interesting diagram, showing the increase in the size of blast furnaces. The first one is about fifty feet high, and had a capacity of 5076 cubic feet, and in it about twenty-nine-hundredweight of coke was consumed per ton of iron produced. By increasing the size to a height of eighty feet, and the capacity to from 12,000 to 30,000 cubic feet, the consumption of coke has been marvellously reduced—viz., from twenty-nine or thirty-hundredweight to about twenty-two-hundredweight per ton, the other conditions remaining the same. There is a model and also a large diagram of the Regenerator gas furnace, which was mentioned in the paper. In the diagram the arrangements are shown all on one plan, in order to make the action clear. Then there is another model of Mr. Head's furnace, an ordinary furnace, working with coal, in which the air is supplied by a steam jet. The whole of the arrangements are of cast-iron. In this model a regenerating gas furnace is shown as the work is actually arranged in practice; the regenerators being under the furnace, and the reversing valves and flues, which are shown on the diagram below the furnace, are placed at the back. It is really on the principle as the hot-blast stove, which was first introduced, and the application of it to blast furnaces came afterwards. In fact, there are four hot-blast stoves on a small scale under each furnace. Here again is a model of the coal-dust furnace of Mr. Crampton, as applied to puddling. In this furnace there is a chamber lined with oxide of iron, open at one end and enclosed at the other; it has a removable flue piece, by means of which the charge is introduced and withdrawn. In actual work, as I explained before, the jet of coal-dust and air is blown in at the centre of the flue piece, and the flame passing to the chimney escapes round the jet, and so goes to the chimney.

The PRESIDENT: I rise now to propose a vote of thanks to Mr. Hackney for his interesting communication, which I am sure you will accord.

I will now call on Mr. Preece for his address on Electric Telegraphs—a subject of vast interest and great detail. I fear Mr. Preece will not have as much time as we should like to give to it, but I have no doubt he will make it interesting.

ELECTRIC TELEGRAPHS.

Mr. PREECE: Mr. President, ladies, and gentlemen, telegraphy is the art of conveying the first elements of written language to distances beyond the reach of the ear and the eye. When electricity is used to effect this operation we have the electric telegraph, and inasmuch as there are many fundamental phenomena of electricity which are used as the bases of these telegraphs, so we have these telegraphs divided into different systems. I propose to show you how these systems of telegraphy have grown by a principle of evolution from the first incipient idea to the present almost complete form of perfection by stages of growth which will be represented by instruments displayed in this noble collection. The invention of the telegraph cannot be claimed by any one, for the simple reason that the very first idea conveyed by the notion of electricity and its action at a distance must have given the first idea of conveying information to a distance. Electricity, or that form of electricity which is used—namely, voltaic electricity—dates from the commencement of this century. In the year 1800 Volta developed that instrument that is called the voltaic pile, and which has been the parent of all our batteries. In the same year, 1800, Nicholson and Carlyle discovered the decomposition of water by the aid of a current, and in less than nine years the combined notion of a current of electricity and the decomposition of water led to the first electric telegraph.

Professor Sömmering of Göttingen—by means of this little apparatus here,—which is the original and identical apparatus made and constructed under his own eyes—by means of thirty-five wires connected between two places, by means of a series of little points each of which represented either a letter of the alphabet or a numeral, and by means of this voltaic pile—the identical original instrument—was able by attaching the ends of two wires to these little brass spots to cause gases—to be evolved from these two points—hydrogen from one, oxygen from the other, hydrogen being evolved in greater quantity ; hydrogen always indicated the first letter sent. Supposing he wanted to send the word "now :" by putting the hydrogen point to "n" and the oxygen to "o," he made a quantity of gas appear from the "n" point and a smaller quantity from the "o" point. And in order to attract attention he devised the first electric alarum—a little bell which you see here, which by the evolution of gas caused a ball to drop and started the mechanism ringing as you see there. Here is also some of the identical wire used on that occasion. This was in the year 1809. Very few years afterwards the attention of the Russian Baron Schilling was attracted to this instrument of Sömmering's, and he developed a form of instrument based, not upon the decomposition of water, but upon the deflection of the magnetic needle in face of the current. This is one of the forms of Schilling's telegraph. It was used in the year 1830; and passing from that we come to the year 1833, when the two great German philosophers, Gauss and Weber, at Göttingen established between the cabinet of the university and the observatory, about a mile and a quarter distant, a telegraph, a copy of which is to be seen in the room downstairs. It is too bulky to bring up here, but it is well worth a visit as being one of the most valuable historical instruments downstairs. From these two instruments, about the year 1836, Mr. William Fothergill Cooke, who was studying at Heidelberg, heard a lecture delivered on the art of telegraphy. He at once grasped the idea. He came to England and associated himself with Sir Charles Wheatstone, and the two together brought out the first needle instrument used in England. This is the identical instrument brought out on that occasion, and here you will see that there are five needles suspended in a line. Each of these needles has two movements—the one to the left, the other to the right ; and when two of these are deflected their line of convergence always points to a letter. When, for instance, this last needle is diverging in this direction and this one in the other, they point to the letter "a," when the first and the fourth are deflected they point to the letter "b," the first and the third to the letter "e," and so on; by that means any letter of the alphabet is instantly indicated to the observer. Now you will observe that this instrument has five needles and requires five wires, but very short practice, very small experience, soon showed that the required letters of the alphabet could be formed by means of four needles instead of five; so that we come now to the four-needle telegraph, which was practically used for the conveyance of messages in this country up to the year 1844. Here is also an actual original piece of the wires that were used on that occasion. This was dug up on the incline between Euston and Camden, and you will see if you examine it that it contains five copper wires covered with hemp, tarred and let into wood. This was buried below the ground, and used in connexion with this instrument. That instrument was employed on the Blackwall Railway, and it happened that two operators at two distant stations on the Blackwall Railway—brothers—found that by making certain preconcerted signals on this instrument, instead of using, as was previously done, one motion to the left and one to the right—by taking two needles and using two and three motions to the right or left, they were able to converse between these intermediate stations. This at once showed to Mr. Cooke that it was possible to reduce the four wires to two. Accordingly, immediately afterwards this instrument was introduced, which is the original double-needle instrument. This identical instrument was used at the Slough station on the Great Western Railway; and there are many interesting records of the messages—strange in those days—which were carried by this wire. In fact this identical instrument has engraved upon its foot-plate a message which was sent in the year 1845 that led to the capture of a celebrated murderer, the Quaker Tawell. One curious point connected with that is this, that neither of these instruments form the letter Q, so that when the word "Quaker" had to be sent there was a great difficulty in supplying it, and it was sent "Kwaker." The clerk at the distant station—Paddington—could not at first conceive what "Kwaker" meant; but after a short time the truth occurred to him, the message was delivered, and the man was captured. Now this, as you see, is a heavy and cumbrous instrument, which only transmitted its message at the rate of about five or six messages an hour, a rate of speed which would at the present day cause all our wires to become congested. Accordingly, the cumbrous form of the double needle was converted into the lighter and more rapid instrument that we have here.

This is the last and latest form of the double-needle instruments used in England. It is still very largely employed on the railways, but it is rapidly making its disappearance. It was very evident at once that if it were possible to form the letters of the alphabet by the combinations of the motions of the needle to the right or left, that it was quite possible to form such an alphabet by means of one needle—in fact, there is no doubt that the very first needle telegraph ever invented, that of Schilling, was based upon a combination of the movements of the needle to the right and the left. Here we have the first form of needle instrument so introduced. We have a single needle which by its deflections to the right or left would form an alphabet: one motion to the left, one to the right, the letter "a;" one to the right and three to the left, the letter "b," and so on; by the combination in that way the letters of the alphabet are formed.

This is a second form of a needle instrument of the same kind, and this is the form that is used in the present day. These instruments are very largely employed at some of our smaller post-offices; and they have one great merit about them—extreme simplicity and the avoidance of the necessity of employing skilled clerks. Very little instruction, very small pay, induce clerks speedily to acquire a knowledge of the single needle.

Now, as we progress in telegraphy, speed has become an essential quality. The first instruments, the four-wire and the double-needle seen here, were made of galvanometer coils of great depth and magnets of great length, their form necessarily became sluggish and cumbrous. The first to make any alteration in that direction was Mr. Holmes, who succeeded in reducing the six-inch needle to this little diamond-shaped form, and by the simple reduction of the size the speed of the instrument was at once increased nearly fourfold. We have by various alterations and suggestions and additions, arrived now at the form of needle which I hold in my hand, and with this needle the motions can be made so rapidly that the eye can scarcely follow them. Now, as you have perceived, these instruments are simply transitory in their signals.

Very early in the history of telegraphy, in the year 1837, a year that was prolific of telegraphic invention, Morse, in America, conceived the idea of *recording* the signals sent by the current. His first notion was to have a style which would make a mark as the paper passed the needle, and his first signals were of this character, representing the numerals. He formed a telegraphic dictionary, and by simply according to each word a certain number, messages were sent by sending a series of numbers. The first message ever sent was "215," "successful;" "36," "experiment;" "2," "with;" "58," "telegraph;" and the first instrument employed registered its signals by a lead pencil. Then, finding that the lead rapidly wore out, he passed to a common pen, which he found great difficulty in keeping supplied with ink. From that he proceeded to the common carbonic paper, and lastly he arrived at a common Morse embosser, where the marks are recorded by the indentation of dots and dashes upon paper. We have here a form of the Morse embosser that was used generally in England up to the year 1853, and you will find that upon this piece of paper are recorded in raised characters the dots and dashes that form the alphabet. Now, in the first place, the pressure on the paper, and the form of these characters upon the paper, means a certain amount of work done by the current. It required pressure; it required force to make these impressions; and, more than that, the constant reading of a white strip of paper, whether merely by a shade or ridge, or by the marks made, became very tedious to the eye, and it was at once apparent that any means which would overcome this difficulty would be a great improvement. Accordingly, an Austrian of the name of John, in the year 1854, produced this instrument that we see here, recording the marks by permanent lines of ink. This very speedily—as soon, in fact, as it reached the prolific house of Siemens—received that finishing touch which genius and skill always impart to works of this class, and we soon had produced the instrument you see here, known all over the world as Siemen's Direct Writing Morse Inker. Here you will find the words recorded in permanent ink upon a strip of paper.

Another Recorder is also on the table—Bains's Chemical Recorder. It differs from the others in recording its signals by the same power which electricity exented in Sömmering's—that is, by the decomposition of chemical materials, and so by decomposing the chemical material in which the paper is steeped, it recorded in a bright colour marks upon the paper. It has a special merit of its own, which in some respects renders it superior to any other form of instrument, and that is, that it is quite independent of magnetism or electro-magnetism. An electro-magnet always means certain force expended, certain time consumed—what we call magnetic inertia ; and we find where speed is essential a considerable advantage is to be found by the use of Bains's Chemical Recorder, and so an instrument which had a fitful existence between 1853, '54 and '56, is now coming again into use, and probably, in England at least, will receive very general acceptance.

Now in all these instruments you will see we require skilled labour. A clerk who has to send the Morse alphabet of dots and dashes has to spend many months in acquiring the art of sending by his hand those intervals of time which require to be recorded. But it very early appeared that advantage would accrue if we could produce an instrument which would indicate directly before the eye of the receiver the simple letters of the alphabet, and in the year 1816 Francis Ronalds designed an instrument which was never tried except in his own house. That was an A B C instrument. In 1841, when telegraphs had become a matter of fact, Cooke and Wheatstone directed their thoughts to the production of an A B C instrument. We have here the original instruments made at that period. Here is one of the first dial instruments. It contains a dial upon which the letters of the alphabet were once depicted. Age has gradually removed them, so that even an opera-glass can scarcely read them. Now here you have this dial with the letters of the alphabet, and here you have an indicator that rotates and stops or hesitates at every letter that it wants to indicate. Here we have the apparatus that was used to send these letters. This instrument itself is not quite complete ; the outer case has disappeared, but there was at the top an index or fiducial spot which always brought the hand up as it were, so that when you wanted to send the letter "K," for instance, "K" would be brought there, and so with any other letter. We have here two dials of the same character, but in one we have a little opening through which the letters appear ; the brass window, as it were, has disappeared from it, but the letter to be indicated was simply shown in front of this window. The other is merely an index as in the first instance. This A B C instrument was also very largely used; it was the first form of instrument that was used by Messrs. Siemens in Prussia in the year 1846. Here is one of the identical instruments that were very largely employed in that country, and they were only after great difficulty and opposition replaced by the Morse apparatus; but these two forms, in England under the hands of Sir Charles Wheatstone and in Prussia under the hands of Mr. Siemens, have been so changed and altered and improved and shaped that probably a more exquisite instrument can scarcely be conceived than the simple A B C apparatus of the present day. This is the simplest form of Wheatstone's A B C. It is so accurate in its working and so simple, that I have known a case at a post-office where an old lady over seventy acquired the art of working this instrument with an hour's practice, and the very next day the office was open to the public, the messages were sent—that is now six years ago—and from that day to this I have not heard a complaint from that office. Children work it, private people work it, and it has really become one of the most valuable instruments for telegraphic purposes. It is rather slow in its operation, so that for the ordinary commercial requirements of the country it is not adapted. The idea of an A B C instrument which indicates the letters of the alphabet to the eye very speedily led to the conception of an instrument which should permanently record the letters of the alphabet upon paper—in fact, that it should print from ordinary type the messages that were sent. The first recorded form of type instrument was made by Mr. Vail in America; and it is a matter of great regret, as far as this collection is concerned, that the Exhibition at Philadelphia should occur at this particular time, for there is no doubt that in America they have many exceedingly interesting historical apparatuses that would have added considerably to the interest of this Exhibition. Amongst the rest, this original type printer, which did not take root in England until the year 1841, when Sir Charles Wheatstone produced this type printing instrument that you see there. It was not successful. It is very pretty as a philosophical toy, but as a practical instrument it never acted, and after having passed through the mill in every electrician's studio in England, on the continent of Europe, and in America, the instrument which has really at last proved itself to be the survival of the fittest, is the type-printing instrument we have here, viz., that of Hughes. There is also an exceedingly beautiful type-printing instrument by Mr. Siemens which is to be seen at work downstairs. This is a type-printing instrument which has been sent over by a French house, and has evidently been sent in such a form as to prevent the possibility of any foreign fingers interfering with it. We have tried to get it in order for this meeting, but unfortunately have failed. However, you will find on examining it, we have a type-wheel on which the letters of the alphabet are raised, and we find as these keys are depressed the letters are printed in bold, unmistakable Roman type on strips of paper.

Now the next system to which I shall call your attention is one dependent upon quite a different organ of the body, and that is the acoustic system of telegraphy ; it appeals to the understanding, not through the eye, but through the ear. Now an alphabet of the Morse type is based, as I told you, on dots and dashes. We can make sounds to represent these dots and dashes. We can have a short sound or a long sound. I am not quite sure that I can make it heard through the whole room, but we will try and make short sounds or long sounds. The letter "e," which is that most frequently used in the English language, can be represented by a short sound. The letter "t," which is also very frequently used, by a long sound. The letter "b" can be produced by one long and three short dots, and so on throughout the whole alphabet. Well, those dots and dashes can also be indicated by bells. For instance, we have here two bells which give out strokes, of different tones, and based upon that was constructed one of the earlier telegraphs suggested by Steinheil. These two little bells that are used are really made to represent Steinheil's Bell Telegraph, and here is one of his coils. The deflexion of the needle simply caused a little hammer to fly out and strike the bell, but in this instrument which we have here this system has been carried to a more practical issue by Sir Charles Bright and his brother, Mr. Edward Bright. This is called the Bell Telegraph. It was exclusively used in England by the Magnetic Telegraph Company, and is now very largely employed in different parts of England by the Post Office. It is an instrument exceedingly rapid in its action, and enables clerks to read with great freedom because their whole attention is devoted to their paper and nothing else. The defect of that instrument is probably its complication, and it requires skilled clerks to work it.

The form of Sounder that has been introduced in America is this little instrument, which we call the "Pony Sounder." This is simplicity itself. It simply consists of a coil, through which the currents flow and an armature which is attracted. There is nothing here to get out of order, and the rate of rapidity with which the instrument is read is something wonderful at times. The rate at which these instruments are worked depends upon the skill of the clerk in manipulating his key. Here is the key that is commonly used. The clerk simply depresses that key, and the rate at which he depresses that key is practically the rate at which the instrument records and the rate at which the clerk receives. But human nature will tire. A clerk who commences to send in the morning at the rate of forty words a minute, in an hour or two descends to thirty-five words a minute, and then gets to thirty words a minute, and in the course of the day his rate is still further lowered. It therefore speedily struck our electricians that an advantage would be obtained if we could replace the skilled labour of the clerk by some automatic apparatus which would send his currents for him—indeed, the very earliest form of telegraphy suggested by Morse in America was an automatic sender. His letters, his numbers, were formed by pieces of type, and these were simply placed in a stick and passed through a machine which sent the currents which made the record at the distant stations.

In the year 1846, Bains conceived the idea of punching these dots and dashes in broad paper, and by passing this strip underneath a spring he thought that he could send to the distant station the signals and have them properly recorded—in fact, the experiment was perfectly successful, and the speed with which messages were sent was something astonishing. In an office of short circuit we were able to attain a speed of about 400 words a minute, and when I mention that the fastest speed previously attained was from thirty to forty words a minute, you will easily understand that the advantage was very considerable. But the conditions of our lines in those days, the character of the instruments themselves was such, that while we were able to get this high speed on short circuits in our own rooms, the instrument totally failed to obtain the same results on our open lines. The result was that punching, as it was called, or automatic sending fell into disuse really because it was not wanted. The instruments we then used were able to transmit messages as fast as the public favoured us with them. The result was that the same desire for fast telegraphy did not exist in the years 1846 and 1848 as it does in the present day. In the present day telegrams have flown in so fast and so thick that the ingenuity of every telegraph engineer is devoted to increasing the capacity of his wires to transmit messages. Forty words a minute, fifty words a minute, sixty words a minute, is not enough. We have to go up to 100, 150, and even to 200 words a minute, and yet the cry is still for more. At the present time the instruments we have practically in use do not exceed 130 words a minute. We have now in work Wheatstone's automatic apparatus, where the messages are punched by means of a little apparatus that is on the table. Clerks are able to punch messages at such a rate that the ear can hardly distinguish differences in the sound, and by such operations these papers are stripped. These punched strips are placed in an instrument, and record the letters of the alphabet upon these strips.

Now there are other forms of automatic senders, one of which I should have liked very much to have shown you. It is in operation downstairs, and it is also the production of Mr. Siemens. There are a series of keys like the keys of an accordion, and the clerk, by moving these keys, is able to raise a species of type which sends the currents in their proper order. The rate at which this instrument works is not so rapid, but for many purposes—accurate sending, duplex working, and so on—it is one which is very likely to receive considerable attention.

There is another form of instrument to which I must allude, but which I am unable to exhibit to you, and that is an instrument called Thomson's Recorder, which is used in connexion with submarine cables. In working a submarine cable we incur difficulties that are not met with on open lines—the currents flow more slowly, more sluggishly, and we are also obliged to use currents of considerably less intensity, less strength, than upon land lines. Consequently the desire of the telegraph engineer is to form an instrument which shall move with the slightest possible current and with the greatest possible velocity. Now, Sir William Thomson has devised two or three forms of instruments, one of which is simply a needle instrument; but, instead of having the needle a mass of matter which moves, the needle is a mere spot of light, or rather the indicator is a spot of light, the needle proper being an excessively minute piece of steel so light that, if I remember rightly, it only weighs about a quarter of a grain. From this mirror instrument he has proceeded to what he calls his Recorder, and there he records upon a strip of paper a thin line of ink, not made by the depression of a wheel or a pen, or any apparatus upon the paper, but simply by a small thin film of ink that is spurted upon it by electrical repulsion. There is no instrument more beautiful in construction than this of Sir William Thomson's, and I am very sorry it is not here for you to see. Probably in a week or two one will be fitted up in the collection, and it will be well worth your inspection.

Now I have rapidly glanced through the different systems of apparatus in use in England and generally throughout the world. In England we really have at the present moment five systems in use, each of which may be called, as I said at first, an example of the survival of the fittest. We have the A B C, which is used at all small post-offices where skilled labour, from its expense, cannot be employed —that is, where messages are few and far between; sometimes where they only reach one a week, others one a day, and so on, where business is slack. It is also used very largely between the merchant and his office, between the hall and the stable, between the counting-house and the drawing-room, and for various purposes of that kind. As business increases and the offices become more important, so we introduce the single-needle instrument, which has this especial merit that it does not require skilled labour. It also has the advantage that a number of instruments can be connected together on the same wire. Sometimes on railways from twelve to fifteen instruments are formed on one circuit. On post-office circuits eight and sometimes ten have been so applied. As the business of the station increases, some proceed from the single-needle to the Morse system, and in the Morse system I include the Sounder. The Morse system is really and truly a relic of the past. Its doom has been sealed so far as England is concerned, and we are replacing the Morse instrument by the more rapid little Pony Sounder. It is rather a remarkable fact that in this instrument, which contains no record whatever of the message which has been sent, we have an instrument which is the most accurate that has ever been produced. The reason for that is simply this :

The sending of the clerk is guided by his ear. If his ear has been trained to read the dots and dashes properly upon this instrument, his hand, which follows his ear, must record the messages he is receiving accurately, and experience proves that to be perfectly true. A short time ago I was visiting a station where some eighteen months since the Morse apparatus had been replaced by the Sounder, and the postmaster told me that although that instrument had been at work eighteen months, not one single case of error had been brought to his notice, whereas, previous to that, scarcely a week elapsed without some mistake having been made in a message. Another great advantage of it is the rapidity, for a clerk having his own attention confined to reading, is in just the same position as you are to me at the present moment. I might be for instance in Southampton, and you here, and yet this little instrument, if your ears were skilled, would have conveyed to you almost as fast as I speak the words I am uttering.

Now as business increases, as circuits become crowded, as messages increase, so the necessity far rapidity arises, and then we fly to Wheatstone's automatic apparatus, which records messages in many instances as fast as they come in ; but even then between very busy places—between London and Glasgow, for instance—that apparatus is not sufficient, and we are now, by means of Bains's Chemical Recorder, still further increasing the rapidity with which the work is done ; so that we hope to be able to keep ahead of the work by improving the rate at which the instruments record.

Now the question naturally arises when considering these various forms of instruments, to whom are we to ascribe the chief merit of a successful and practical invention ? Is it to the philosopher who suggests in a misty future some visionary project ? Is it to the engineer who renders the abstract concrete—the dream a reality ? Is it to the financier or the commercial man who risks his fortune on his foresight, and on his estimate of the value of the philosophical idea and of the engineer's skill and practice ? In all these applications of science to practice these three characters are involved. Take, for instance, telegraphy. We have here the dream of Sömmering, of Schilling; we have the genius of Wheatstone and of Steinheil. On the other hand, we have the practical enterprise of Cooke, of Morse, and of Siemens. We have again the financial foresight of men who are little known—Ricardo, Bidder, Weber, of the great railway companies of this country, of the Government, and of the Governments of Europe—all of whom have lent their assistance in establishing telegraphy on its present great basis. Take again one of the greatest branches of telegraphy—submarine telegraphy. See how our coasts are joined to the Continent, how our country is joined to all parts of the world, enabling us to waft a sigh from Indus to the Pole, or put a girdle round the earth in far less time than forty minutes! Here we have a specimen of the submarine cable which connected our shores with France and Holland. The great genius Wheatstone, as early as the year 1839, I think, formed the idea of a cable connecting England with France. The genius of Faraday and the skill of Siemens succeeded in making submarine cables practicable; but it was the foresight of Crampton, of Carmichael, and others who succeeded in rendering this project successful; so that we see in all branches of telegraphy the philosopher, the engineer, and the commercial man must take their fair share of credit.

Time will not allow me to pursue this matter further. There are many points I have omitted. I have been unable to say anything to you of relays and translators which enable us to speak to distant places; for instance, every day we hold incessant communication with India by means of translators. We have that beautiful system of time apparatus downstairs which transmits time to all parts of the United Kingdom at one o'clock every day, and we have various forms of apparatus used for signalling upon our railways; and before I conclude, I will simply ask this question, which probably has suggested itself to many minds:—If the past shows so much indication of progress, is that progress still in force? Has perfection been reached? I am quite sure it has not. We are still in the infancy of telegraphy. Scarcely a day passes by without some improvement. Some defect shows itself requiring a remedy. Some improvement is suggested increasing speed, and the condition of our lines, the condition of our apparatus, and the various processes by which messages are sent and messages are prepared are daily receiving improvement and accession; and you may depend upon it that if the direction of our telegraphists is maintained as it is at the present moment, if it is succoured by such exhibitions as this, we may expect in a very short time not only to produce greater speed in telegraphy, but what the public want a great deal more—we shall be able to introduce cheaper telegraphy.

The PRESIDENT: Ladies and gentlemen—In rising to propose a vote of thanks to Mr. Preece for his very lucid and satisfactory explanation of the progress of telegraphy, I can but bear testimony to the great difficulty with which the subject is surrounded. We have seen that telegraphy dates from a comparatively short time ago, but that its progress has been very rapid, involving researches in science as well as the practical work of the engineer to an extent which is perhaps unrivalled by the development of any other invention. The

---

Note as to the origin of Submarine Telegraphy.

Submarine telegraphy in 1851 was deemed by most engineers and the public to be visionary, if not impracticable.

Its extension over the whole world since its first practical introduction in 1851 has been immense, and the advantages to the world at large incalculable. It was established in the following manner:—

Various propositions were from time to time put forth to effect the object, but few people were prepared to take the risk until a Company was formed having most influential men on the direction, who advertised in the usual manner for subscriptions. Such, however, was the want of confidence felt in the scheme, that only about two per cent. of the necessary capital was subscribed, and this money was consequently returned to the applicants. Notwithstanding this apathy of the public, some of the directors and their friends did not cease to entertain a full conviction of its possibility; and they subsequently consulted Mr. T. R. Crampton, C.E., on the subject, and offered to assist towards providing the funds if he felt sufficiently confident of ultimate success. Mr. Crampton undertook the entire charge and responsibility of the form, construction, and laying of the cable; also took upon himself rather more than one-half the pecuniary risk, the other half of the money being found by Lord de Mauley, Sir James Carmichael, Bart., Messrs. Davies, Son, and Campbell (the Solicitors of the Company), the Hon. Frederick W. Cadogan, and Mr. Haddow. The cable was in the same year (1851) successfully laid by Mr. Crampton between Dover and Calais.

The great risk the parties ran can be appreciated from the fact that three successive attempts to establish submarine cables between England and Ireland by other parties occurred soon afterwards, and all failed,

The above-named gentlemen were also instrumental in laying the next successful cable between Dover and Ostend, which was constructed in a similar manner to the original one.

No great improvement has yet been effected in the form and mode of construction of heavy cables, thus proving satisfactorily that the first type of submarine cable laid upwards of twenty-five years ago is practically right in principle.

subjects here touched upon are very various, and it would be impossible to do full justice to all their branches. For instance, submarine telegraphy has been perhaps rather slightly touched upon by Mr. Preece, though it is in itself a very large and important branch of the whole. I am quite sure you will join me in a vote of thanks to Mr. Preece for his very able paper.

The next paper on our list, Mr. Henrici's, will not be brought forward to-day, and therefore there is a short time for discussion if any one wishes to continue the subject.

If not, I have, in closing the meeting, to close this section of the Conference. We have had before us important communications on all the leading branches of applied science, and I think that we may congratulate ourselves upon the manner in which these subjects have been brought before us, and have been illustrated by the models from the collection downstairs, a collection which is unrivalled of its kind. The Conferences will be followed by a series of lectures or explanations of the exhibits themselves, because it was impossible at these general meetings to enter into the merits of the particular exhibits, and time not being sufficient for the purpose. It may be objected that we have not done justice to many of the most interesting machines which may be found downstairs, but if we had attempted to discuss the merits of particular exhibits, we should certainly have failed to give a bird's-eye view, so to speak, of the riches contained in this exhibition, and by having obtained this bird's-eye view, we shall be better able to turn our attention to particular exhibits, and range them amongst the whole. I myself would have liked to have given an explanation of one or two apparatus in which I feel particular interest, but for the same reason I have abstained from doing so. I hope we shall all separate with the conviction that these discussions have not been without profit.

END OF VOL. I.

Printed in the United States
By Bookmasters